Market-Oriented Technology Management

Innovating for Profit
in Entrepreneurial Times

Springer
Berlin
Heidelberg
New York
Barcelona
Hong Kong
London
Milan
Paris
Singapore
Tokyo

Fred Y. Phillips

Market-Oriented Technology Management

Innovating for Profit in Entrepreneurial Times

**With 92 Figures
and 43 Tables**

 Springer

Prof. Fred Y. Phillips
– Department Head –
Oregon Graduate Institute of Science and Technology
Department of Management in Science and Technology
20 000 N.W. Walker Road
Beaverton, Oregon 97006
USA
e-mail: fphillips@admin.ogi.edu

ISBN 3-540-41258-1 Springer-Verlag Berlin Heidelberg New York

Library of Congress Cataloging-in-Publication Data applied for
Die Deutsche Bibliothek – CIP-Einheitsaufnahme
Phillips, Fred Y.: Market oriented technology management: innovating for profit in en-
trepreneurial times: with 43 tables / Fred Y. Phillips. – Berlin; Heidelberg; New York;
Barcelona; Hong Kong; London; Milan; Paris; Singapore; Tokyo: Springer, 2001
    ISBN 3-540-41258-1

Springer-Verlag Berlin Heidelberg New York
a member of BertelsmannSpringer Science+Business Media GmbH

© Springer-Verlag Berlin · Heidelberg 2001
Printed in Germany

The use of general descriptive names, registered names, trademarks, etc. in this publication does not imply, even in the absence of a specific statement, that such names are
exempt from the relevant protective laws and regulations and therefore free for general
use.

Cover design: Erich Kirchner, Heidelberg

SPIN 10734570   42/2202-5 4 3 2 1 0 – Printed on acid-free paper

To my students

Preface

This book takes a decision-making orientation to technology management. Its view is more operational than strategic. However, I have tried not to adhere too closely to this orientation when doing so would prevent the reader from approaching an issue in a creative or integrative way.

While much that has appeared in recent print under the banner of technology management has been reworked material on manufacturing engineering or R&D management, this volume addresses an originally motivated collection of topics drawn from sociology, government and policy, history of technology, marketing, entrepreneurship, and other disciplines. Twenty-five years as a manager, consultant and educator have convinced me that these address the real problems of industry and the real needs of students. I hope their presentation here will help lead to a distinctive identity for technology management as a field of study while maintaining respect for the worth of manufacturing engineering and R&D management as related, if more traditional, fields. Some instructors may want to use readings from those fields to supplement this book's emphases on technology cycles, diffusion of innovation, marketing, and U.S.-Japan comparisons.

Although the topics of competitive analysis and of managing researchers, developers, and their relationship with marketers may also be considered part of technology management, these topics are not heavily emphasized in this book. Problems of maintaining an innovative organizational atmosphere are mentioned but not pursued in depth. This is simply because these topics are dealt with in courses other than the technology management course here at Oregon Graduate Institute, and in order to keep the book at a reasonable length for use in a one-quarter course.

The book can be used as a primary text in a graduate course on technology management, or a supplementary reading in an operations management, technology marketing, or integrative management course. No calculus is needed, but a basic familiarity with spreadsheets is prerequisite; the text reviews some specific math and spreadsheet skills that are needed to negotiate the book's concepts and chapter-end problems. The mathematical material is thus not beyond the reach of undergraduates, but the text presupposes a familiarity with current trends in business that is more likely to be shown by a motivated M.S. or MBA student or an experienced manager. I hope that the book will also be enjoyed and used by managers who are not enrolled in university courses.

Principles and issues in technology management are the primary focus here. This reflects my wish to make technology management scientific (or at least systematic) while continuing to ground it solidly in real industrial examples and relevance to current affairs. The book emphasizes problem solving. Quantitative problems comprise about half of the chapter-end questions. Students must use rhetorical skills to address the remaining problems successfully. This is in accord with industry's need for managers who can communicate and persuade. While the book does not utilize cases in the Harvard University sense of the word, a few less-structured problems allow readers to exercise creativity in building plausible strategies and decisions in complex environments with incomplete information, and scenarios of business-technology futures. (Note: Instructors desiring an electronic file of answers to selected chapter-end problems should write to me on their university letterhead, c/o MST department, Oregon Graduate Institute of Science and Technology, 20000 NW Walker Rd., Beaverton, Oregon, USA 97006, noting the email address to which the file should be sent.)

Theory should not only help advance science, but make science easier to learn. One can ingest a great deal about technology management by reading journals, but the piecemeal tips, techniques and cases described in journal articles can be difficult for a student to assimilate. This book aims to use theory (diffusion of innovation, developmental learning, economics, etc.) to give the student a framework for making sense of the tools and examples given in the literature, without burdening the student with excessive theoretical detail. Ideally, students completing the book and its exercises will be able to devise their own tools (e.g., instruments for determining the critical hurdles in a technology transfer) to supplement the tools given herein, by virtue of having mastered underlying principles.

The book's material has been tested over three years of teaching high-tech marketing at the Austin Technology Incubator (in a course for MBA students at The University of Texas at Austin), plus four years teaching Issues and Trends in Technology Management in the executive M.S. in Management in Science and Technology at Oregon Graduate Institute of Science and Technology (OGI) in Portland. The book also incorporates some material from my U.T.-Austin MBA course in Managing the Product Cycle, some from my Software Commercialization course at OGI, and some material generated by the Japan Industry and Management of Technology program (AFOSR grant #F49620-92-J-0522) at U.T.-Austin. I hope the book's concepts are sharpened by the frequent recourse to U.S.-Japan comparisons of technology management practices.

The emphasis on marketing that will be found in these pages is also a distinctive feature of the book. Every insertion of a new technology or technology product must be "sold," whether the insertion is purely intraorganizational or whether a vendor company is selling to a buyer company. People on the originating and receiving ends work together to evaluate, test, and apply the innovation. Though others may use different words, I call that marketing. Students should be encouraged to imagine applying each of the principles in this book first as a "seller," and then again as a "buyer." Instructors are encouraged to call on prominent local technology entrepreneurs to serve as living cases, as described in the appendix to Chapter 1.

My thanks to the students who contributed article summaries to this book project; to Cenquest, Inc., for kindly allowing me to use several figures they rendered for the distance learning version of my course (and from which, unfortunately, I had to remove the color); to my wife Hyonsook, for her sharp eye for typographical errors; and to the faculty, staff, and administration of Oregon Graduate Institute of Science and Technology for allowing me the freedom to complete the book. Readers will have my thanks for communicating errata and suggestions to me at fphillips@admin.ogi.edu.

Table of Contents

PART I

TECHNOLOGY LIFE CYCLES

"People will buy something new. They will not buy something revolutionary."

Marketers' proverb

"A technology can be called 'revolutionary' if it changes people's physical or mental capabilities by a factor of ten."

Feigenbaum, McCorduck & Nii, *The Rise of the Expert Company*

1 Introduction: Revolutionary Technologies

1.1 Revolutionary Technologies: Why Bother?

For many of the world's needed innovations we must thank inventors who are motivated only by the joy of working with advanced technology. They are aided by venture capitalists who see the opportunity for an extraordinary return on investment. Not to mention sales executives who wish to ride growing markets, especially when there's an equity kicker. Still other entrepreneurs enjoy having a vision, seeing it actualized, and making a lot of money. But socially desirable benefits emerge from technological innovation and its successful diffusion, regardless of the motives of the actors. The benefits are improved productivity, increased exports, reduced waste and pollution, and the creation of jobs.

Why are these important? Remember the grim Reverend Malthus, who observed in 1798 that human populations grow exponentially while their food sources do not. If we were to freeze all technology at 1995 levels, the food and other resources available to the world would increase, but only slowly. (Continued exploration with current tools would, for example, reveal a few new oil deposits.) Human reproduction, which remains a low-tech activity for most of us, would continue rapidly, of course. It is easy to anticipate, as the Club of Rome did in the early 1970s, that available resources per person could decline to a point where civilization could not be maintained.

Technological innovation and its diffusion results not only in the discovery of new resources, but in the more efficient use of currently available resources, and, indirectly through increased affluence, voluntary decreases in birth rates. Put simply (though grandiosely), innovation contributes to the sustainability of human civilization on earth. What is the alternative? When the resources available to a population diminish beyond a certain point, the result is war or plague that sharply decreases the population.

It has been pointed out that war-plague cycles have characterized human history for millennia, and it is only recently that technological advance has even more effectively destroyed the traditional lifestyles of many human societies. This indictment of innovation is worth thinking about. But it is ultimately irrelevant, for two reasons. First, the farflung societies that have benefited from modern technology will not willingly give it up. Where technology has produced, say, reduced infant mortality, people will not willingly return to a regime of high infant mortality. Even people who have not yet experienced such improved public health benefits nonetheless know that technology can make it possible − because they've seen it on television! There is a strong demand for beneficial technologies. Second, we have begun to move beyond the technologies of the industrial revolution. These "mass" technologies tended to homogenize consumers. The new technologies of information and customization allow for and even encourage diversity of application. These adaptable technologies will be used to differentiate people and societies rather than level them, and so will be more interesting and exciting, more readily accepted and less politically inflammatory.

1.2 Why This Book Was Written, and for Whom

I hope this book will serve the needs of technology entrepreneurs, corporate innovators and intrapreneurs, and graduate students contemplating a career in technology management, technology marketing or technology entrepreneurship. The book grew out of the author's experience as an R&D manager and consultant in the market research industry, and later as a teacher (at The University of Texas at Austin and at Oregon Graduate Institute of Science and Technology), and his desire to "make sense of it all" in a way that is appealing to those just starting on any of the paths just mentioned.

The IC^2 Institute at U.T.-Austin, where the author was Research Director, is well known for a variety of areas relevant to the technology manager. These include the link between technological innovation and economic development; technology entrepreneurship and incubation; the "technopolis"; international technology business studies; and industrial competitiveness. This book tries to bring some topics from these areas together in a way that is relevant to the student and the working technology manager.

Finally, the book was motivated by the shortage of books that are specifically useful to the technology marketer.

It is fairly easy to identify the people in an organization who do accounting, or who do finance or manufacturing. But who does technology management? Certainly CTOs (chief technology officers) and CIOs (chief information officers) manage technology. So do entrepreneurs, presidents, engineering managers, quality managers, strategic procurement managers, and many other individuals within the firm. Outside the technology company, industry analysts and venture capitalists are involved with important aspects of technology management.

One might conclude that, unlike the study of accounting, which leads clearly to a career track in an organization, the study of technology management does not, and therefore has only more diffuse benefits. We will not worry that point here, though. Even if the benefits are diffuse, they are significant. And everyone who reads this book will find tools and ideas that enable them to do their job more successfully and profitably.

We will take the view that the technology manager's job includes some or all of the tasks below, no matter what their titles may be. Technology managers...

- obtain needed technologies and technology products from outside the firm;

- manage the generation of needed innovations within the firm;

- decide what kinds of innovation should be fostered within the firm and what kinds are best sourced externally;

- maintain the ability of the firm to do both of the above;

- combine internal and external innovations to develop viable products;

- market and sell the products;

- do all the above in a way that contributes to a growing revenue stream for the firm; and

- do it in an ethical and fulfilling way, with an understanding of the role of technology management in the economy and society.

1.3 What the Book Covers (I)

Later sections of this chapter make it clear that the great majority of entrepreneurial opportunities lie in providing well-established solutions to customers' routine problems. (If you have looked ahead to Figure 1.8, you know that we will call these solutions Type I products.) Growth opportunities abound in Type I markets for the entrepreneur who can provide slightly better efficiency and service. But many students in graduate engineering laboratories, many engineers in advanced corporate laboratories, and many entrepreneurs in new business incubators are committed to selling new solutions to novel problems – i.e., those Type IV products that provide the smallest proportion of profit opportunities – and hope to build a business around them. Their concerns, and the related concerns of makers and users of Type II and III products, will be the primary emphasis of this book.

The book will also look at niche strategies. A revolutionary product may be too advanced for most market segments, yet only slightly daring – and easier to sell – to a specialized group of customers with special needs. Conversely, the product may be old hat to many segments, but totally new, and perhaps a hard sell, to a set of customers that have not used such products previously.

In marketing, it is elementary that awareness precedes information gathering, and information gathering precedes the purchase decision. In high technology markets, a customer must internalize a great amount of information about a product that may have a very short life cycle. Time, timing, and lead times become important parts of marketing thinking and marketing action. Information must be released sufficiently in advance of the product, and a plan must be in place for the inevitable leveling off of demand. We will also look at time and technology cycles.

Management science has advanced to the point that we know what questions to ask about type IV products, about niches, and about timing. We know *how* to ask the questions. This is no mean feat! But we still don't know the answers to the questions. There has been little systematic gathering of data. But we can look at cases, anecdotes, and the opinions of those experienced in both success and failure. This is what we will do in the following chapters. This is not all we will cover, though. But we must introduce a few needed concepts before continuing with "What this book covers, part II."

1.4 Theory and Practice

The book is organized to let theory and practice interact. There are four theory areas that are highly relevant to technology marketers. These are the marketing theory of the product life cycle, the sociological theory of the communication and diffusion of innovations, the learning theory of developmental psychology, and the theory of the ecological niche. Chapters 2, 3 and 8 review these with an eye toward what technology managers can learn from them. The intent is to give you a memorable framework for mentally organizing your technology management tools and strategies – *not* to bog you down in theory.

Life cycle theory (which is introduced in part in the next section) helps us understand the timing of investment, development, sales and obsolescence, and how to forecast these. Diffusion theory deals with the kinds of customers that will be responsive and unresponsive to our products at various times, and how to overcome customers' and co-workers' resistance to change. Developmental psychology gives a complementary viewpoint on how people react to information technology products, user interfaces, and complexity.

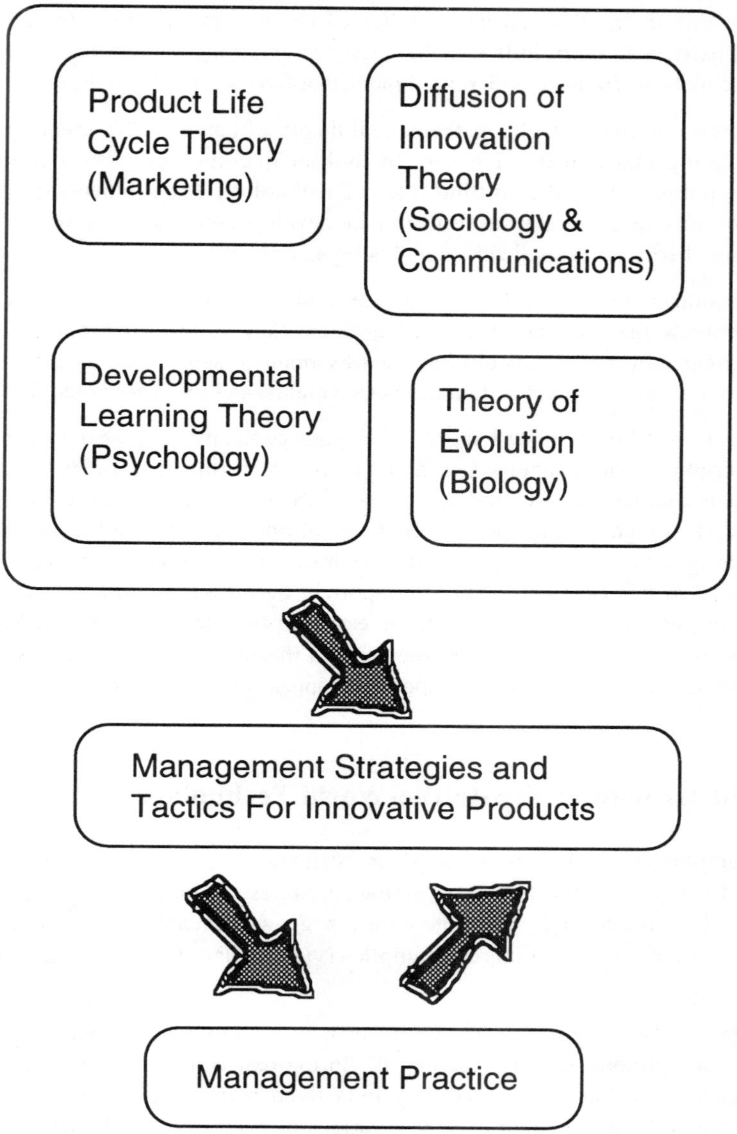

Figure 1.1 Conceptual basis of this book

A fourth theory that entrepreneurs will find meaningful is the theory of the ecological niche. Businesspeople can quickly grasp the analogy between growing and protecting brands, and the process by which separated biological populations specialize, speciate and co-evolve with their environmental niches. The book will develop this theme through case studies of companies that have found niches, erected barriers to entry in that niche, then grown strong enough to venture forth from the niche to compete with other brands that have similar "metabolisms."

A fifth relevant theory is the mathematical theory of games. This theory provides a fascinating and constructive way to look at negotiations with a supplier or alliance partner, or at the shifting tides of technology entrepreneurship as large companies try to entice startup firms to take development risks. We will not delve into game theory or its applications in this book, however.

The remainder of the book details strategies and tools. We will take a look at how these embody the four theories, to what extent they validate the theories, and to what extent they are (or should be) used by management. Each strategy is best suited to a certain time and certain market segments, and these are noted for each.

Chapter 10 touches on what happens after your company evolves out of the Type IV description, and Chapters 5, 6 and 11 provide further context by comparing entrepreneurial technology marketing in the U.S. and Japan. Chapter 7 provides a framework for gathering the market information you will need as the product cycle progresses. Some of the useful tools of technology forecasting are introduced in this chapter. They are important because they may lead you to the *vision,* discussed in Chapter 11, that is essential to entrepreneurial marketing of high technology. Short cases are interspersed throughout the chapters, and some longer ones, written by student collaborators, appear in Chapter 11.

1.5 The Growth of New-to-the-World Technologies

The cumulative market penetration or "diffusion" of a new product can be pictured using any of a number of s-shaped curves, including the logistic curve, portrayed in Figure 1.2. Fast penetration growth means new customers are acquired at a rapid rate. Figure 1.2 implies (via its "first derivative") the graph of Figure 1.3.

Mathematically, the penetration can hover near its maximum attainable level for many years without actually reaching it. In the real world, obsolescent products experience no further sales. (Really long-lived technologies, for example the wheel, are no longer thought of as innovations or even as technologies!) So there is a limit in time to the adoption of a technology, and graphs like Figure 1.3 should have a cut-off point to the right.

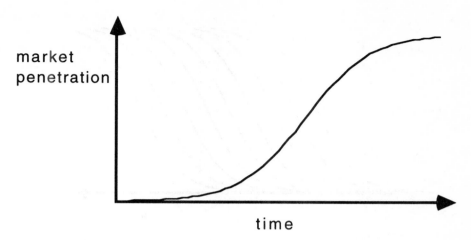

Figure 1.2 The time path of market penetration

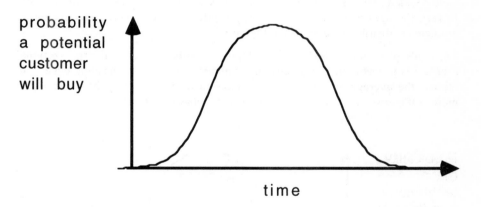

Figure 1.3 When will a customer most likely buy?

Suppose in each of several years, several new technological products reach the market. In the graph below, there is a curve for each of the years we are considering. The curve portrays the "average" diffusion parameters for all the year's innovations.

Now suppose that in year t, a company commences a search for a product to solve a factory automation problem. The search, naturally, considers all currently extant technologies relevant to the manufacturing problem. In other words, the search considers all products represented by the curves that intersect the vertical dotted line in Figure 1.4.

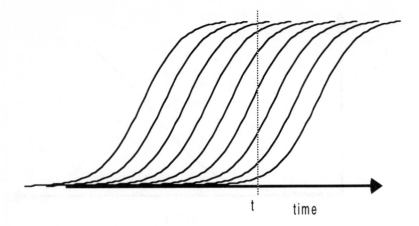

t   time

Figure 1.4 Penetration curves for innovations introduced in successive years

The dotted line intersects few technologies that are very new. And, given what we have decided about obsolescence, it intersects few technologies that are very old. In fact, the ages of the technologies applicable to the company's manufacturing problem are distributed as shown in Figure 1.5.

Thus, it's plain that a technology adopted to solve the typical organizational problem is probably not very new, and is almost certainly not "leading-edge." In reality, the average age of a newly purchased technology may be a decade or more. Of course, it is "new" to the company purchasing it for the first time.

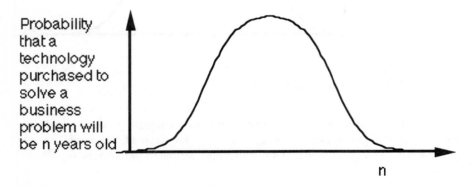

Probability that a technology purchased to solve a business problem will be n years old

n

Figure 1.5 The most likely age of the technology that solves the customer's problem

We have just used theory to validate an item of managerial folklore, namely, "I can solve 90% of my problems with well-known technology. I only need leading-edge stuff for the other 10%."

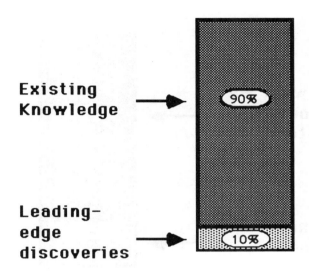

Figure 1.6 Where do operational improvements come from (I)?

Executives also say, "80% of my company's potential for improved profitability will be realized by improving the efficiency of our current activities. Only 20% of our profit increases for the next few years will come from truly new activities." Figure 1.7 represents that statement graphically.

It is important to realize that both kinds of improved profitability come from innovations, because the changes will be new to that company (even if such measures are "old hat" to other companies.) A company may, of course, derive more than 20% of profit increases from new products it develops. The company has, though, a tried-and-true procedure for developing new products. The 20% in Figure 1.7 refers to activities the company has not engaged in before, which require changes in the way (parts of) the company are organized.

It is impossible to resist cross-footing Figures 1.6 and 1.7 into a "management matrix" (Figure 1.8).

You should be aware that some writers divide the vertical axis in Figure 1.8 into three segments: continuous innovations, meaning that the customer continues his habitual behavior when using the new product; discontinuous innovations, where the product causes a near total dislocation in the customer's behavior; and a middle category called "dynamically continuous," for products somewhere in between.

Another researcher speaks of (i) products that use a different technology to provide a familiar benefit at similar cost; (ii) products that provide a familiar benefit with significant new economies, efficiencies or other advantage; and (iii) products that provide a new, unfamiliar benefit.

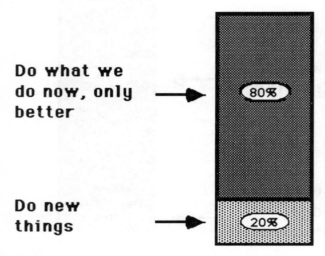

Figure 1.7 Where do operational improvements come from (II)?

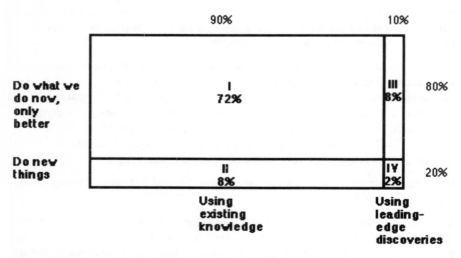

Figure 1.8 Where do operational improvements come from (III)?

Thus, there are four kinds of technological products.

I.   Those that use well-established technologies to help a company improve the efficiency of current operations.

II.   Those that use well-established technologies to help a company move into new kinds of operations.

III.   Those that use leading-edge technologies to help a company improve the efficiency of current operations.

IV.   Those that use leading-edge technologies to help a company move into new kinds of operations.

1.6 Type IV Products

The leading edge moves quickly. Fortunately, this means that type III products quickly turn into type I products (and type IV products into type II products), perhaps in the space of two years. It's surviving those two years that's the hard part[1].

The quotation that opens this chapter defines a *revolutionary* technology as one that increases productivity by an order of magnitude. As most manufacturing or other organizational systems are not stable under order-of-magnitude variations in one of their components, it makes sense that a revolutionary technology requires the organization to change, to do things differently. Revolutionary technologies, when packaged and marketed, are necessarily type IV products.

The percentages shown in Figure 1.8 are very rough. Booz, Allen and Hamilton (*New Product Management for the 1980s,* New York, 1982) claimed that 10% of new products then appearing each year were "new to the world." These, of course, include consumer products like stove-top stuffing and rollerblades. So the proportion of new-to-the-world industrial technology products must be signficantly less than 10%.

A central point of this book is that makers of type IV products have to deal not only with the usual questions of adding value and reducing risk for the customer,

---

[1]   Geoffrey Moore (author of *Crossing the Chasm*) maintains that *after* you have survived the first two years, the hard part is just beginning. He's right too. Perhaps not coincidentally, two to three years is the planned tenancy of start-up companies in new business incubators.

but also with all the problems of organizational culture change that attend the insertion of a revolutionary technology into the customer's company.

Nor is *that* the full extent of the type IV entrepreneur's problem. He/she must have a vision of *how* the customer firm can be reorganized to absorb the new technology, that is, what its new organizational chart and information flows will look like. Communicating this vision to the potential customer will be a significant part of making the sale.

1.7 What the Book Covers (II)

Thus, the type IV entrepreneur must attend in special ways to issues of

- the buying decision process
- technology marketing
- channel management
- designing the communication mix
- management of change in organizations
- competitive advantage
- policy environment

- industrial marketing
- government marketing
- pricing
- futurology
- corporate culture
- global scanning
- growth and hypergrowth.

In this book we will try to touch briefly on each of these areas, looking at whether conventional management thinking is applicable or extendable to type II, III and IV products.

Type IV entrepreneurs, as we've noted, have some advantages to offset the added difficulties of blazing new paths. A few of these stem from the lessened competitive pressures on a truly original innovation. The type IV entrepreneur may have the luxury of making mistakes or schedule slippages during product development. Because markets are fragmented for mature, conventional technologies, mature products must be targeted at small segments, limiting their opportunities to recover investment. A type IV product, on the other hand, *may* be attractive to a large market.

A rather different consideration is that of social and governmental constraints. Conventional technology products must comply with detailed regulations and well-articulated public opinion. Of course, if your new-to-the-world product is a drug, it will be subject to FDA review. If it is polluting, you will face citizen opposition. But there may be advantages to introducing a new-to-the-world product that is hard to categorize: Potential government and special interest group reaction may be slow. This is not to encourage you to market a product that is harmful to people or the earth, but only to make the point that even beneficial new products are often misunderstood; and bureaucracies do not react quickly to what they cannot classify.

1.8 Looking Ahead

What are the barriers to adopting a revolutionary technology? They are:

- perceived learning burden
- perceived risk
- fear of change
- lack of vision
- lack of knowledge of how to implement change in an organization
- and other costs of switching from the product that is currently used.

These can be summarized as the "FUD factor" – Fear, Uncertainty and Doubt. They will be discussed in more detail in Chapter 8. But it's not too soon to start thinking about them.

Business uses of the Internet were slow to take off because of *perceived learning burden*. The Internet was seen as a complex, academic device quite distant from the well-known ways of engaging in commerce. Consumer resistance to irradiated food is an example of *perceived risk*. Use of local-area computer networks have been cited as a case in which *fear of change* slowed what might have been a fast and smooth transition of business offices to networked computing. The NeXt-Step desktop operating system, widely seen as a superior OS, was not adopted in companies due to the companies' *lack of vision*. Microsoft Excel was slow to replace Lotus Corporation's 1-2-3 spreadsheet product because the costs of installation, training and converting existing files constituted *high costs of switching*.

A maker of self-reporting utility meters chose to sell them to rural electric cooperatives. These co-ops tended to be resistant to change, and the vendor, while technically competent, *lacked knowledge of how to implement change in the customer organization*. The experiences of the Magellan Group's Rosetta product[2] and American Express Corporation's "Authorizer's Assistant"[3] suffered in initial sales because *customers were not organized to feed needed information to the new product.*

As many of these examples are drawn from the computer industry, let's take a more detailed look at that industry's practices in moving technology products into customer organizations. We'll then close the chapter by nailing down the definition of an advanced technology product, and by discussing technology fusion, technology convergence, and technology mapping.

1.9 Computer Industry Booms, But Where's the Marketing?

The American Electronics Association and other observers see the computer industry as a link in a "food chain" of mutually dependent U.S. industries. Losing any of these to foreign competitors would further weaken our nation's competitiveness in the world economy. In this section of the chapter, we'll look at the marketing problems of the personal computer (PC) industry – now in its early maturity – as an example framework for marketing other innovative and high tech products.

A circa-1990 Congressional committee estimated that a full 25% of world GNP is spent on information technology. But marketing blunders had left the U.S. with a retarded share of this (now demand-driven) market. Let's analyze some marketing mistakes and opportunities.

• Calling the machines "computers" was the original marketing sin. Today, most computer users don't compute; they process words or access databases. "Workstation" is a dismal label too. Does "an exciting career in data processing" sound like an oxymoron? "Desktop publishing" is slightly better, but it doesn't excite.

---

2   This tool for measuring relative productivity of chain (or franchised) retail outlets is a powerful and useful analysis. But it requires cost data for each outlet that is sometimes difficult for the chain's headquarters to produce. It has been found that much of the data needed for assessment is lumped into headquarters' General and Administrative expense and cannot be expressed for each outlet under many clients' current accounting practices.

3   This artificial intelligence tool was designed to give a rapid response to loan applications, by tagging bad loan signals in real time during the application process. The first release of the tool proved to be a good discriminator of good loan risks from bad risks, but only if the application data were supplemented by data from credit agency and other references. Because the latter took time to acquire, the 'real-time' feature of Authorizer's Assistant – which was its principal advantage as a product – was defeated.

• "Diskless PCs" died a well-deserved death in the 1980s. They were slaves to networks; there was nothing "personal" about them. Like "horseless carriage," this negative nomenclature did nothing to convey the unique benefits and promise of the new technology. The mistake is about to be repeated with the net computer: Try getting a customer excited about a "thin client." Likewise "artificial intelligence" is misleading to some consumers, offensive to others.

• Terminals and PCs are housed in unsightly boxes that few executives want on their desks.

• How are these machines advertised? With stupefyingly technical spreads, featuring pictures of the ugly boxes. Commercials that avoid this trap, on the other hand, often have been so odd as to be incomprehensible.

• Complaints abound that one computer company's high prices for upgrades have shafted the hobbyists that supported the company in its early years; that another company, using manipulative tactics to lock in customers, sells them conservative or backward technology.

• In 1989, the *Wall Street Journal* interviewed executives at IBM and at the industry's top market research firms. The conclusion? IBM's PC sales might have been growing at 40% per annum, but no one really knew. High-end machines were fueling the leading PC companies' growth, according to the *Journal.* "That scares me," said an industry executive, "Are we trading higher dollars for fewer customers?" Why didn't he know?

When hardware manufacturers don't understand what businesses and individuals use computers for, they can't clearly communicate their product to their ad agencies. The agencies then fail to understand uses, features, or benefits. The result is a tendency to sell MIPS and megaHertz rather than the ways computers make work more efficient and humane.

Ed Thomas, former general manager of CompuAdd, likens the computer to the automobile, which in its early versions could be driven only by a mechanic. Thomas notes the advent of the automatic transmission made automobiles practical for families that could afford the car but not the driver. The computer industry is now, he said, looking for the automatic transmission. This could mean advanced interaction modes, such as voice input/output, are the marketing key. Or perhaps the key is integrating applications, or finding the right balance of personal convenience and privacy versus connectivity.

Selling hardware alone sharply limits attainable market share. One thoughtful response has been the "usage concept" providing broad application power within a limited framework. But even the better-defined usage concepts have fuzzy edges, making marketing difficult. Word processing shades over into page layout, CAD/CAM into CIM. Less sharply defined usage concepts, like Executive Information Systems or Personal Information Managers, result in diffuse marketing strategies. Half-formed usage concepts are rushed to market before prices decline.

The majority use computers for a limited set of repetitive tasks. But a substantial niche of buyers buys computing tools with which to build idiosyncratic applications. A minority of these applications prove to be solutions to someone else's problem. But lack of marketing sophistication prevents many leading-edge users from turning these ideas into entrepreneurial ventures.

Bundling hardware with software that has been optimized for it has worked especially well for peripherals manufacturers. Supermac Technologies has been a paragon of this approach, offering spooling and backup programs that are exemplary in their attractiveness and ease of use. This solves the consumer's business problem while differentiating the hardware from similar competing products.

CompuAdd's Ed Thomas remarked further (in 1988) that the coming battles over features would center on architectures rather than processor speed. But until the advent of home networks in the late '90s, an architecture's capability for networking (its most important feature) was of consequence only to IS managers, not to individual users. Marketing has not successfully addressed the distinction between personal computing and group/distributed computing.

Nor is there an accepted best answer to the question of open architecture. Closed architectures make a focused marketing statement easier. Open architectures mean more flexible machines and a larger potential pool of buyers. Apple decided to license its closed architecture, a move that was widely regarded as having been made too late, and then reversed course and bought out its cloners' agreements. What is the right timing for opening an architecture?

What is the role of standards in computer marketing? Standards imply cheaper manufacturing, greater reliability, and wider capabilities for connections between machines – but also a stifling of innovation and restricted choice for consumers.

U.S. marketers' greatest success has been in mastering complex chains of distribution (OEMs, VARs, consultants, retailers, manufacturers' and retailers' direct salesforces, direct mail/phone) and purchase (recommenders, approvers, purchasers, users). One downside of this is that market research becomes a nightmare.

Another is the threat of complacency. Industry executives consider this complexity an entry barrier to the Japanese – "They've stymied us with their complicated distribution systems, now we'll stymie them with ours." Nonsense. Although there have been American successes recently, the Japanese dominated the early laptop market. In the late 1980s, U.S. companies once again began by taking only the high end (Grid Systems, in the case of laptop computers), leaving the mass market for the Japanese. The later U.S. rebound might have been due more to currency fluctuations than to marketing acumen.

Sometimes we succeed in spite of ourselves. Market researchers IDC and Dataquest report that first-quarter PC sales in 2000 were 15-20% above sales in 1Q1999. The growth was in consumer PCs; corporate sales were sluggish.

1.10 What Makes a Technology or a Product "Advanced"?: Innovative Products and Revolutionary Products

The section heading above emphasizes that there is a difference between a technology and a product. We can make this point graphically, in Figure 1.9: A device is a physical realization of a technology, and performs a limited number of functions. It may not perform these functions reliably or conveniently. A device may function adequately in a limited range of environments (e.g., on a laboratory bench) when used by experts. A device may also demonstrate the validity of a technology or the feasibility of a product.

In contrast, a product has (or should have) a clear positioning for the marketplace, an unambiguous message that its sellers intend for it to convey to customers, an attractive and ergonomic physical form, an appealing and protective packaging, an instruction sheet or manual, and systems in place for distributing, supporting, maintaining, repairing, and recycling the item. It should be tested for reliable operation in all the environments in which the customer is expected to operate it (snow, desert, underwater, with dirty power supply, etc.)

Thus, developing a product can require five to ten times the time, expense and effort required to produce a device. It also requires a market orientation that is especially important for the scientist, the engineer and the inventor/entrepreneur to attend to.

Well, then: Whether it's a technology, device, or product, what makes it "advanced"? The Austin Technology Incubator was the early home of a company, Partnerwerks, that uses the tools of communication theory and sociology to teach other firms to build and utilize cross-functional teams. Another ATI company, MBA Practice Management, provides business management services for small medical practices. Neither company's product involves any technology beyond the usual office telephone and PC. Yet both Partnerwerks' and MBA's products require some behavior change on the part of the customer, and increase the customers' productivity. These companies do not sell *technological innovations,* but they do sell *organizational innovations.*

Other innovations do not involve products per se, but rather ways of making products. These procedures or *process innovations* show the customer how to arrange, organize, or use the customer's technology better. But it is difficult to point to a particular machine and say, "that is the process innovation." History's most dramatic example of process innovation may be Eli Whitney's concept of interchangeable parts for guns. Before Whitney, guns (and all other assembled products) were individually "crafted" rather than made identical. Broken items had to be returned to the craftsperson for repair. After Whitney, untrained persons could assemble products from manufactured pieces at low cost – and when an item needed repair, a spare part could be slapped in from inventory. This innovation was truly *revolutionary* in that it changed people's physical or mental capabilities (or in measurable terms, their productivity) by a factor of at least ten.

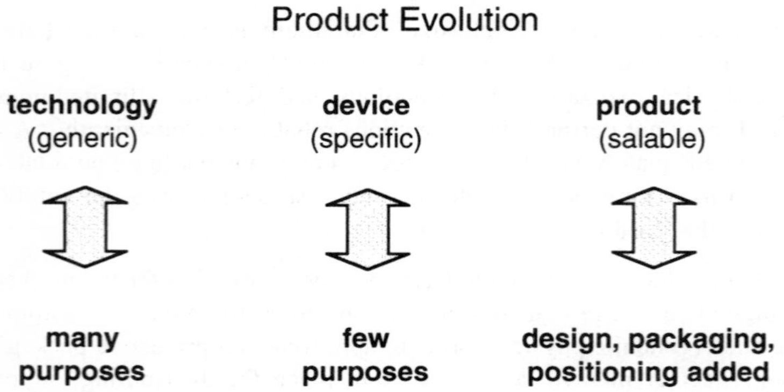

Product Evolution

technology	device	product
(generic)	(specific)	(salable)

⇕	⇕	⇕

many purposes	few purposes	design, packaging, positioning added

Figure 1.9 Consumers are not interested in technologies or devices.

Revolutionary products can be identified with the Type IV products of Figure 1.8. Elsewhere in this book, we use the phrase "new-to-the-world product." Indeed, the principles discussed in this book are applicable to revolutionary products and to new-to-the-world products. But they can also be applied to products that are new to a market segment, even if they are not new to the world. To continue clarifying terms, *high technology* products are so named because they are in the introductory or early growth stages of their life cycle. Winchester drives ("hard disks") are familiar products in the mature life cycle stage, but newer models require constant product and process innovation because they compete intensely on the dimension of storage density (bits stored per square inch of disk surface). They might be called an *R&D-intensive* product. *Technology* products like radios, hard disks, or zippers are in the mature stages of the life cycle; they no longer strike customers as newfangled. This book primarily addresses high technology, rather than just "technology" products.

In the final analysis, we can say the following chapters deal with marketing *innovative and high tech products*. These include technological innovations, organizational and social innovations, and process innovations. They may arise either in the entrepreneurial or intrapreneurial setting. Some of these may be revolutionary innovations. Others may be Type II or Type III products in the definition of Figure 1.8. These help the customer do new things or conduct existing activities with new methods. Often, they require considerable attitude change or behavior change on the customer's part before the customer will purchase and use the product. The marketer's task is to help the customer overcome the barriers of attitude and behavior change. The product designer's task (among many others) is to design products that minimize, in the first place, potential buyers' perception that they will have to change at all.

1.11 Technology Fusion

In days past, isolated research laboratories could develop expertise in any needed discipline. Advances were linear extensions of the lab's prior knowledge. Today's phone, fax and Internet connections mean that no lab is isolated. Nor could today's budgets support facilities devoted to wide expertise. The result is that profitable advances can be made cheaply by combining knowledge from different fields, different companies, different industries, and different geographies. This combination process is called technology fusion. The current base of scientific knowledge is such that combining existing ideas and technologies can generate profits for decades to come even with no new advances in basic science. Inter-company alliances are formed more and more frequently for purposes of technology transfer to feed the fusion process.

Today's technology companies are "networked" for this purpose, and, as we shall see, are pressed to determine which scientific/engineering subfields to specialize in in-house, and which to source externally and "fuse." Indeed, in most industries, so many commercial advances are due to fusion that no single company can maintain the wide breadth of expertise required to compete. Alliances are here to stay.

In class discussions, the author's students have been unable to name any technological advances that are *not* the result of fusion, and this casts doubt on whether technology fusion is really a deep concept. Yet there is no doubt that it is a more widespread and critical phenomenon than in earlier times, and Kodama (see box below) draws some important management principles from the idea of technology fusion.

Technology fusion should not be confused with technological convergence. Technological convergence occurs when different technologies, which previously provided disjoint sets of user benefits, evolve to the point where they can provide an identical user benefit at comparable cost. Thus, convergence must be described in terms of two or more technologies, and one or more benefits.

For example, in the near future we will be able to say that:

• Movies can be delivered to the home by broadcast, by private satellite, by coaxial, via power lines, or by telephone line. This demonstrates convergence of the modulation/carrying capability of the various transmission modes.

• These movies can be viewed on a TV or a PC. Similarly, the Internet can be surfed on a PC or a TV. Thus the convergence of TV and computing technology for a wide variety of endusers' purposes.

• Friends can converse at a distance by POTS[4], cellular, PHS, CB, or via other radio bands that until recently were reserved for ham operators. This indicates the convergence of telephony and radio for mass-market voice communication.

Technology Fusion and the New R&D

Fumio Kodama's HBR article focuses on how companies define and use research and development in order to maximize success. He articulates two R&D possibilities: 1) the "breakthrough" approach of R&D that replaces an older generation of technology; or 2) the technology fusion approach that focuses on combining existing technologies into hybrid technologies. Kodama states that companies need to use both kinds of technology strategies in order to succeed by blending technological advances from a variety of industries. He also states that many Western companies still rely almost exclusively on the breakthrough approach, while leading high-tech companies in Japan are succeeding through use of technology fusion. According to Kodama, there are three basic principles essential to technology fusion.

1. The market drives the R&D agenda, not the other way around. Technology fusion begins with a new understanding of market needs, a step Kodama calls "articulating demand," in which market data is translated into a product concept, then broken down into a set of development projects. Kodama's example is Fanuc's development of the computer/numeric control (CNC) machines, in which Fanuc identified a market need that was not being met and broke the problem down into a series of projects. One of these benefited by Fanuc's close connection with Fujitsu, giving Fanuc access to technology that led to the development of a lucrative new product based on these fused technologies.

2. Companies need intelligence gathering capabilities to keep tabs on technology developments both inside and outside the industry. Companies need to gather

---

4   POTS: Plain old telephone service; CB: Citizens' band radio; PHS: Personal handyphone systems, first introduced in Japan, are known in the U.S. as Personal Communication Systems (PCS).

intelligence across a broader spectrum than the limited channels that have been generally considered to date – and on more than the obvious competition. In most cases, obvious competitors use similar technologies and production systems, so the information gathered has only limited value. What Kodama calls "invisible competitors" are companies that are unfamiliar and possibly unknown, and may in fact be companies outside the industry that possess technology that could become a threat if turned to new markets. An example is the threat to small form-factor disk drive makers from flash memory semiconductor companies.

3. Technology fusion grows out of long term R&D ties with a variety of companies across many different industries. Kodama calls this principle cross-industry R&D, and uses as examples Japan's high-tech companies that are aggressively diversifying their technologies. For example, Asahi-Kasei, a textile producer, is applying its fiber technology to building materials and medical applications. In order to succeed, cross-industry R&D must be both substantial and reciprocal: substantial in that management makes a commitment to support the R&D work through the end of the project; and reciprocal so that both parties hold equal responsibility and authority for the project's success. Kodama's example is the evolution of Japan's fiber-optics systems, developed by a partnership of Nippon Sheet Glass and cable maker Sumitomo Electric Industries. This partnership resulted in stronger fiber-optic cables with improved transmission quality.

Kodama F (1992) Technology Fusion and the New R&D. Harvard Business Review July-August. Summarized by Cheryl Coupé.

1.12 Technology Mapping

As an easily understood "gozinto" diagram, a technology map (in its "antecedent" varieties – see Figure 1.10) is an ideal way to represent technology fusion graphically. A technology map can communicate complex developments to non-technical audiences. It can help in competitive analysis, guide early-stage market research, help structure alliance decisions, clarify technology transfer mechanisms, and help identify a firm's best core technologies. Figure 1.10 shows a classification of technology maps.

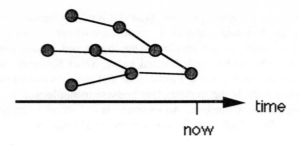

(a) Long time scale "Antecedent-Historical"

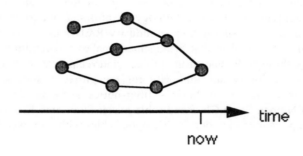

(b) Short time scale "Antecedent-Current"

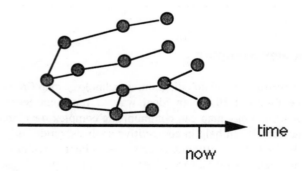

(c) Long time scale "Descendant-Historical"

(d) Short time scale "Descendant-Current"

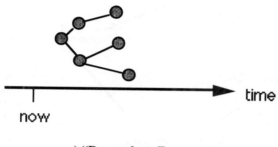

e) "Descendant-Forecast"

Figure 1.10 Classification of technology maps: Each node ⬤ is a technology.

Figure 1.11 is an historical/antecedent map showing the evolution of the electronic computer; for the full, fascinating story of these historical linkages, see James Burke's book, *Connections,* on which his public television series was based. Technology fusion was a part of this early history, but hundreds of years were required to achieve what in today's fast and globally competitive world must be done in months.

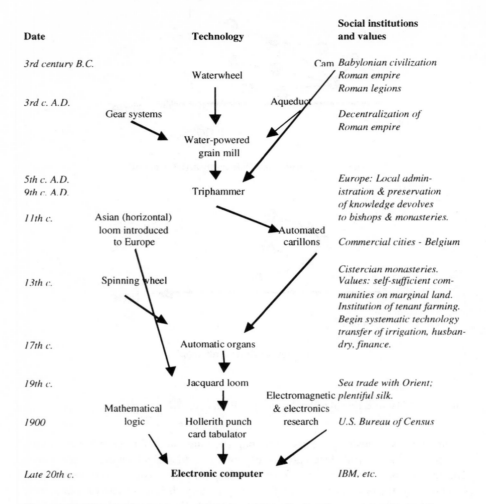

Figure 1.11 Historical technology map (Adapted from Burke, *Connections,* chapter 4)

One of these links, the triphammer (or water-powered cam) was a basic technology for its time, leading to several advances affecting many industries. Figure 1.12 is an historical/descendant map listing the triphammer's consequent technologies in rough sketch.

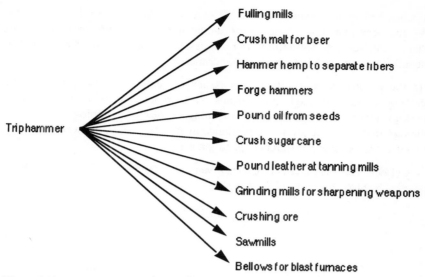

Figure 1.12 Descendant-historical technology map: The triphammer

---

The Dendrimer

A three-dimensional, man-made molecule, called a dendrimer, was developed by Dr. Donald Tomalia at Dow Chemical. In the early 1970s Dr. Tomalia began researching dendrimers, but had a hard time persuading his bosses at Dow to support his research. Quitting his job at Dow, Dr. Tomalia continued his research and formed a company in 1992 to commercially develop and market dendrimers. A Dow official confirmed publicly that "Dr. Tomalia did indeed discover the dendrimer technology." Dow Chemical decided dendrimers did not fit with its business, but acquired a 10% stake in privately held Dendritech.

Unlike a buckyball (another 3-D molecule that is – unless filled with a drug or other payload – an empty spherical shell of carbon atoms), a dendrimer is made of bonded atoms branching in all directions from a central point. Branching ceases at the maximum radius when the branching atoms are packed tightly. Many immensely beneficial dendrimer-based products seem possible. Dendrimers have properties that could let them, for instance, carry genetic material into human cells to repair birth defects. Dendrimers are believed to break down inside animal cells. In other application areas, an IBM researcher says, "Dendrimers have opened up the possibility of constructing electronic devices at the molecular level."

Dow Chemical is accustomed to making chemicals in bulk, but it takes seven weeks to form just a few pounds of dendrimer. Although "a kilogram of dendrimer can generate $50 million in revenues," Dow would have had to change its manufacturing processes to produce dendrimers.

While the dendrimer story illustrates issues of technology strategy and the generation of breakthrough innovations, it is placed in this discussion of technology mapping to illustrate the meaning of *basic* technology, that is, a technology that can generate many products, many companies, and even possibly many industries in several different application areas.

Naj A K (1996) Persistent Inventor Markets a Molecule. The Wall Street Journal Feb 26. Summarized by Mary Bourret and Liem Nguyen.

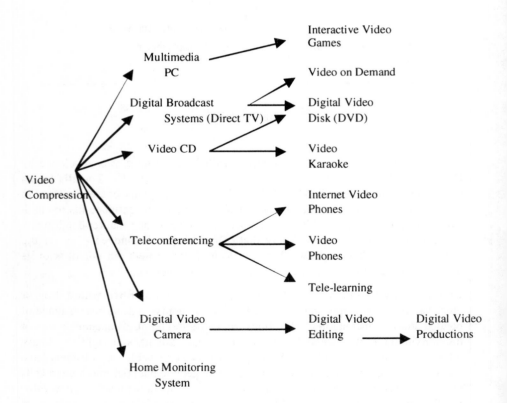

Figure 1.13 These descendants of video compression technology are either currently available, or will be available in the very near future. Contributed by George K. Chen.

Figures 1.13 through 1.14 show how variations on the basic technology map types of Figure 1.10 can be useful for a variety of business purposes. In particular, the forecast maps are sometimes called *roadmaps,* and are used to formulate targets for engineering milestones[5]. Further variations may note the companies involved in each product area, or even the people who have moved from company to company as an industry has developed.

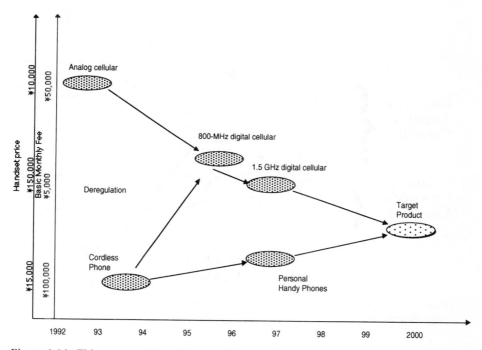

Figure 1.14 This map combines features of current and forecast maps, and incorporates dates, prices, and targets. It also notes a regulatory event affecting the evolution of the product. Data from Nomura Research Institute

A final figure, Figure 1.15, summarizes the main points of this chapter: The technology manager combines competencies from inside and outside the firm to produce new technologies, new devices, and ultimately new products that contribute to the firm's growth. Technology fusion occurs first when the existing competencies are combined to form new technologies. Also possible, but not shown in the figure, is external acquisition of technologies, of devices, and of

---

[5] SEMATECH's roadmap for semiconductor manufacturing technology can be downloaded at http://notes.sematech.org/NTRS/RdmpMem.nsf/Lookup/ntrs94/$file/ntrs94.pdf.

products (components) that occasion further fusion and find their way into the product development stream.

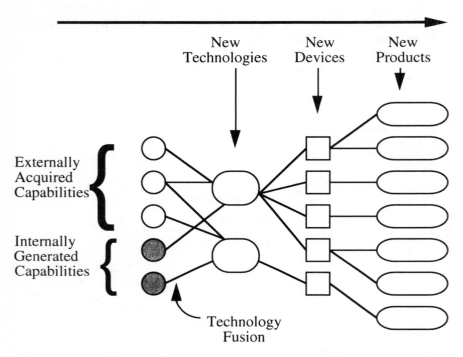

Figure 1.15 Summary of technology generation, acquisition, fusion and evolution into profitable products

1.13 Notes

"Reverend Malthus." Samuelson P A (1970) Economics. McGraw-Hill New York (8th Edition) 28

"Club of Rome." Meadows D and D Meadows (Ed) (1972) The Limits to Growth: A Report for the Club of Rome's Project on the Predicament of Mankind. Universe Publications June

"Another researcher divides innovations..." Goldberg A I (1997) A Structured Approach to Market Feasibility Studies for High-Tech Products. Proceedings of PICMET'97: Portland International Conference on Management of Engineering and Technology (IEEE Engineering Management Society) July 892-895

"computer industry as a link in a 'food chain'..." Young R (1994) Silicon Sumo. IC^2 Institute, University of Texas at Austin

"James Burke's book Connections." Burke J (1995) Connections. Little Brown and Company (revised paperback)

"IDC and Dataquest report that first-quarter PC sales in 2000 were 15-20% above..." Hamilton D P (2000) Growth of PC Market Continued In First Quarter: Research Firms Find. Wall Street Journal (April 24) A3

1.14 Appendix: The IC^2 Institute and the Austin Technology Incubator

The IC^2 Institute of The University Of Texas at Austin is a research center for the study of innovation, creativity and capital – hence IC^2. The Institute was approved by the University administration in 1977 to study, analyze, and report on the enterprise system through an integrated program of multidisciplinary research, conferences, symposia, and publications.

The Institute grew out of a recognition by selected faculty members of the College and Graduate School of Business that the United States was undergoing a fundamental transformation. America's economic infrastructures, social institutions, and international competitiveness were changing. These changes were affecting the relationship of the public and private sectors, and were having an impact on the way business, government, and academia interacted with one another.

Understanding the scope, nature, and direction of these changes required a new type of research organization – one that could rigorously assess the impact of two key societal drivers – technology and ideology; bring new analytical methods to bear on problems affecting the nation, individual states, local communities, academia, and business firms; and evaluate issues relating to the viability of emerging industries, growth and survivability of business enterprises, and the role and purpose of private-public sector institutions.

Six characteristics distinguish the IC^2 Institute as a unique research organization: Dealing with unstructured problems. Developing multidisciplinary teams. Going beyond functional boundaries. Linking theory and practice more effectively. Providing opportunities to think anew within a university environment. And transferring research results to other institutions.

The IC^2 Institute is a self-governing body of scholars reporting to the Office of the President of The University of Texas at Austin[6].

---

[6]   This description of the IC^2 Institute is quoted from an early annual report of the Institute.

The IC^2 Institute now serves as an interdisciplinary research center for U.T.-Austin, as a staging ground for interaction among academics and professionals from business and government, and as an interdisciplinary think tank on issues of business, technology and society. Much of the Institute's research is focused on the innovation and commercialization processes, that is, on how to accelerate the transformation of a scientific advance into a widely accepted product that will benefit consumers, businesses, or local or national economies.

But IC^2 is not just a think tank, it is a "do-tank." In the latter, activist role, the Institute has created a number of professional societies and other kinds of institutions that carry out experiments in innovation. Among these are the Austin Technology Incubator, the Austin Software Council, and the Entrepreneurs' Council. Each of these organizations provides case material for this book. Indeed, all three embody what has come to be known as "The Austin Experiment" and the "IC^2 Model" of technology-based economic development. They have been central to Austin's recently acquired reputation, bestowed by the national and world press, as a "technology hot spot."

The Austin Technology Incubator currently nurtures about 25 start-up technology firms, and has "graduated" about sixty companies that are now in fast-growth mode. Companies apply to enter the Incubator by submitting a business plan. Companies applying for entry include (i) the small start-ups owned by self-employed inventor/entrepreneurs, (ii) start-ups consisting of technologies and former employees that have "spun off" from large companies; (iii) new divisions and business units of non-Texas companies; and (iv) technologies and personnel that have "spun out" from U.T. The spin-offs and spin-outs usually involve license agreements with the parent entities that hold patents on the new company's technology.

ATI provides its tenant companies with reasonable rental rates (to date, ATI has not taken equity in tenant companies) and shared receptionist and office equipment. But most tenants say that the true value delivered by ATI is the expert business assistance provided by ATI's staff and consultants, the introductions to the Texas business establishment and to the industrialists that visit IC^2 from many nations, and the opportunity to commiserate daily with others in the entrepreneurial predicament.

Further value – for entrepreneurs, students, and faculty – comes from the interaction of students and entrepreneurs. Students work as interns at ATI, and faculty teach courses at ATI. ATI entrepreneurs drop in on the academic courses to learn and to teach. They serve as "living cases" for the courses. Students learn from working on projects important to the incubated companies, and the companies benefit (more often than not!) from the students' involvement.

1.15 Questions and Problems

Short Answer (Answers should be two or three sentences.)

S1-1. (a) Name some current innovations that might help preserve civilization, and others that might help destroy it.

(b) Are hard disks a hi-tech product? A tech product?

(c) Which of the following are likely to be a continuous innovation? A discontinuous innovation?
- a new model of an existing product
- a new product in an existing product category
- a new product category

S1-2. The chapter lists four types of technology products.

(a) List at least three products of each of the four types.

(b) What role can each type play in the ongoing profitability of a company?

S1-3. The chapter lists seven barriers to the acceptance of an innovative technology. Provide four more examples of innovations or products that faced barriers to acceptance. For each, which barrier was most significant? Who or what group was the resisting party? Was the barrier finally overcome?

S1-4. Draw a picture of a purchase probability vs. time curve that shows the cessation of sales of an obsolescent product.

S1-5. How do "technological convergence" and "technology fusion" differ? Give an original example of each.

S1-6. (a) Are there circumstances in which breakthrough R&D processes are more appropriate than technology fusion? What are they and why?

(b) How would you go about identifying and researching the "invisible competitors" discussed by Kodama?

(c) What would be some of the difficulties or issues involved in cross-industry R&D?

S1-7. Consider the laser to be a "technology." Name a "device" and a "product" that have resulted from laser technology. What physical and informational packaging distinguish the product from the device?

S1-8. (a) A technology that starts an entire industry (and thus many product categories and a plethora of individual products) is called a basic technology. (This use of the word "basic" is different from its use in the phrase "basic research.") Name a basic technology, and indicate the industry/product growth that it has generated.

(b) Name a technological advance with less far-reaching impact, i.e., a non-basic advance.

Discussion Questions (Answers should be one or two paragraphs.)

D1-1. (a) Dow Chemical rejected early research on dendrimers, which turned out to be a truly basic technology. Why? Was Dow wearing blinders that did not really relate to batch speed? If so, what are they?

(b) What industries were spawned by nylon, polyethylene and plexiglas?

D1-2. The world population is increasing with great rapidity. Is it realistic to think that any advances in technology – even assuming fast, efficient diffusion/commercialization – can make a difference in the resources-per-capita equation? That is, can technology industry really ease human misery on a global scale? Distinguish myopic (exploitative) gains from sustainable gains.

D1-3. If you are working in high-tech, remark on whether you entered the field due to idealistic motives; by accident; in order to get a well-paying job; to work with interesting technologies; to make money in an entrepreneurial venture; or for other reasons. Talk this over with classmates, and write down some observations about the uniformity or the variation in their expressed motives.

D1-4. Can you think of any non-incremental innovations that are not the result of technology fusion? Why does tech fusion present different problems to firms today than in earlier times?

D1-5. The OECD (Organization for Economic Cooperation and Development, to which the "first-world economy" countries belong) defines a "high-tech" company as one with an R&D budget exceeding 3% of sales. Using public databases, find the R&D expenditures (as a % of sales) of a chip manufacturer (e.g., Intel); a computer manufacturer (e.g., Dell); and a major consumer goods manufacturer (e.g., Procter & Gamble). How does the OECD definition compare with current common usage of the designation "high-tech"? What are the limitations of OECD's definition? How do the three companies' R&D ratios support your answers to these questions?

Miniproject (Miniprojects require 1-4 pages of graphics and narrative.)

M1-1. Draw a technology map (preferably a current or forecast map) showing the technological antecedents or descendants of a product that is of interest to you. Try to include 10-15 antecedent/descendant technologies. We have classified tech maps in the text; what kind is yours? Does it help you understand the technology business area to which it applies? If so, how?

"Revenue generating is not the most important consideration in the market development phase. The most important activity in this stage is *learning*. You need to find: what product configuration is most in demand, who the best customers are, how customers actually use the new product, how they perceive the product's benefits, [and] how to best distribute the product."

J.W. Taylor, *Planning Profitable New Product Strategies*

"You will."

AT&T

2 Technology Life Cycle and Market Segmentation

The picture of the technology cycle in Figure 2.1 begins at a point in time (marked "technological innovation") where a device exists. Other companies have complementary devices; for example, one firm may have set-top home multimedia boxes, another firm having movie-on-demand databanks, and a third company having to-the-home cable infrastructure. At this point, each firm begins a low-level investment[1] to investigate the possible profitable alliances and the potential market size. This potential is called "latent demand" because there is no demand in the economic sense for a product that does not exist, and consumers cannot articulate their desire for something they have never heard of (more about this in Chapter 9). But the wise innovator begins to "warm up the audience" even before the product development cycle begins. This is what AT&T has done with the YOU WILL campaign. It is interesting that this kind of advertising is different from traditional corporate image advertising. Image advertising differs from product advertising in that it promotes the company as a whole, and by extension, its entire current (and perhaps future) product line. The YOU WILL campaign, in contrast, prepares the customer for a particular technological direction the company will take, though the form of the product and the identities of the companies in the alliance that will produce it are still unknown.

When a specific product is envisioned, significant development investment begins. This investment is not recovered until a significant interval of time following the

---

[1]   Intel devotes "serious resources" to technology development five years before a resulting product will go into manufacturing. The long leads, according to Dr. Ken McQuhae of Intel, allow the company to study the supply chain and customer chain, and remedy any capacity bottlenecks.

introduction of the product to the market. By this time, although it is not shown in the Figure, development has begun on the second generation of the product. During the development cycle, further market research elucidates consumers' desire for the innovation, and estimates demand – demand being the combination of desire and ability to pay.

The market cycle begins when the product is launched. At first, customers respond to the producer's marketing messages ("technology push"). Later, in a "demand pull" phase, customers buy because they have seen earlier customers buy. After initial demand growth begins to level off, the product may become a commodity with little differentiation from competitors' products. Success in this phase, if there is to be no repositioning or further differentiation, lies in dominating the distribution channels for such products.

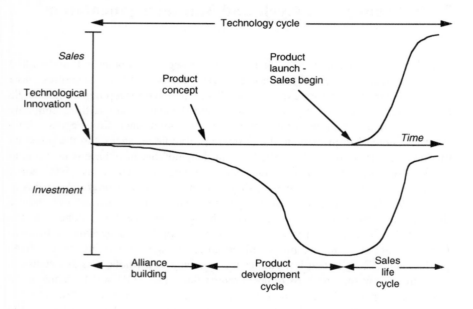

Figure 2.1 Some phases and milestones in the technology cycle. Not shown are the eventual decline in product sales, and the overlapping development of subsequent products from the same core technology.

The cumulative penetration curve introduced in Chapter 1 is ubiquitous. Examples can be seen in the business press almost every week, and each follows the familiar S-shaped curve: microwave ovens; internet connection sites; cable TV installations; color televisions. Here are two examples:

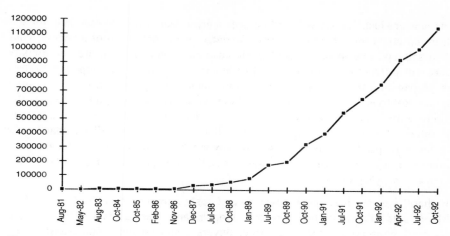

Figure 2.2 Early growth of connected internet hosts (Source: SRI International)

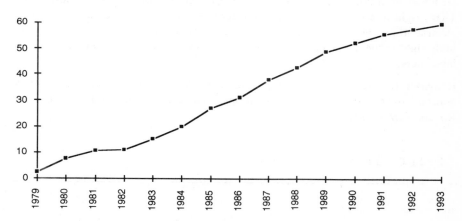

Figure 2.3 Early household penetration growth, microwave ovens, in % (Source: International Microwave Power Institute)

2.1 Stages in the Life Cycle: Buyer Segments

Each stage of the demand life cycle corresponds to a discrete buyer segment. In a strange twist of terminology, the earliest buyers are not the so-called *early adopters*. The earliest customers for a new-to-the-world product are called *innovators*. This customer segment has a high interest in the benefits brought by the new product, and are willing to devote a great deal of time to searching for and learning about these benefits. This, of course, is the definition of a "nerd"; your first sales will be to people within the customer company who are not socially active or influential. Although word-of-mouth and imitation are important factors in propagating trial purchases between the later adopter segments, there is little buyer-to-buyer communication to help you bridge the gap between the "innovators" and the next segment of triers, the *early adopters*. It is a unique problem of new-to-the-world products that they are more likely than other technology products never to sell to anyone but "innovators." Because the innovator segment is tiny, they cannot sustain your product, or your hopes for your company's growth.

If there is a gap between innovators and early adopters, there is (as Geoffrey Moore puts it) a chasm between early adopters and later adopters. The latter are far more conservative than the earlier buyers – and far more numerous! They represent the true profit opportunities. It is difficult, though not impossible, for an unknown entrepreneurial startup to sell to them. In the exercise at the end of the first chapter, my students thought American Innovations Corp.'s difficulties stemmed from not knowing how to change the culture of their target customers. That's right in a way, but it would be truer to say that the chosen market – large electric utility companies – is so conservative that AI Corp. was in effect trying to jump right to the late adopter stage of the life cycle without first enjoying sales to innovators.

Remarkably, the proportion of people in each adopter segment has proven constant over many kinds of innovations. Table 2.1 is adapted from Everett Rogers' *Diffusion of Innovations:*

Table 2.1 Characteristics of Adopter Segments

Segment	% of population	Description	Associated LC Curve Interval
Innovators	2.5%	"Venturesome"	Introduction
Early Adopters	13.5%	"Respected"	Early Growth
Early Majority	34%	"Deliberate"	Take-off; Fast Growth
Late Majority	34%	"Skeptical"	Maturity
Laggards	16%	"Traditional"	Late Maturity; Decline

Innovators are younger on average than the later adopters, and make the most use of impersonal information sources outside their immediate circles of colleagues. Early Adopters have the highest social status, the greatest opinion leadership, and the greatest contact with change agents within their immediate circles. The later adopter segments are increasingly conservative in their approaches to change, and have less and less contact with individuals outside their immediate neighbors, with the laggards being extreme in these regards.

Most college marketing texts lay out the different strategies that are appropriate for selling to the successive customer segments which correspond to the life cycle stages. Later in this book, especially in Chapter 8, we will focus on the segment strategies and tactics that are unique to advanced technology marketing.

Free PCs

In 1998, a company called free-pc.com launched a program to give away ten thousand PCs. Not to charities, but to ordinary households. That was also the year in which U.S. household penetration of PCs was just equal to 50%.

These were not state of the art PCs, nor were they completely undesirable machines. Winners of free PCs agreed to give screen space to advertisements and to pay for an Internet Service Provider (ISP) contract at a price that was not really the lowest available on the open market. Ads were for traditional consumer products, not for far-out dotcom services. Users' Internet experience would be somewhat controlled by free-pc.com, made something more like watching television.

Table 2.1 shows that households in the second 50% to acquire PCs would be quite conservative, delaying the purchase of a computer – which is generally the most expensive tangible item a consumer household can buy, after a house and car – until comfort levels are very high. free-pc.com used price (you can't beat free), familiar ads, and the television metaphor to achieve this comfort level.

Did FreePC.com's strategy effectively generate sales for its advertisers? That is, did the late adopters engage in online purchasing? Was the price of the ISP service exploitative, or was it a fair exchange for making the online world comfortably available to late adopters? While we debate these questions, new free PC offers proliferate. There is no doubt that free-pc.com cleverly applied the market life cycle principles to pioneer a creative marketing strategy.

Holland K (1999) (Ed) The Great PC Giveaway. Business Week (February 22) 45; Martinez B (2000) Lender Finds Free PCs Have Potent Allure. Wall Street Journal (May 18) B1

2.2 Modelling New Product Growth

Many s-shaped curves, like the logistic model shown in Chapter 1, can roughly capture the diffusion (cumulative growth) of a new technology or a new product. The Math Review at the end of this chapter gives a comprehensive look at growth models and their underlying concepts. But the model most often used for technology based products is the Bass model and its variations. The model is

$$x_t - x_{t-1} = [p + qx_{t-1}][1 - x_{t-1}].$$

The number of new buyers at any time t is equal to the cumulative penetration (total first-time buyers) at t, minus the cumulative penetration at time t-1. The Bass model says that the number of first-time buyers of the new item at time t is a certain fraction of 1-x(t-1), the number that have not yet bought. What is that fraction? It is [p + qx(t-1)], a quantity which we will now deconstruct.

The parameter p is a coefficient of persuasion, reflecting the degree to which potential buyers are swayed by advertising and other non-peer influences. The other parameter, q, is a coefficient of imitation, reflecting the degree to which potential buyers want to "keep up with the Joneses," that is, the extent to which they are influenced by their peers' prior purchases of the item. Clearly, the effect of p is independent of how many people have purchased at any given time. But the effect of q is greater when *more* peers have already purchased. For that reason, q is multiplied by x(t-1) in the fraction we are looking at. The fraction of the remaining population who will become new adopters at each time period t is thus a function of the total persuasion effect p, and the total imitation effect

$$qx_{t-1}.$$

It has been argued that qx_{t-1} is not a sufficient measure of imitation effect in electronic communcations products. After all, the first person to buy a telephone had no one to talk to on the phone! It is simplistic to say that 100 phone subscribers gave that person 100 people to talk to, because conference calling is available, and because all subscribers benefit from the more rapid communications made possible by the phone conversations of others. So we might say each additional subscriber makes subscribing to the telephone look more attractive by a factor of more than one. If attractiveness is to increase supra-linearly with the number of existing subscribers, the imitation effect might look like

$$q[x_{t-1}]^2$$

The resulting diffusion/adoption curve is still s-shaped, and has been claimed to fit some telecomm products better than the original Bass model.

All the references to "purchases" in the above paragraphs are equally applicable to the other signs of technology adoption, like first use, signing up, signing off, and signing on – and the Bass model is a valid reflection of each of these.

2.3 Are There Technology "Products" Any More? [2]

A recent demonstration at a large high technology company was perhaps more revealing than intended. The capability we were shown provided clear benefits for customers, was currently being sold in a shrink-wrap box containing a circuit board and some software, and had a brand name. But one marketing manager called it a "set of products," another called it a "technology," and a third remarked, perhaps inadvertently, that it was "going to be a product." After the demonstration, one of us gently asked a marketing executive, "If the marketing group doesn't agree on whether this is a product or not, how will the customer possibly comprehend it?" The executive replied, "Yes, but by the time we all agree on it, the next release will be out, and we'll be back where we started."

Time is required to conduct one or more new technologies through the intermediate stages that precede a marketable product. Product life cycles have been and continue to be getting shorter than ever before, particularly for high technology products. Compression of sales cycles (and the attendant compression of development cycles) is driven by rapid introductions of successive releases, and new and substitute products. It is not because of a short span of enthusiasm on the part of customers. Indeed, technology may change much faster than consumer psychology and needs.

This is why some technology marketers have begun to market earlier in the technology cycle, often before a product has been specified concretely. Among the less salubrious results is "vaporware." Rather more constructive is the "You Will" campaign of AT&T. The AT&T ads feature a leading question concerning future ways of communicating ("Have you ever attended a meeting from the beach?") superimposed on a machine with indeterminate features. The tag line, "You Will," is meant to warm customers up to the idea of communicating in this

---

[2] Material in this section is adapted from Phillips F, Ochs L and M Schrock (1999) The Product is Dead; Long Live the Product-Service. Research•Technology Management July-August 42(4):51-56 (The authors gratefully acknowledge the creative input of Joe Ellertson of Tektronix, Inc. We benefited also from conversations with Dr. Deb Chatterji of the BOC Group and Mr. Peter Coldrey of Sola International, Inc. An earlier version of this work appeared in (1997) PICMET '97: Proceedings of the Portland International Conference on Management of Engineering and Technology July)

new fashion – even though AT&T clearly does not have a product ready for market.

Traditional Implementation/Commercialization Path

Competence (general)	Technology (generic)	Device (specific)	Product (salable)
Domain knowledge and integrative skills applicable to scientific, engineering, and business processes.	Knowledge, skills and machines applicable to many engineering and product areas.	A machine that performs one or a few functions, in the lab or shop.	A device made salable by addition of reliability and customer testing, design, packaging, positioning, and, increasingly, service.

Evolution of Competitive Advantage

Figure 2.4 Superior products and marketing were once seen as the only sources of competitive advantage. Now, the firm's competencies – which lead to expert service, institutional flexibility, and a continually updated product line – are recognized as the true source of competitive advantage.

Almost all tangible products have a service component. For high tech products requiring much user learning, the service component is a significant fraction of the total package. That package includes the physical product, its packaging and advertising message, its distribution, its setup, its after-sale support, and much more. The increasing role of the service component is evident to consumers, for example, in the form of telephone support for software and PC products. It is evident to industry executives, for instance, in the outsourcing of data processing services.

The perceptive marketing executive quoted earlier acknowledged that the whole question of product identity could be bypassed if the benefit were delivered directly to industrial customers by knowledgeable service consultants. Indeed, the brand could then be associated with the benefit rather than with the particular circuit board, giving the company greater leverage from the brand name. The obstacle to this was the shortage of knowledgeable service consultants. "The average street-corner VAR[3] ," said the executive, "is not sufficiently knowledgeable. The big six accounting firms [specifically, their systems integration divisions], while knowledgeable, are too few and too much in demand as resellers..."

If there were no shortage of service consultants, the growing service component of the high-tech tangible product could be allowed to dominate the tangible

---

[3]   Value-Added Reseller.

component, from the enduser's point of view. We could then proclaim, "The product is dead." Because product is one of the most venerable concepts in marketing[4] , this is a true watershed event in business management. Yet dispensing with the idea of product is one of the few remaining ways – if not the only way – to compress the technology cycle still further.

It has been said that customers do not need quarter-inch drills; they need quarter-inch holes. With the death of the tangible product, customers can have "quarter-inch holes" when and where they need them. The holes – that is, the benefits once provided by "products" – are delivered by professional service people who are conversant in the product technology and in the customer's business. We call these professional service providers/application domain experts "Expert Service Providers" (ESPs).

This section presents further evidence for the death of the high tech tangible product, and explores the implications of the thesis. The evidence is based on trends seen in the literature, and on a survey conducted by the authors. The implications include the benefits to producers and customers of eliminating the tangible product, the economics of the ESP professions, and imperatives for educating these consultants.

To say "The product is dead" is of course somewhat hyperbolic. Companies still think of themselves as vendors of products. But the balance of tangible vs. service offerings is now tilting toward services more markedly than ever, and this is true in a wider range of industries than ever before (particularly mid-range industrial products and consumer technology products) – justifying, we think, the dramatic headline.

2.3.1 Evidence for the Death of the Product

The shift is making itself evident in the vocabulary of industry salespeople. Videoconferencing provider PictureTel speaks of its "desktop solution" to distinguish it from the company's "full-size offerings." Gross and Coy note the growing use of the word "architectures" to mean grand schemes for integrating many products into smoothly functioning systems. Words like "platform," "standard," and "environment" are often heard as substitutes for "product."

---

[4]   *Kohler's Dictionary for Accountants* defines "Product" as "A good or service resulting from an operation or series of operations." Neither Cunningham and Cunningham's nor Lazer's marketing textbooks formally define "Product," although Lazer notes descriptively that "A product is an entire offering that a company makes to the marketplace, as perceived by customers.... [Products are] 'attractive propositions.'" Surprisingly, a search of current texts in *Production and Operations Management* and the *Penguin Dictionary of Economics* turned up no alternative definitions. Lazer's definition foreshadows our argument that the product as a discrete set of features is dead.

The same authors note, "To help counter [illegal copying by consumers], software publishers know they must give customers something more than a platter of bits. Makers of mainframe software are stressing customization or 24-hour maintenance agreements." Smaller software entrepreneurs, lacking sufficient marketing resources to keep their unique packages salable, instead use them as internal resources to back niche consulting businesses. As for hardware, says Roger N. Nage, deputy director of Lehigh University's Iacocca Institute, "Those who offer attractive packages can virtually give away their hardware.... Tomorrow's factories will sell customer gratification, not things." Notable in this regard, but hardly unique, are the cellular phone companies using the "Kodak strategy" to give away phone instruments and make money on air time, paging services, and calling plans.

Mentor Graphics (Wilsonville, Oregon) sells approximately 250 software tools that are used in the design of integrated circuits and printed circuit boards. Mentor also integrates numerous third-party tools. Several years ago, according to one manager there, Mentor produced over 100 "release events" in one year; the customers cried "Uncle!" He says, "We have since backed off, but customers still have trouble keeping up with the new versions." One of Mentor's responses has been to develop a Professional Services Division, which includes consulting, training and custom design services. Since its beginning less than five years ago, PSD's revenue has grown to $51 million in 1995 (about 13% of total revenue). It is one of the fastest growing segments of Mentor's business.

At Mentor Graphics, a concept called "The Whole Product" goes far beyond what is traditionally meant by "product." It includes not only the tangible product, training, hotline support, marketing, sales, etc., but also the many things that constitute the relationship with the customer. Phil Robinson, Vice President and General Manager of the Customer Support Division, believes what Mentor provides to customers is the very ability for the customers to build their products. This includes ongoing tool and database compatibility, building coalitions among vendors to solve customers' problems and the knowledge to recommend products and services, even if they are competitors' offerings.

2.3.2 Reasons for the Death of the Product

Several converging trends herald the death of the product. These trends involve life cycle effects, heightened competition, and advances in information technology.

• The commoditization or maturing of electronics and semiconductor technology, and the resulting squeeze on margins, mean that the quest for "value-added" has migrated to expert service.

• The basis of competitive advantage has shifted from the individual product to its underlying architecture, and a firm's ability to flow new technology into its subsystems and rapidly tailor them to the needs of specific customers. Today the battlefield is product platforms and not single products.

• Joe Ellertson, a manager at Tektronix Inc., remarked to the authors that many companies in technology based industries choose to use the product as an organizational principle; the structure of their business is put in place to most effectively support the product. Yet evidence accumulates from many industries that companies that focus and organize around the customers' needs are more successful than companies that cling to the internal, product focus. Most commonly, marketing managers are given responsibility for major accounts or market segments whose requirements are met by a mix of products and services.

• By increasing their emphasis on relationship marketing and alliances, companies aim to lock customers into streams of products. A recent study of industrial purchasing suggests that satisfaction with the purchase process – which arises from excellent service – has more of an influence on the likelihood that the customer will buy future products from the same supplier than does satisfaction with the product itself.

• Mass customization, using computers and automated production to provide individually tailored products at costs consistent with mass produced products, enables companies to differentiate their offerings. But if every customer receives something different, how can a "product" be advertised?

• Object-oriented technology means that software components can be assembled rapidly into an enormous variety of functional capabilities, making it harder for any one of these functionalities to take on firm identity as a product.

• The short market life cycles of individual products lead producers to emphasize brand over product in order to economize on corporate communications and to avoid overloading customers with information about product changes. As one observer noted, "This industry was built on product advertising." Nevertheless, a recent survey indicates an increased emphasis on brand marketing within the high tech sector.

• The evolution of knowledge-based business, according to Davis and Botkin, means that "businesses will come to think of their customers as learners and of themselves as educators." Businesses, say Davis and Botkin, profit from making smarter products (like self-diagnosing fuel injectors, and home appliances that communicate via the Internet) that better help customers. Smart products "both oblige and help" customers to learn. If the delivery of the needed knowledge is not yet automated, it must be delivered by human service providers.

2.3.3 Summary of Survey Results

We surveyed 41 managers and engineers at technology firms in the "Silicon Forest" corridor near Portland, Oregon, an area currently outstanding in the export strength of its semiconductor, electronics, software, and transportation/shipping industries. The opportunity sample respondents all held positions with responsibility for technology management, though in various senses of that phrase, and all were asked to respond with reference to "a particular product and its successive releases and upgrades." The survey's purpose was to confirm and measure the momentum of the trends that indicate the death of the product in high-tech industry.

56% said successive releases were coming out with increasing frequency, and 35% felt this rate was "faster than the rate that would let the customer fully comprehend the feature changes and added benefits." 30% said they were commencing marketing activity for each release earlier in the development cycle than was the general case three years ago – though 23% said "later in the cycle," perhaps indicating that early marketing effort is now focused on the brand or standard rather than the individual product release. When asked, "Does your company attempt to compress the time between releases by redesigning the service component for successive releases, while minimizing actual hardware/software changes?" 40% responded "yes," and 26% said "sometimes." 71% of both groups said this is happening more frequently than three years ago.

Three years ago, 68% of respondents used ESPs to help customers use products; today it is 82%. 42% agree that "More than was true three years ago, the identity of my product is now defined by the service component rather than the tangible component." 60% place more importance (now relative to three years ago) on brand awareness than on a customer's awareness of a given product, although 66% report that the status of product managers relative to brand managers at their company is "about equal."

62% of the products were industrial products (half of these were self-rated "high priced"), and 38% were consumer products (60% of these "high priced"). Primary customers were OEMs (23%), large industrial endusers (23%), small/medium industrial endusers (38%), and individual consumers or households (15%). 55% of all respondents market through VARs, but none use VARs without also using other channels. Nor do any use retail stores as their only channel. 31% use major account teams, and 19% use only customer direct (mail, phone, online) channels; 50% use more than one channel.

Use of ESPs was highest among producers of low-priced industrial products, consistent with the effect of commoditization, and was highest among those selling to industrial endusers. Presumably, OEMs employ specialists to understand the components they are buying.

2.3.4 The Experiences of Other Companies

Sola International, Inc., is a maker of spectacle lenses, many of which combine new materials with advanced lens design. Sola's VP of R&D remarks that "The company runs the risk of overwhelming customers with a wide range of new products. At trade shows, customers sometimes don't even realize a new product has been launched. While we have two or three major products that we have been able to communicate well, in general we are introducing products faster than they can be effectively promoted." Netscape's chief technology officer, referring to the recurring three-to-four month upgrade cycles of his company's web software, says, "So much, so fast, with whizzy-cool features, has left customers with battle fatigue." In 1998, Sun Microsystems will slow its Java upgrade cycle for the same reasons. Microsoft's Windows NT Version 5.0 will not be released until it is fully debugged and tested, and "customers, partners, and software vendors have reacted positively to the move." Sola, Netscape, Sun and Microsoft must decide whether to slow their product introductions (risky, if competitors learn to satisfy customers' new needs faster and with less confusion), or acknowledge the death of the product, de-emphasizing the identities of individual releases and building the profitability of product-service packages.

General Electric will realize 62% of its income from services in the year 2000, compared to 38% now. The Wall Street Journal reports that CompUSA, OfficeMax, and Tandy Computer City find it pays to help consumers and home businesses install and upgrade their PCs. It pays $80 to $100 per visit, in fact – with a profit margin of up to 50% – and customers return fewer computers. A service venture of Xerox and OfficeMax earns "several million" annually serving these markets. Other electronics retailers emphasize training classes, phone help lines, or free installation of upgrades in the store.

2.3.5 Implications for Managers

Clearly, we do not overwhelm customers with product releases in order to serve customers better. By considering the idea of the death of the product, those of us who are producers can examine our motives: Are we overloading the customer due to fear of competition? As an attempt to accelerate revenue or establish industry standards? To compensate for a shortage of trained ESPs? Is this the right way to go, how will customers respond, and what are the long term consequences?

Buyers of extremely expensive industrial equipment have always expected high levels of service. Our survey shows that this expectation or need now applies also to mid-range industrial products, and also to some consumer technologies.

Our survey respondents rated on a 5-point scale (1=strongly disagree, 5= strongly agree) a number of potential benefits to the producer of bringing the product's benefits directly to the customer via ESPs. These were "reduce the need to communicate (to the customer) the features and benefits of new releases"; "reduce the time needed to prepare a new release"; "reduce the need to crisply define the 'product'"; and "reduce the time needed to recover the investment in a new release." Agreement and disagreement were evenly split among all these (means between 3.00 and 3.04) except the last, for which the mean rating was a slightly higher 3.34.

ESPs help implement the expanded concept of product that is necessary for today's technology company. ESPs should be trained to:

• Increase customer satisfaction. At Mentor Graphics, ESPs have contributed to the company's high ratings on customer satisfaction surveys.

• Convey good ideas for new products. Significantly, Mentor's ESPs bring customer ideas and needs back to the software designers, and are active members of the integration teams.

• Create good ideas for new products. Exposure to the customer's environment gives the ESP ideas for new products to solve the customer's problems.

• Create new applications within the customer's environment that will increase pull for the tangible product(s).

• Cushion the risk of missing a market window. ESPs provide the customer with the benefits of the earlier and the later release of a product, regardless of the formal release date of the later version.

To fulfill these roles, an ESP must be creative, flexible, and skilled in the industry's research and application. Issues still to be addressed satisfactorily at Mentor and other firms include finding and training good ESPs, managing the knowledge they hold, making sure the knowledge is shared with fellow ESPs, automating this established knowledge so the ESPs can leverage their time to stay on the leading edge of new knowledge, and maintaining the application domain knowledge that eventually erodes among ESPs that have been hired from domain firms into Mentor. Few engineers leave college knowing how to do customer service; this is an opportunity for universities to develop courses in "customer engineering."

There are three broad implications for R&D managers we would like to highlight. They are, first, the growing scope of product and technology development; second, the risk of overwhelming customers with product releases; and finally, the potential benefits of ESPs.

The scope of product and technology development is expanding, driven by the increasing value of customer service in the total product offering. Indeed, as R&D managers we now must focus on and actively shape our firm's "value delivery process", a process that encompasses the traditional product development process and its underlying technologies, along with the firm's customer service and support processes. This demands excellent cross-organization ties and teaming, close working relationships with customers, knowledge of the customer's key business processes, and application domain expertise to deliver distinctive customer value in the area of service. It necessarily demands engineering customer service with the same degree of attention that we've historically paid to product capabilities and performance specifications. R&D must instruct customer service representatives not only about what is currently offered, but also about what is possible.

Moving new capabilities quickly to market has been an imperative for R&D in recent times. But if rapid commercialization confuses the customer, it is not good for the company. This is true even, or perhaps especially, if your competitors are confusing customers with rapid product upgrades. In such cases, it is better to transfer the innovation to the brains of the ESPs than to burden the customer with it directly.

Through much of this century, we have sought to replace skilled labor with less expensive factors of production. Qualified engineers are indeed in short supply, but analysis of the technology cycle shows that time is even scarcer. The pendulum swings back, and we see the relative advantage of relying on ESPs.

"Product," in the past, meant something that a company makes, puts in a box, ships to a customer and, barring defects in material or workmanship, never sees again. What is proposed to replace "product" is a new meld of service component and tangible component, the latter having a feature set with a fuzzy definition. What the customer receives is everything necessary to be successful.

2.4 How Does "High" Technology Become Just Plain Technology?

Figure 2.5(a) below shows the life cycle curve of a product that was "revitalized" after its initial market had matured. This is by no means an unusual life cycle pattern. In fact, revitalization may occur several times over the LC, resulting in the "scalloped" LC curve shown in Figure 2.5(b).

There are three classic strategies for revitalizing a mature product:

- new niches
- new technologies
- new uses.

Apropos of the "new niches" strategy, in October, 1993, the *Wall Street Journal* reported "Compaq Computer Corp. has a novel idea for selling personal computers: pitch them to women."

For advanced technology products, we might add a fourth strategy: the creation and observance of industry standards.

The PC market is now in its first maturity. Revitalized take-off cannot occur without one or more of the above strategies. Multimedia computing represented a new use; handheld PCs are a result of newer, miniaturized technologies that will lead to new uses. Can these, and other revitalized products, be called new-to-the-world products in the sense of this course?

Is the second take-off in Figure 2.5(b) similar to the first? In other words, are the marketing challenges the same? In general, they're not. The first take-off is that of a new-to-the-world product category. No one has heard of the product, nor heard of anything *like* the product. The second, post-maturity take-off is usually a new product within a familiar product category, and so does not present the same marketing challenge or target the same avant-garde innovator market. However, the differentiated product may be *so* differentiated that its take-off is similar to that of a new category. The PC category has spawned Personal Digital Assistants (PDAs), but I think PDAs will be one of these exceptions that must be marketed as a new-to-the-world product.

If we allow the mass market to qualify as a new "niche," then Figure 2.5(c) is a good representation of the LC of VCRs. Recall that VCRs were first targeted only to studios and were not intended for the home market. The post-revitalization portion of the curve accounts for far more sales than were possible in the first growth stage. This shape is not unusual for an LC curve, either. New distribution channels might qualify as a fifth strategy for revitalization. For example, a maker of auto accessories sold through auto parts stores might become a supplier of original equipment to the auto maker. This could easily result in a curve like that of Figure 2.5(c).

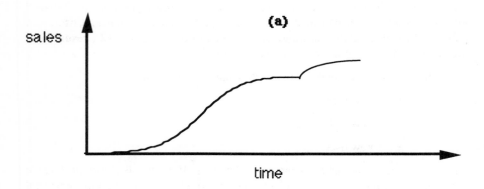

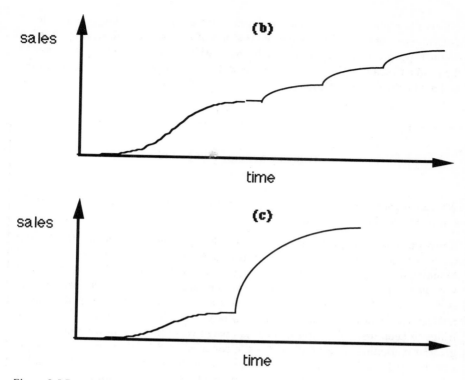

Figure 2.5 Renewing/revitalizing the market life cycle

The general shape of the life cycle is inexorable. The progression of buyer segments and their characteristics (innovators, early adopters, early majority, late majority, laggards) is unchangeable. But the specific slope of the curve for your product – how much *time* you must spend transiting each of the LC stages – is to some extent under your control. An inspiring example is provided by Japanese makers of consumer electronics. By providing advanced or revolutionary functionality in small bites – one machine, one function – and by paying exquisite attention to aesthetic design and to ergonomics, they are able to make new-to-the-world products palatable to the conservative mass of buyers almost immediately. These Japanese firms have almost broken the shackle imposed by the life cycle curve.

As this chapter's leading quotation indicates, the entrepreneur's most important activity in the first two stages of the life cycle is *learning*. Let there be no mistake: By the time the rapid growth stage of the cycle is reached, investors want their payback and production must be financed – the most important thing then is to *make money*. But, as Pat Flanagan of Reveille Technologies told my students, people working in startups like to learn. The learning is a payoff that offsets the higher risk of a startup. They learn a lot because they do a little of everything. There are two dangers. First, that if they sell a high-learning-burden product, people who like to learn will underestimate their customers' aversion to the product's learning burden. And second, that after fast growth is attained, they will not be the right people to continue working at the company.

Saga of a Sodium Compound

We can illustrate two of the principles of this chapter by looking to a humble compound, sodium bicarbonate.

Known as baking soda, $NaHCO_3$ is used in baking, and many people brush their teeth with it. How could sales of this commodity chemical be increased? Manufacturers convinced consumers that baking soda, sprinkled on a carpet and vacuumed up again, will freshen and clean the rug. This sold additional product, and encouraged the manufacturers to attack a household unspeakable: the smelly refrigerator. Consumers now open a full box of $NaHCO_3$ and place it inside the fridge to keep it smelling fresh. This campaign was followed by what may be the greatest marketing coup of all time – the conquest of the odiferous drain. People now buy baking soda, take it home, and pour it down the kitchen sink! Baking soda manufacturers kicked this mature product into new growth by finding *new uses*. And this was not the end of it. Baking soda chewing gum is now on the market!

Mixed with aspirin and dissolved in water, $NaHCO_3$ makes a quick-release remedy for pain and stomach upset. A story is told about a brainstorming session at a Madison Avenue advertising firm. How to increase sales of these tablets? The discussion went on and on, without progress, until one shy novice ad exec spoke up. "Would it hurt," she asked, "to take two?" What famous advertising campaign resulted? Sales increased because of customers' use of *two units,* where one had been the expected purchase prior to that point.

The Long Cycles: Computer and Dynamo

The last two decades have seen rapid innovation in computer technology. But over the same twenty years we've seen disappointingly slow gains in measured productivity. This quandary has been called the "Productivity Paradox. The paradox is not without historical precedent. If we take a long view and examine the beginnings of the electric age (i.e., the electric dynamo) we can more readily understand the productivity paradox of the computer age.

The common notion today is that computer technology will swiftly lead society to vast improvements in productivity. But despite the infusion of computers in our lives the potentialities of rapid productivity growth as a result of computers have not been realized:

• Computer technology accounts for approximately one-half of the United States gross investment in equipment.

• "The recent boom in office automation and the rise of computer-intensity of service industries have not been accompanied by surging output per hour in those activities," according to Stanford University economist P.A. David.

• Economic growth since the mid-70's has slowed. According to OECD statistics, total factor productivity advance has slowed.

The paradox may be explained by noting that the pace of realized improvements (i.e., technical change resulting in increased productivity) is not tightly tied to the rate of innovation. However, in our rush to satisfy our high expectations of technology, we tend to lose sight of the intricacies associated with changing from one technical paradigm to another. With this "telescopic vision," our technological future seems closer at hand and possibilities seem magnified. We are prone to "technological presbyopia," concentrating on the arrival rather than the journey.

• People are inclined to concentrate on the future and hold onto the prospect of dramatic improvements.

• Technological presbyopia may be a predisposition of industrial democracies – to direct energies toward the conquest of science and its commercial exploitation.

"Forward-thinking" does not take into account the historical, social-political and economic complexity involved with migration from technological regime to another technology paradigm transition.

By looking at the story of electric dynamo we should be able to avoid immoderate hopefulness and impatience on our journey into the information age.

• As industrial countries went from industries based on steam to industries built around electricity, rates of productivity declined.

• Productivity suffered for decades while at the same time there was rapid innovation in electric technology.

• The full transformation to the new electric technology was drawn out and uncertain.

• Since the first electric dynamo in 1870, engineers saw the potential for electric technology to be revolutionary, but it took 40 years before there was widespread electrification of American facilities.

• Part of the delay was due to the durability of old steam powered technology. Steam had staying power much like paper-based methods continue to co-exist with computer-aided procedures.

• When it finally came, the payoff of electric technology was big. The measures of productivity and total growth soared in the 1920's once electric technology became common place.

David P A (1991) Computer and Dynamo. Stanford University Center for Economic Policy Research Palo Alto CA (Reprint No 5). Summarized by Kevin Craine.

2.5 Technological Substitution

A better way to do something almost always comes along. But the better way does not replace the old way instantaneously; it diffuses through the potential user population according to an s-shaped growth rule. While the new technology grows, the old technology declines in usage. The growth of the new at the expense of the old is called *technology substitution*.

The Fisher-Pry substitution model for successive generations of a product or technology is:

$$\log (s_{n-1} / s_n) = kt,$$

where s_{n-1} is the market share of generation n-1 (the old technology), and s_n is the market share of generation n (the new technology). k is a parameter, and t is time. The Fisher-Pry was the first influential model of technological substitution, and is still a good conceptual basis for understanding substitution. But due to its rigidity (models with only one parameter usually cannot take on enough different shapes to encompass all the situations the real world can offer), the Fisher-Pry model is now used only rarely. For a more useful but more difficult model, see Norton J A and F M Bass (1987) A Diffusion Theory Model of Adoption and Substitution for Successive Generations of High-Technology Products. Management Science Sept. 33:1069-1087.

2.6 Technological Substitution in the Market Research Industry[5]

It is easy for manufacturers and retailers to know how much product moves out of their doors. It is more difficult, but very necessary, for businesses to know what kinds of people buy what kinds of goods, and with what frequency, what degree of brand loyalty, etc. Only in this way can the free market, that is, the *laissez faire* process of manufacturing what people actually want, work.

The other side of *laissez faire* is manufacturers' freedom to attempt to create and influence demand. Although public relations and other corporate functions play a role, the primary vehicles for doing this are promotion and advertising. So the triple problem of market research is: What do people want and buy? What advertising and promotions have they been exposed to, and how do they respond to them? The key to answering these questions is survey research.

2.6.1 Purchase and Audience Panels

Modern survey research (as opposed to other kinds of market research like focus groups, test kitchens, and so on) adopted its analytic methods from statistics. These methods allow making inferences about large populations based on the observation of small samples. But the interesting technological substitution issues are in the ways in which the sample data are collected, and in the interplay of technological, business, and social changes. The four players in this drama are manufacturers, consumers, retailers, and media.

Regardless of whether advertising "works," manufacturers accept that an advertising presence is needed, and will pay more for repeated exposure to larger and more desirable audiences.[6] This was the impetus behind the first radio audience measurement service in the 1930s (see Table 2.2). In 1941, the Market Research Corporation of America (MRCA) commenced door-to-door interviews with sample households to ascertain their demographic characteristics and buying habits. In general, MRCA questioned housewives (to use the term that was common in that era) about what they had bought that week. On occasion, the company conducted "cupboard inventories" in the household, with the owner's permission. Cupboard inventories determined, for example, that a "Stock up on Soup!" ad campaign would persuade consumers to carry larger inventories of soup in the home – relieving distributors and retailers of the cost of this inventory. In a

---

[5]   The author acknowledges the help of Dr. David Learner in clarifying some points in this section.

[6]   As a 30-second commercial spot during the 1998 Superbowl cost a manufacturer $1.5 million, companies want to know that is money well spent! Similarly, gaining a single market share point in a large market, like breakfast cereals, can mean millions in added revenues.

tacit division of the research market, MRCA conducted purchase behavior surveys, and Nielsen concentrated on media audience measurement.[7]

By the 1950s, increased urbanization meant that housewives, often newly moved to the city and less trusting of intruders, were less likely to respond to interviewers. MRCA adapted by changing over to mail surveys, using a pre-printed diary form in which consumers recorded their purchases for the week. The A.C. Nielsen Company, recognizing the growing presence and influence of television, began measuring television audiences. This was done with a mixture of mail diaries and in-home devices that recorded the time of day and what channel the TV was tuned to. Notably, the devices could not measure who was watching the TV, their degree of attentiveness, or, indeed, whether anyone was in the room with the TV.

The 1960s saw the mathematical revolution in market research. Mathematical modelers took advantage of the large survey data sets that had been assembled (at that time, MRCA's consumer purchase information constituted the largest proprietary database in the world), and created new algorithms for predicting brand shifting, purchase frequencies, and other marketing phenomena. The advent of commercial database software in the 1970s made this easier[8].

But the 1970s also saw changes that would transform the panel survey business completely. Increasing affluence and increasing divorce rates meant more households were being formed. These households were smaller (often single-member) and displayed new buying patterns. The households rarely had a stay-at-home member, and in general people became or perceived themselves to be more busy than in more traditional times. Single-member households were often younger, and young people had never been enthusiastic survey respondents. These influences decreased cooperation rates for market research surveys. Nonetheless, while it might have seemed that people would be reluctant to fill out the very detailed MRCA and Nielsen diaries, MRCA's data were validated by comparing their totals with factory shipments. The MRCA and manufacturer data series, while not coinciding (MRCA did not report, e.g., institutional sales to

---

[7]   We focus here on Nielsen (now a subsidiary of Dun & Bradstreet) and MRCA because of their leadership in surveying the same set of sample households on repeated occasions, a technique called panel sampling. Many other firms perform *ad hoc* sampling, which involves surveying a sample of households only once. But panel sampling results in lower error rates when trends (in market shares, market penetrations, etc.) are measured. And because these trends are of prime importance to large consumer goods manufacturers, the panel sampling firms had a uniquely prominent role. Also worth noting is that these two firms measured contemporaneous consumer behavior. Despite the measurement problems described above, this is much more accurate than trying to measure consumers' attitudes, beliefs, intentions, or recollected behavior – though many firms try to do that, too.

[8]   and gave rise to private databases that quickly exceeded MRCA's in size.

restaurants or the military), did mirror each other's upward and downward trends in a reliable fashion.

On the technological front, laser scanners could now read standardized bar codes. These codes, originally designed for distribution control, could be used to record purchases of an item at a super-market checkout. Nielsen and some start-up companies attempted higher-tech means of recording TV viewership. Next-generation set-top boxes had buttons for each household member, and members were requested to push their own button upon entering and leaving a room where a television was playing. But the cooperation rate for button boxes was not satisfactory. Medallions containing personalized radio frequency devices were then introduced – but as the styles of the 1970s passed, people were loath to wear medallions on chains. Set-top boxes with heat sensors were the next attempt – and the measured TV audience was augmented by dogs, infants, space heaters and toaster ovens.

2.6.2 The Rise of Scanner Panels

At this time, few stores used checkout scanners. In 1979, a start-up with an audacious plan to revolutionize consumer surveys raised enough IPO[9] capital to *give* scanners to every supermarket in a half-dozen "pod markets" throughout the U.S., in return for rights to the checkout data. Each pod market was a small city with demographics mirroring those of the U.S., an isolated grocery shopping area, and an isolated cable TV market. In each pod market, a sample of households were recruited, asked to fill out a paper questionnaire on household demographics, and issued an I.D. card with a unique bar code, to be swiped at the checkout stand prior to scanning the grocery purchase. For the first time, supermarket purchases could be automatically recorded and linked to households with known characteristics. Because this seemed "objective" and eliminated some key entry tasks, manufacturers were excited about the prospect of more accurate data.

Nor was this the limit of excitement about this scheme. Arrangements with the cable company enabled manufacturers to air two versions of a commercial, with each version cablecast to a different sub-sample of the panel households. It was then possible (or so went the claim) to measure the differential effect of the ad copy on subsequent purchasing! A remarkable passion had been aroused in the mature and conservative consumer goods industry; after fifty years, the "old-fashioned" paper-and-pencil diary questionnaire was to be supplanted by a high-tech solution, and the way seemed clear to answering the question of advertising efficacy. MRCA started to lose clients to the upstart, IRI (Information Resources, Inc.).

---

[9]   Initial Public Offering.

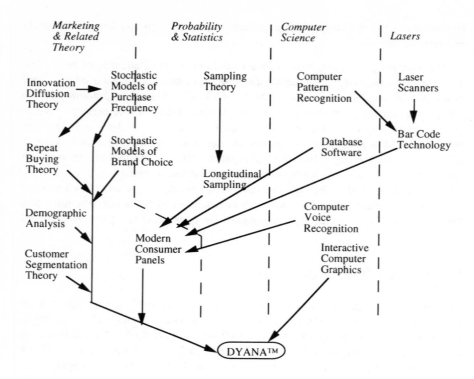

Figure 2.6 Antecedent-current technology map: The development of the DYANA™
interactive market research tool at MRCA Information Services in the early 1980s.
Technologies from four scientific areas are used.

What went wrong for IRI? At first, a lot:

• Not all manufacturers used their allotted UPCs to uniquely differentiate their
products. Without the supplementary information given in paper diaries, the
specific product purchased often could not be identified.

• There are regional and urban/rural differences in tastes and available brands.
Even the balanced demographics of pod markets could not produce nationally
projectable purchase data.

• IRI panel members could easily forget to take their I.D. cards to the supermarket.

• The location of the pod markets were well known, and split-cable ad tests could
hardly be kept secret. Competitors could and did issue coupons and air opposing
ads in order to sabotage other companies' ad tests.

• Scanner data could, in principle, show not just the price of a purchased item (paper diaries could do this just as well), but also the prices of competing products that were not bought at the time the panel member was shopping. It could show the shelf location of products in the store, or whether they were on aisle-end display. But the task of processing scanner databases to extract useful information for decisionmaking was, initially, too difficult. Until companies and programmers learned to turn the databases into useful reports, the consumer goods manufacturers could not get full benefit from the data.

• Only supermarkets were given scanners. But people buy food items at convenience stores, K-Marts, gas stations and department stores. MRCA's diaries captured purchases of foods from all retail outlets – a critical advantage in the eyes of manufacturers.

• Scanner databases suffered from their own, unique key-entry errors. Through the mid-1990s, studies reported that up to 9% of prices shown at the scanner checkout differed from the prices marked on shelves or packages, or were otherwise in error.[10]

It took manufacturers several months to see that these problems compromised the actionability of their market research data. They then began to re-subscribe to MRCA's service.

2.6.3 Substitution of Scanner Data for Diary Data

MRCA, a privately held company, did not wish to dilute ownership in order to raise capital and risk it in a "me too" scanner panel offering. But MRCA took IRI's early success to heart, examined its core strengths, and decided to compete not on the basis of data collection technology but on the basis of reporting and customer service technology. As already noted, scanner data processing was a less than mature art. MRCA capitalized on its expertise in processing diary data, implementing the industry's first interactive report generator, DYANA™. DYANA's ability to quickly answer clients' questions extended the life cycle of diary panels (see Figure 2.6).

---

10 A 1998 study by the Federal Trade Commission (CNN Interactive, http://cnn.com/US/9812/16/price.scanners.01/, found that the wrong price is scanned in one out of every thirty transactions.

Table 2.2 Technological Substitution in the Market Research Industry and Attending Business and Social Changes

---

(Technological changes are in plain text, business changes are in *italics*, and social changes are in **boldface**.)

1930s   • Statistical sampling theory.

1936   • Arthur C. Nielsen, Sr., licenses device from MIT to record the stations to which a radio has been tuned; *in 1942 starts Nielsen Radio Index.*

1941   • *MRCA*
        - door-to-door.
            + purchase interviews.
            + cupboard inventories.
        - manual punchcard DP.

1950s   • **Urbanization; less response to door-to-door interview sampling.**
    • *MRCA moves to mail diaries.*
    • Nielsen TV viewership measurement.

1960s   • *Trash audits.*
    • Digital computers; proprietary languages for DP.
    • Mathematical models for analyzing market data.

1970s   • Commercial database software used by market research industry.
    • Laser scanner technology.
    • **More women in workplace.**
    • **More divorces; expanding economy; average household size shrinks.**
    • "Advanced" TV viewership measurement.
        - button boxes.
        - medallions.
        - infrared.

1979   • *IRI business plan & IPO; $200 million raised.*
        - pod markets.
            * isolated grocery trading area.
            * isolated cable reception area.
        - give away scanners.
        - split-cable experiments.
    • *IRI starts to take customers from MRCA.*

1980s   • *IRI expands from pod markets.*
    • *Nielsen emulates IRI model;* uses pattern recognition algorithms to recognize what commercial is being received.
    • DYANA™.
    • *Customers return to MRCA due to DYANA, shortcomings of scanner data.*
    • **Increase in unlisted phone numbers**; random digit dialing.
    • Cheap microcomputers, microprocessors, microcontrollers.

• But reliable voice recognition technology still not cost-effective.
• Survey-on-a-disk for computer industry market research.
(Sawtooth, Intelliquest).
• Automated call centers.

1990s   • *IRI, Nielsen overcome most technical difficulties with scanner*
                 data; recapture market share from MRCA.
            • **Home and free time more valued by working adults.**
            • **Further internationalization of technology markets.**
            • Home scanners, home scanner panels.
            • *IRI $6 million syndicated advertising effects study fails.*
            • Wide diffusion of fax machines; fax surveys.
            • E-mail surveys.
            • Exploding Internet use; interactive WWW questionnaires.
            • Data mining.
            • Successful voice recognition technology for census data.

1996   • *ABC, NBC, CBS, Fox place joint ads in trade press criticizing*
            inaccuracies in Nielsen TV measurement data.

1990s and Later...
            • Nielsen plans to use embedded codes in digital TV to identify
            incoming programs.
            • Image recognition computers to recognize individual TV viewers.
            • Tracking web page hits; "cookies."
            • Integration of TV and WWW...

Nielsen, by this time an enormous company, adopted scanner technology to ease the collection of their "store audit" data service, which involves tracking movement of product through stores (without regard to who buys it). Nielsen then moved into the scanner panel business by issuing hand-held scanner wands to a sample of households. The devices use the household's telephone to upload data to Nielsen's computers during nighttime hours. By the early 1990s, most stores selling food items had bought their own checkout scanners, and about 60% of retail food product movement passed across checkout scanners. IRI was able to expand its store base (and enter the store audit business) by buying scanner data from stores in cities beyond its original pod markets. IRI and Nielsen had thus overcome a few of the difficulties that beset the startup of scanner panels, but home scanners had their own error problems, and store data purchased by IRI represented fewer than 1% of U.S. counties. It was the convergence of the technologies for store audits and consumer panels that finally drove out diary panels; IRI and Nielsen began to bundle store audit scanner data with scanner panel data, giving the latter to their clients essentially at no extra charge. This drew legal scrutiny (compare it to the question of Microsoft bundling web browsers with its operating systems), and clients knew scanner panel data were not really as accurate as diary panel data. But the price was irresistible. Scanner panels became the manufacturers' data source of choice for consumer package-goods purchase information, essentially driving diary panel services from that market.

Scanner panels experienced the interrupted take-off life cycle shown in Figure 2.7. (Life cycles of this shape may also result from faulty or recalled products; or from initial sales to VARs, OEMs, and third-party developers that are not followed immediately by enduser sales.) Diary panels are still in demand, however, for tracking sales of items that are not checked out using standard codes or scanners. This includes many consumer goods such as clothing, auto supplies, shoes, jewelry, and home furnishings. Diaries are still best for tracking the consumption (as opposed to the purchase) of foods.

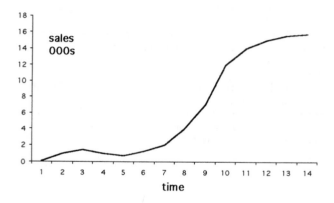

Figure 2.7 Interrupted growth cycles can be due to distribution pipeline loading, faulty product, or (in the case of scanner panels) technological convergence.

It is worth noting that supermarket chains rapidly learned that information is power. They used their own scanner data to compute the profitability of every foot of shelf-facing in every store. This enabled them to negotiate with manufacturers about the shelf space allocated to each of the manufacturers' products, and to estimate the profitability of new product offerings from given manufacturers. In some cases, this led to the levying of "slotting allowances," payments from the manufacturers to the store to allow the display of new products. Scanners had indeed shifted power from the manufacturers, where it had been traditionally, to the stores.

Power is also shifting to individuals and to local advertisers, as the range of electronic entertainment options skyrockets. *Business Week* notes that in the 1960s, our choices amounted to NBC, CBS, and ABC. By the 1990s we had added UHF, cable, and direct satellite options, as well as VCRs for timeshifting programs and viewing recorded tapes, for a total of about 75 "channels." The choices, according to the magazine, will expand to 300 in the 2000s and then to 1,000 in the 2010s as HDTV and Internet-based virtual reality "channels" become widespread. Today, US cable ad revenue is $8 billion per year, and has been growing 15-17% each year, while broadcast ad revenues have leveled off at $20 billion per year as audiences decline. The resulting audience fragmentation means that the number of people viewing (in the future we may say "participating in") a given channel may be small, and the margin of error in measuring this number will be large. The consumer wins because of increased entertainment choices, but advertisers and media have already begun to strike back. "Digital ad insertion" allows cable operators to send different ads to different neighborhoods, during the same commercial break. Tele-Communications Inc. (TCI) says it will have digital ad insertion systems in place for 90 percent of the 24 million households in its service areas by the end of 1998. Eventually, they will be able to "serve different ads to a teenager watching The X-Files in her bedroom and her parents catching Mulder and Scully in the den." According to *Wired News,*

> In traditional coaxial cable architecture, one signal is sent to all of the tens of thousands of homes in a service area from a single "head end," the industry's term for the transmission source. But now, as the industry moves to a hybrid fiber coaxial architecture, which allows for two-way communication, its systems include many more head-ends to transmit signals to smaller nodes. This means a different set of signals can be sent to each node, serving as few as 500 homes.

> When a cable operator pulls down a program from a satellite to distribute across its network, national advertising is already inserted into some of the commercial breaks, with spots left open for local ads. In the old days, those gaps had to be dubbed in from tapes. Now local ads are stored digitally on servers provided by companies like SkyConnect. And now that the hybrid fiber coaxial architecture has more head-ends, it is easy to plug different ads into the signals sent to different nodes.

The *Wired News* article reports that Kraft Foods has signed a multi-million dollar deal with TCI to develop services including the first household-addressable TV ad. But, it goes on, "Targeting could be particularly attractive, for example, to a pizza delivery company that knows which neighborhoods account for the majority of its business." Digital ad insertion can make local advertising (which is both important for cable company revenues and impossible to measure under older technologies) better-targeted and more effective.

2.6.4 Technological Change and the Death of "Quality Research"

Interactive computing led to other advances in survey research in the 1980s and 90s. On-screen questionnaires eliminated the confusing "If you answered yes to question 9, go to question 11b" instructions often seen on paper questionnaires. Phone surveys could be completely automated using random digit dialing, voice/audio databases, and push-button telephone tone responses to questions.

Fax and email offered new channels for collecting market research information. As faxes and email addresses were not uniquely identified with particular households, offices or individuals, the ideals of random, demographically representative statistical samples began to fall by the wayside. Phone survey firms had used callback protocols to maximize the probability of reaching a household that had been chosen for a sample. Now, with people using answering machines and caller I.D. to filter calls, researchers considered it lucky to reach a household at all. "Opportunity samples" ruled the day. Ideal statistical sampling was particularly difficult on the World Wide Web, as a culture of "alternate personae" (that is, lying about one's identity, age and gender) had already taken hold among WWW users. Data mining, the use of automated statistical tests and pattern recognition algorithms to find regularities in large databases, also violated traditional rules of statistical inference – but became common and even necessary in many businesses.

Managers, rather than moaning about the demise of traditional, "high-quality" techniques for collecting and analyzing market research data, should instead think about how best to use the newer technologies to assist good decision making. Some have done so. One result is WWW advertising billed on a "per click" basis, indicating the prospect not only saw but responded to the ad. Digital interactive television, combined with cameras and image-recognition systems in set-top boxes, may finally tell researchers which household members are facing the TV at any time.

But adjusting to the new is rarely easy. Many companies, mistakenly viewing the WWW as "the next television," became concerned with measuring the audiences of websites. As mentioned, web surfers often use the WWW under false demographic pretenses. In addition, it was hard to tell whether a hit on a WWW page was a human or a crawler, 'bot, or search engine. Individuals use access accounts belonging to others, and may routinely erase the "cookies" left on their hard drives. Nielsen's early claims to have mapped the demographics of Web users were widely questioned, and finally scientifically discredited; Nielsen has since introduced a new, improved methodology.

Will "share and ratings" numbers for the Web be perfected? It doesn't matter. There is little point in answering old questions about new technologies and media. The real challenge is figuring out what new, relevant questions the new technology lets us ask and answer. What are users' expectations regarding interactive media like the Web? Are Web surfers as susceptible to suggestion as TV viewers? Or do their feelings about control and creativity, as they navigate hypermedia[11], change their attitude to advertising? How much personal data do

---

[11]   "Push technology" on the Web was an egregious example of ignoring the interactive nature of the Internet in order to netcast content to a passive, couch-potato surfer – a psychographic that probably does not exist. This author looks forward to the day

they wish to share, and what compensation do they expect for this? These are a few of the new questions, actually new opportunities, that are opened by the new media.

2.7 Why Such an Emphasis on Cycles?

Technologies, products, product categories, and industries all experience life cycles, and obviously model cycles are nested within product cycles, which in turn are nested within category cycles, and on and on in a potentially very confusing way. Moreover, the meaning of cycle compression and life cycle psychographic segmentation differ according to whether yours is a new product (technology) in an old industry; an imitation product in a new industry, etc.

The several cycles and their related concepts (technology cycle, product development cycle, sales cycle, economies of scale, learning curves, demand elasticities, etc.) can be analyzed jointly with a lot of higher mathematics and great difficulty. But managers don't tend to do or use analyses of this complicated order. Rather, they analyze single cycles using the tools of this chapter, and comprehend the epicycles only as conceptual tools.

At a practical level, why is this worthwhile? Here are four reasons.

1. Compressing the cycle allows you to deliver more up-to-date technology to the customer. As product development projects proceed, it gets more expensive to make changes in the design (Figure 2.9). Especially in tightly coupled designs (in which changing one design element necessitates changing several others), ECOs (engineering change orders) are dreaded.

In the 1980s, General Motors total design cycle for a new model car was ten years. Honda's was four years. Let us suppose that two automotive innovations became cost-effective and available in quantity near the middle of the decade. (In Figure

---

when advertisers recognize the "upward" flow of content on the Internet, possibly even supporting individuals' efforts to publish web content.

Privacy and personal data issues arise because, unlike MRCA and Nielsen volunteer households, HDTV and WebTV viewers involuntarily reveal their web navigation history (and hence perhaps their lifestyle and product preferences), and the technology leaves "cookie" files on the viewer's hard drive. In a letter to The Oregonian on (ironically) July 4, 1998, the CEO of International Diversified MacroTechnology Inc. defended these practices as making "a visitor's experience more useful and enjoyable." One can argue the ownership of the data, but not the fact that the website owner has made unauthorized use of the user's disk space. At best, it seems comparable to walking into the user's home and borrowing his/her stereo set without permission.

2.10, for ease of thinking, we use anti-lock braking systems and fuel injectors as examples of such innovations, although these were not the years these particular inventions came to market.) Due to the cost of mid-project ECOs, it would have been difficult or impossible to incorporate thes innovations in GM's 1990 Buick – whose development cycle started in 1980. It was easy to incorporate them in the 1990 Honda because its development cycle began in 1996. This made the purchase choice easy for a consumer who was safety-conscious, technology-conscious, or convenience-conscious. Indeed, it was in the 1980s that Honda and other Japanese makers took big bites out of the market share of GM and other American manufacturers who had not yet seen the value of compressing development cycles.

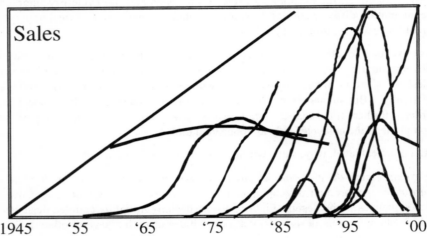

Figure 2.8 Cycles within cycles: Life cycles of individual computer models, processor types, and classes of computers from mainframe to micro underpin the total life cycle (shown as an approximately straight line) of the still-growing computer market.

2. *Although their timing is hard to predict, life cycle phase shifts are inevitable – don't be caught unawares.* In 1987, at the fastest-growth phase of the PC industry, I visited a computer company. The young people in their market research department struck me as the most arrogant businesspeople I had ever met (and I had met a lot of arrogant businesspeople!) Each of them believed that the fast growth of the company was due to their own personal brilliance. In fact, they had spent their entire working careers, since leaving school, in the fast-growth phase of the cycle. It was all they had known as working people; if they had learned about bell-shaped LCs in school, they had decided it was just another academic fantasy.

Arrogance and complacency are inappropriate. All products launch, grow, mature, and die. Their LCs show their introduction, early growth, take-off, fast growth, levelling off, and decline. We don't necessarily know when a growth curve will level off, but we know that IT WILL.

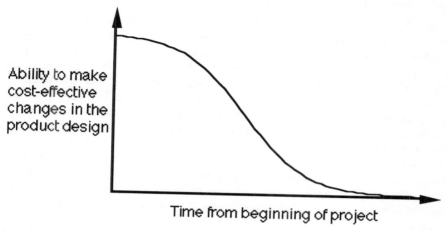

Figure 2.9 As a project progresses, the ease of making design changes decreases (the cost of making changes increases).

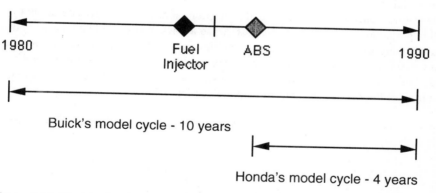

Figure 2.10 Shorter development cycles enable a manufacturer to bring the latest innovations to consumers – more cheaply and earlier than competitors.

3. *Whether a company manufactures or sources your product's components, better predictions of sales can reduce its inventory costs.* When life cycles are short and component prices are decreasing on a monthly basis, this is particularly important for profits and competitiveness. Kurawarwala and Matsuo combine moving-average estimates with Bass-model life cycle equations to achieve better forecasting for individual PC models. Figure 2.11 shows the actual sales of five

PC models. Their life cycles follow a roughly bell-shaped curve, but also show seasonal effects and end-of-quarter bunching of corporate orders. Kurawarwala and Matsuo's work successfully models these complex curves, and has given the sponsoring company a substantial cost advantage over its competitors.

4. Be able to diagnose companies. The self-reading utility meter company we met in Chapter 1 was trying to sell a new-to-the-world concept to a very conservative market segment. We now know why this idea had limitations, and could advise companies that are in a similar situation. We can presume AT&T, by advertising "You Will," is looking for alliance partners. Hewlett-Packard, a company started by engineers to sell products to other engineers, responded to investors' need for more growth, by moving into products for the office market – and faced challenges that were predictable from life-cycle segmentation theory. Some years ago a chemical company executive told me, "We need to find new uses for titanium dioxide." You can use LC theory to pinpoint that company's position on its product life cycle. A software executive considering the "product is dead" phenomenon said it "made us seriously re-examine the role of customer service in this company," and provided a way of explaining and justifying the needed changes.

It's one thing to estimate your product's total market potential. But it's essential to your cash flow and ROI planning to know *when* this potential will be realized, and in what amounts. The s-curve gives you the time path of revenues. This is especially relevant when pitching your product or company to an investor. Investors want to see your financial projection, and want to know you did not pull this projection out of thin air, or just "trace a hockey stick." The s-curve and its underlying generators (i.e., external and imitative influences) give your financials a reassuring rationale. And by estimating the time when demand for generation n of your product will begin to level off, you are aided in deciding when to introduce the next generation (n+1) to the market.

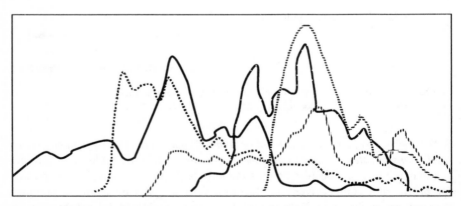

Figure 2.11 Sales (market cycles) of five consecutive models of PC from one manufacturer. Total time shown is three years.

Technology Colonization

We have to resist "media imperialism," according to media mogul Barry Diller. This is the tendency to colonize, or to define a new technology in terms of the old. After all, the automobile proved to be much more than a horseless carriage!

New media can empower and liberate business leaders if the leaders let themselves imagine untried business models and paradigms, Diller says. Edward R. Murrow used television as an eye to take the viewer into other people's homes. Murrow was among the first to fuse television and the imagination, showing that TV creates intimacy.

Diller believes we should be convergence contrarians, willing to challenge conventional wisdom, yet able to explore other possibilities, treating a new medium on its own terms. For example, Adolph Zukor began to create the infrastructure for what would become the movie industry. The motion picture industry then began to break free of the constraints of the Edison-controlled Motion Picture Trust.

Finally, he says the fear of displacement is misguided. The arrival of cable did not signal the end of the networks. The networks, while stressed by cable channels, have allied with companies offering a wide range of media options, including Internet, making a richer viewing experience and creating cross-marketing opportunities.

Diller B (1995) Don't Repackage – Redefine! Wired February. Summarized by Annie Leong.

2.8 Notes

Table 2.1 adapted from Rogers E M (1962) Diffusion of Innovations. New York: The Free Press 169-189

Product life cycles have been and continue to be getting shorter than ever before... Gaynor G (1993) Exploiting Cycle Time in Technology Management. McGraw-Hill New York; Rosenau M (1988) Speeding Your New Product to Market. Journal of Consumer Marketing Spring 5:23-36; Stalk G (1988) Time: The Next Source of Competitive Advantage. Harvard Business Review July-August 66:41-51

Almost all tangible products have a service component... Lazer W (1971) Marketing Management: A Systems Perspective. John Wiley & Sons New York 239

Kohler's Dictionary for Accountants... Cooper W W and Y Ijiri (1983) Kohler's Dictionary for Accountants. Prentice-Hall Englewood Cliffs NJ 398 (6th Ed)

Cunningham and Cunningham... Cunningham W H and I C M Cunningham (1981) Marketing: A Managerial Approach. South-Western Publishing Co. Cincinnati

Lazer's ... Lazer op.cit

Penguin Dictionary of Economics... Bannock G, Baxter R E and R Rees (1972) The Penguin Dictionary of Economics. Penguin London

customers do not need quarter-inch drills... Levitt T (1983) The Marketing Imagination. The Free Press New York

Gross and Coy note the growing use of the word "architectures"... Gross N and P Coy (1996) The Technology Paradox. Business Week February 23 (online edition)

internal resources to back niche consulting businesses... See Petzinger T Jr (1995) The Front Lines. Wall Street Journal November 3 (second front page)

companies aim to lock customers into streams of products... Peppers D and M Rogers (1996) As products get smarter, companies will have to focus on relationships. Forbes ASAP Supplement (Feb. 26) 69; Donath B (1994) Sell 'subscriptions' instead of 'products.' Marketing News (Sep. 12) 28(19):12

excellent service...has more of an influence on the likelihood that the customer will buy... Tanner J F Jr (1996) Buyer Perceptions of the Purchase Process and Its Effect on Customer Satisfaction. Industrial Marketing Management 25:125-133

Mass customization... Pine B J II (1993) Mass Customization: the New Frontier in Business Competition. Harvard Business School Press

to emphasize brand over product... Farquhar P (1994) Strategic Challenges for Branding. Marketing Management Summer 3(2):8-15; Pettis C (1995) TechnoBrands: How to Create & Use "Brand Identity" to Market, Advertise & Sell Products. American Management Association New York

"This industry was built on product advertising..." Paustian C (1994) Marketers rebuild their brand muscle. Business Marketing (Sept.) 79(9):B-3, B-4

an increased emphasis on brand marketing within the high tech sector... Kosek C (1995) Product vs. brand image debate. Business Marketing 80(9):A4-A-18

"businesses will come to think of their customers as learners and of themselves as educators..." Davis S and J Botkin (1994) The Coming of Knowledge-Based Business. Harvard Business Review September-October 165-170

General Electric will realize 62% of its income... Smart T (1996) Jack Welch's Encore. Business Week (October 28) 154-160

CompUSA, OfficeMax, and Tandy Computer City... (1995) Wall Street Journal Business Bulletin November 9 (page 1)

Netscape, Sun, and Microsoft items are from... Andrews P (1997) Internet time slows to a virtual crawl in '98. The Seattle Times (Dec. 28) C3

Today the battlefield is product platforms and not single products... Meyer M H (1997) Revitalize Your Product Lines Through Continuous Platform Renewal. Research•Technology Management March-April

"value delivery process..." Watson G H (1994) Business Systems Engineering (especially chapters 4 and 11)

Kurawarwala and Matsuo... Kurawarwala A and H Matsuo (1998) Product Growth Models for Medium-Term Forecasting of Short Life Cycle Products. Technological Forecasting & Social Change March 57:169-196; See also... Kurawarwala A and H Matsuo (1996) Forecasting and Inventory Management of Short Life Cycle Products. Operations Research 44(1):131-150

The MRCA and manufacturer data series... did mirror each other's upward and downward trends in a reliable fashion... Wind Y and D B Learner (1979) On the Measurement of Purchase Data: Surveys vs. Purchase Diaries. Journal of Marketing Research (February) XVI:39-47

Checkout scanners... If you wish to try processing scanner data, here is a source: Over seven years of weekly store level data for the Dominick's Finer Foods chain is available at the University of Chicago web site,

http://gsbwww.uchicago.edu/research/mkt/MMProject/DFF/DFFHomePage.html

Complete documentation is available on the web. Data include weekly sales, prices and cost information for each of some 85 DFF stores. Data are available in some 26 food and non-food categories. The data are available in PC SAS format as zipped files.

raised enough IPO capital to give scanners to every supermarket... See also Levin S G, Levin S L and J B Meisel (1987) A Dynamic Analysis of the Adoption of a New Technology: The Case of Optical Scanners. The Review of Economics and Statistics 12-17, who use proportional hazard models to model supermarkets' adoption of scanners in the nation at large.

DYANA™... F Phillips (1985) Advanced DSS Design in Consumer and Marketing Research. DSS'85: Fifth International Conference on Decision Support Systems. Anthologized in Sprague R and H Watson (Eds) (1986) Decision Support Systems: Putting Theory into Practice. Prentice-Hall

Fox place joint ads in trade press criticizing inaccuracies in Nielsen TV measurement data... Associated Press (1996) Networks join forces for anti-Nielsen ads. The Oregonian Dec. 25

Image recognition computers to recognize individual TV viewers... Aust E W (1996) Television Ratings. Scientific American November 127; See also... Nielsen Media Research (1987) What TV Ratings Really Mean. A.C. Nielsen Co. Northbrook IL

in the 1960s, our choices amounted to NBC, CBS, and ABC... Business Week (1998) The Entertainment Glut. (February 16) 88-95. (Audience measurement in the age of entertainment glut is addressed from the viewpoint of solving the measurement problem, not from that of technological change, in ESOMAR (1998) Electronic Media and

Measurement Trends: On a Collision Course? European Society for Opinion and Market Research Amsterdam

broadcast ad revenues have leveled off at $20 billion... Randolph Court (1998) "TV Ads Tailored to You" Wired News February 10 http://www.wired.com/ news/news/email/other/busin reported these figures from the Cable Advertising Bureau.

Mulder and Scully in the den... Wired News op.cit

Nielsen's early claims to have mapped the demographics of Web... Kline D (1995) Market Forces: Nielsen Gets Nervous. Hotwired October

2.9 Appendices

2.9.1 Spreadsheet Skills

Figure 2.12 illustrates a simple method for forecasting the growth of a new product's sales using the Bass model, a spreadsheet program, and the first few week's sales figures for the product.

The task is much easier if you are familiar with your spreadsheet software's "Fill Down" command. You would do well also to master the idea of "relative addresses" and "absolute addresses." (You will usually want to refer to the locations of your model's parameters as absolute addresses.) You will need the program's "Regression" feature as well. Please refer to the spreadsheet program's manual or online help to learn these topics.

The growth curve we wish to estimate is not a straight line; it is an s-shaped curve. But this section's promised simple method uses only the simple x-y linear regression feature of your software. This is possible because of the logic of the Bass model: It says that the probability of a new buyer buying for the first time in time period t is

$p + q$ [fraction of possible buyers that have already bought].

The number in brackets is, assuming we have a good idea of the total possible number of buyers (market size), a known quantity. For this reason, the probability of a new buyer buying for the first time in time period t is a linear function of the unknown parameters p and q. Given a few period's observations of sales, we can use linear regression to estimate q and p.

A statistician might object to this "quick and dirty" method, for reasons having to do with the "distribution of the error term." However, this author takes the view that in a changing market, with only a few initial data points in hand, more formal statistical niceties will not result in a materially better estimate.

In the example below, the total expected market size is 10,000 customers. Note that [column 5] = [column 4]/(10,000 - [column 3]), and [column 6] = [column 3] /10,000.

Table 2.3 New Customer Data for Use in Estimating the Bass Model

(1) Time Period	(2) Cumulative Customers	(3) # of Old Customers	(4) # of New Customers	(5) New Customers as Fraction of Remaining Mkt	(6) Old Customers as Fraction of Total Market
0	0	0	0	0	0
1	80	0	80	0.008	0
2	240	80	160	0.01612903	0.008
3	490	240	250	0.02561475	0.024
4	870	490	380	0.03995794	0.049
5	1460	870	590	0.06462212	0.087
6	2285	1460	825	0.09660422	0.146
7	3435	2285	1150	0.14906027	0.2285
8	4860	3435	1425	0.21706017	0.3435
9	6410	4860	1550	0.30155642	0.486
10	7810	6410	1400	0.38997214	0.641

Again, according to the Bass model, the probability of a new customer at time t equals p+q [fraction of customers that have already bought]. In the columns on the left of Table 2.3, the "# of new customers" is our observed data. Columns 2, 3, 4 and 6 are derived from the fifth column; be sure you understand how. Let's graph the relationship between columns 5 and 6:

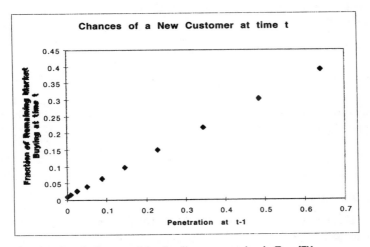

Figure 2.12 Estimating the Bass model using linear regression in Excel™

According to Excel's regression feature, the Intercept (=p) is 0.01109, and the slope (=q) is 0.59496. Plugging these values of p and q into the Bass equation

$$X(t) = X(t-1) + [p + qX(t-1)][M - X(t-1)]$$

lets us graph the smoothed s-shaped curve of "Customers" vs. "Time Period" in Figure 2.13.

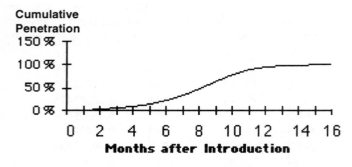

Figure 2.13 Estimated shape of Bass curve for the data in Table 2.3

2.9.2 Mathematics Review

2.9.2.1 Simultaneous Equations in Two Variables

We will use temperature conversion as an example. I can never remember how to convert degrees Fahrenheit ($^\circ$F) into degrees Centigrade ($^\circ$C), or vice versa. But I do remember that

$$0^\circ C = 32^\circ F \qquad \text{(freezing point of } H_2O\text{);}$$

$$100^\circ C = 212^\circ F \quad \text{(boiling point of } H_2O\text{); and}$$

the relationship is linear, i.e., it can be expressed in the form $y = ax + b$, where a and b are (unknown) constants.

So...

$$32 = 0a + b, \qquad \text{and}$$

$$212 = 100a + b$$

From the second of these, we have b = 212 - 100a. *Substitute* this into the first equation to get 32 = 0a -100a +212. Thus, a = 180/100 = 9/5.

Then 212 = (9/5)(100) + b = 180 + b, and

b = 212 - 180 = 32.

Thus the conversion formula is

$$^{o}F = (9/5)^{o}C + 32.$$

We will use this substitution method in later problems involving the Bass Model of new product growth and the Fisher-Pry model of technological substitution.

2.9.2.2 Net Present Value (NPV) and the Cost of Capital

If the cost of capital is 10%, then:

• The value to you of getting $100 one year from now is $100/1.10, since you could invest that amount and have it come to $100 in a year. Similarly, the value to you (now) of a promise of $100 in two years is $100/(1.10)(1.10).

• Perhaps strangely, you must also divide by a power of the cost of capital to compute the net present value of an investment that requires future payments. Suppose you agree to pay $5,000 up front plus exactly $5,000 per year for four years on a new machine (some of the latter may be maintenance, etc.). The *present* value of the total expenditure is

$$\$5K + 5K/1.1 + 5K/(1.1)^{\wedge}2 + 5K/(1.1)^{\wedge}3 + 5K/(1.1)^{\wedge}4$$

because you would be able to bank the latter amount and use the interest to pay off the purchase/maintenance schedule in contemporary dollars when the payments are due.

• Conversely, the cost to you of paying off a $100 loan a year from now will be $100x1.10 – assuming you and the creditor have agreed on a 10% interest rate. Paying off the $100 in two years really costs you $100(1.10)(1.10) in present-day dollars.

Thus, if you have the option of putting $5,000 down on a new machine and paying two additional annual payments of $1,000 (after one year and two years, respectively), or the option of paying $6,750 up front (assuming you have that much cash) . Which is the better deal? Use the fact that if you put cash in the bank, you could earn 10%/year interest.

$$NPV \text{ (mortgage)} = \$5,000+1,000/1.1+1,000/(1.1)^{2} = 5,000+909+826 =$$
$$\$6,735 < \$6,750,$$

so taking the delayed payment plan is better for you.

2.9.2.3 Growth, Growth Rates, and Limits to Growth

Companies based on technological or other kinds of innovation are attractive to investors because they are growth propositions. It is advisable for the entrepreneur/intrapreneur to understand the dynamics of growth, in order to make a reasoned case for her company's prospects. The analyst also needs this understanding, to make reasoned recommendations to investors. Individual new products are also expected to grow in sales, to an extent that covers their own costs, subsidizes other, failed new product projects, and yields a profit. At the macro level, the growth of industries and their by-products are constrained by the carrying capacity of the natural environment. For these reasons, we study growth dynamics in this section.

2.9.2.4 Kinds of Growth

This section refers to the graphs on the next two pages. In these, the symbol t in expressions like "x(t)" denote a time index; they do *not* mean multiplication of x by t. Next to each graph is an equation in *iterative* form, showing how each x(t) is calculated based on the previous value of x, which would be denoted x(t-1).

Figure 2.14 shows *linear* growth. In this kind of growth, each value of x is equal to the previous value of x, plus a constant..

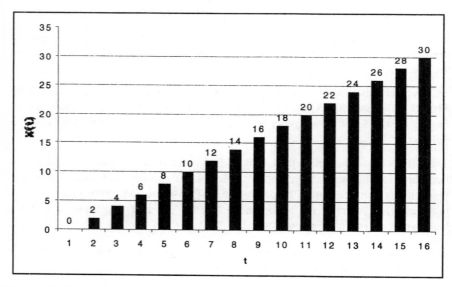

Figure 2.14 Linear growth. $X(t) = X(t-1) + k$. Here, $k = 2$.

Linear growth is usually associated with unchanging technology. If you continue manufacturing your product at the same rate of productivity and the same levels of material and labor inputs, your total cumulative production will grow linearly. If you plow another acre of soil of equal quality, plant soybeans and cultivate them just as you do your other acres, your harvest will increase linearly.

Figure 2.15 shows *exponential* growth. x(t) is equal to x(t-1) plus a multiple of x(t-1).

Exponential growth characterizes the unchecked growth of a reproducing population. If all citizens of a nation were to marry and have four children, each couple would have four offspring; the offspring would have eight of their own (after marrying the offspring of another couple); and so on. Each generation would have twice the numbers of the previous generation, and the total population would grow exponentially until land or food ran out. If a single bacterium undergoes mitosis and becomes two, and they split into four, and so on, they increase exponentially until energy resources run out. A population of algae on a pond increases exponentially until it reaches the edge of the pond.

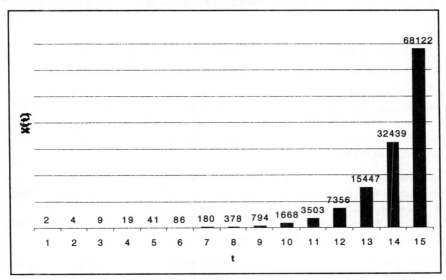

Figure 2.15 Exponential growth. $X(t) = X(t-1) + kX(t-1)$. Here, $k = 1.1$.

In Figure 2.16, the increment or "delta" in x(t) is not a fraction of x(t-1), but rather a fraction of [M-x(t-1)], that is, a fraction of the total amount of potential growth remaining, where M denotes the total potential population.

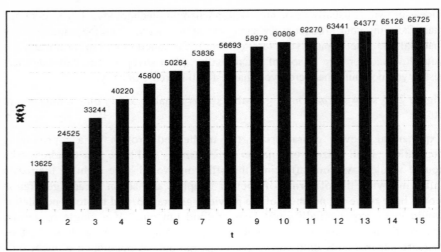

Figure 2.16 Exponential growth with ceiling. $X(t) = X(t-1) + k[M-X(t-1)]$. Here, k = 0.2, and the population ceiling M (or in marketing terms, the maximum market potential) is 68,000.

Because the remaining amount of potential growth decreases at each iteration, the delta becomes smaller as we move to the right end of the horizontal axis.

This kind of growth equation recognizes the "untils" noted above. This curve characterizes the growth rates of some new products [Fourt L A and J W Woodlock (1960) Early Prediction of Market Success for New Grocery Products. Journal of Marketing October 31-7], and was used for some years prior to the first publication of the Bass model.

Note that it assumes the *same* fraction of the remaining potential buyers will buy the product for the first time in each period -- regardless of how many total people have actually bought it already. In the context of new product introduction, this fact is usually interpreted to mean that there is some kind of external force, like advertising, that a certain fraction of a population reacts to.

$[1-x(t-1)]$ is a fraction of the remaining growth potential if $x(t-1)$ is itself a fraction (of total potential buying population or of total potential sales). Then "1" or "100%" is the maximum. If, instead, M is the total amount of product you expect to sell (e.g., expressed in tons or number of packages) and $x(t-1)$ is the amount sold so far, then $[M-x(t-1)]$ must be substituted in the equation where $[1-x(t-1)]$ appeared before.

Figure 2.17 combines the exponential growth effect of a delta that is a fraction of last period's population with the limiting effect of a delta that is a fraction of the remaining population. In this equation, the fraction of the remaining population that adopts, buys, or is affected by the innovation is kx(t-1), that is, k times the cumulative population that adopted in or before the prior period.

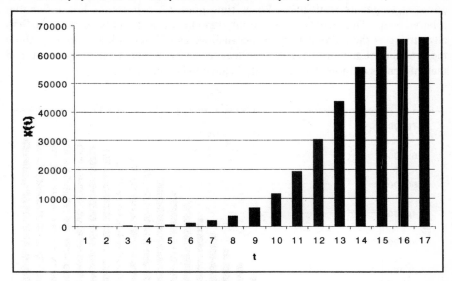

Figure 2.17 Exponential growth with imitation effect. X(t) = X(t-1) + kX(t-1)[M-X(t-1)]. Here, k = 0.8 and M=68,000.

This fraction kx(t-1) is bigger if x(t-1) is bigger, so in the early periods the deltas are larger and larger. (In the later periods, the remaining non-adopting population becomes small, and so the deltas become small also.) This is in contrast to Figure 2.16, where the fraction remains constant for all periods. The equation for Figure 2.17 is therefore usually interpreted to mean there is a bandwagon effect, or word-of-mouth, or keep-up-with-the-Joneses. In this view, people buy the product because they see other people buy the product, not (as in Figure 2.16) solely because of external (advertising) influence. Figure 2.17 shows what marketers call "imitation effect."

The Bass model combines external and imitative effects. The model is

$$x(t) = x(t-1) + [p + qx(t-1)][1 - x(t-1)].$$

Here, in each time period the probability that a member of the non-adopting population will adopt is $[p + qx(t-1)]$. p, showing the external effect, is called the "coefficient of external influence." q, representing the imitative effect, is called the "coefficient of internal influence." This model has been extremely robust, accurately tracking the growth of many technology products.

If you mentally track how the model will work in each period, you'll note that in the very first time periods, x(t-1) is still tiny – so q has very little effect, and adoptions in those periods are dominated by the value of p. In later periods, x(t-1) grows, so q has some work to do – you might say, a growing population of adopters to get its teeth into. So as time goes by, p becomes less influential relative to q. This does seem to be the way the world actually works: At first, advertising is the dominant effect, because we as customers know nothing about the product. Later, we have seen colleagues and coworkers use the product, and that influences us more than the vendor's advertising messages.

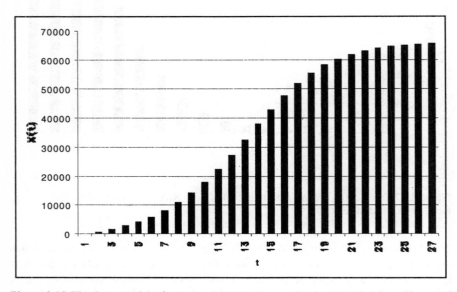

Figure 2.18 The Bass model of new product growth contains both advertising effects and imitation effects. X(t) = X(t-1) + [p + qX(t-1)][M - X(t-1)]. Here, M = 68,000; p = 0.1; and q = 0.3. All X(t) are buyers, not fractions of the potential market.

Range of q and p for real products: According to Sultan, Farley, and Lehmann, the range for p values "usually" has been 0.00002 to 0.23 and the range for q values 0.00003 to 0.99 for the Bass model estimates appearing in the literature. However, there is no mathematical restriction on p and q values, and other authors report that the range of p is 0.000021 (black-and-white TV) to 0.03297 (color TV) and that of q is 0.2013 (room air conditioner) to 1.67260 (b/w TV). In particular, the Easingwood, Mahajan, and Muller search of earlier diffusion studies identified q values larger than 1.

Sultan F, Farley J U and D R Lehmann (1990) A Meta-Analysis of Applications of Diffusion Models. Journal of Marketing Research (February) 27:70-77

Easingwood C J, Mahajan V and E Muller (1983) A Nonuniform Influence Innovation Diffusion Model of New Product Acceptance. Marketing Science Summer 2:273-295

2.9.2.5 Differential Equations

All the growth equations shown above are in *iterative* form. You may be more used to dealing with such relationships in *differential equation* form. Since we are dealing with discrete time intervals, we will only have to worry about the simpler cousins of differential equations known as (*finite*) *difference equations*. These give a value for the ratio (change in dependent variable)/(change in independent variable).

Take the linear growth equation $x(t)=x(t-1)+c$ as an example. x is dependent, and t is independent. If, e.g., $t = 3$, then

> change in dependent variable $= \Delta x = x(t)-x(t-1)$ and

> change in independent variable $= \Delta t = 3 - 2 = 1$.

In fact (and obviously), from one time period to the next, Δt is always 1. So $\Delta x / \Delta t = \Delta x = x(t)-x(t-1)$. Thus, the difference equation form of the linear growth relation is:

> $x(t)-x(t-1) = c$.

2.9.2.6 Analytic Equations

The analytic form of the linear equation is

> $x = ct$ if $x(0)$ was zero, and

> $x = ct + d$ if $x(0)$ was equal to a nonzero value d;

or, in general,

> $x(t) = ct + x(0)$.

This is a convenient form, because if you know a value of t, you can always compute the corresponding value of x, that is, $x(t)$, without having to know the value of $x(t-1)$.

However, getting to the analytic form is not always easy, especially for complicated functions, and you won't have to do it in this book except in some optional problems.

To get to it in the linear case, remember that the difference equation tells us $x(1)-x(0)=c$, and similarly, $x(2)-x(1)=c$. Substituting the first of these into the second, and rearranging, we see that

> $x(2) = 2c + x(0)$.

The principle continues for each higher t. In each case, $x(t)=tc+x(0)$.

2.9.2.7 Cumulating and Decumulating

We have been using x(t) to represent the *cumulative* number of buyers up to and including time period t. Unless we choose to consider product returns as "negative sales," x(t) cannot decrease as t increases – even if no one buys in period t, then x(t) remains equal to (not less than) x(t-1). But how many people *do* buy in time period t? That is, of the cumulative market in time period t, how many *new* buyers appeared in the most recent time period?

If we are given a finite difference equation representation, where Change in Dependent Variable = Δx = x(t)-x(t-1), then the (uncumulated) number of new buyers is simply Δx.

But we have been viewing the Bass model in terms of the cumulated fraction of potential buyers (i.e., the market penetration),

$$x(t) = x(t-1) + [p + qx(t-1)][1 - x(t-1)].$$

Here, the incremental penetration in period t must be calculated as the penetration at t *minus* the penetration at t-1:

$$(\text{Incremental penetration at t}) = x(t) - x(t-1) = [p + qx(t-1)][1 - x(t-1)].$$

2.9.2.8 Honesty in Graphics

For convenience, in this text and many others, growth models are displayed as line graphs:

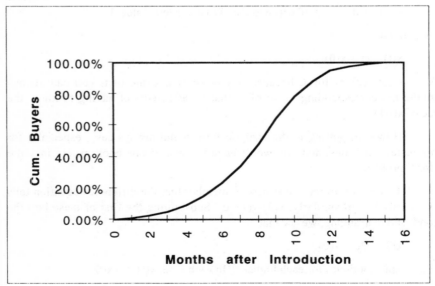

Figure 2.19 Line graph presentation of new product penetration

But line graphs imply that you can find a value for the growth curve at any value of t, even non-integer values. Often the available data will be structured only in whole months (weeks, or other "chunks"), so that "cumulative customers at t=4.5" has no meaning or is not measurable in the real world. For these reasons it is more honest to portray product sales growth via a bar chart:

It is easier to read the exact number of total customers for any single month from the bar chart.

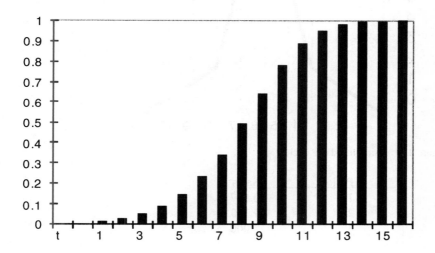

Figure 2.20 Bar graph presentation of new product penetration

It is true, though, that the bar chart does not convey the overall shape of the growth pattern with the high visual impact that the line chart does. And if (as we do in some of the homework problems and in Figure 2.22) we approximate a bell-shaped growth curve with a "triangular" model, it is easier to see the triangular nature if we use a line chart. It is all right to use a line chart as long as you keep in mind that interpolated values are just estimates and may not be verifiable.

How can we recover month-by-month sales from a chart like the Figure 2.21? First, note that the bell-shaped curve decreases after period 6, so it cannot be a cumulative representation; the chart shows period-by-period sales. But how does it show them? Calculus students will be inclined to worry immediately about the "area under the curve." But if this chart were generated from data like the following, and graphed "dishonestly" as a line chart, it would be reasonable to look at the chart and conclude that sales in period four were six units.

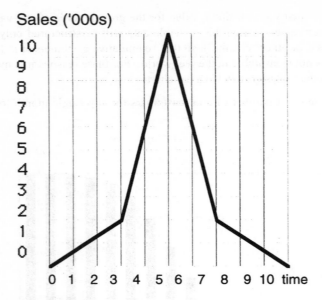

Figure 2.21 A simplified life cycle

Period	Sales
1	0.667
2	1.333
3	2.000
...	...

In another of the homework problems, a graph like the following is used, with the suggestion that w = 1 year and total sales for the first year are (0.5)mw units:

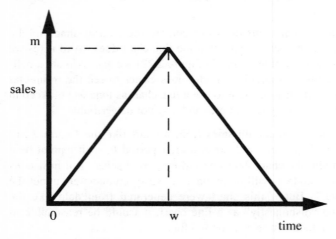

Figure 2.22 A triangular approximation to the life cycle

Because (0.5)mw is the area of the half-triangle on the left, the implication is clear that the graph shows sales in infinitely divisible units of time, and the sales over any *interval* of time are the area under the curve for that interval. For example, how many units were sold in month 5? We must "integrate" the curve for the interval t=5 months to t=6 months. Total sales through t=6 months are $(0.5)(0.5w)(0.5m)$. Total sales through t=5 months are $(0.5)[(5/12)(w)][(5/12)(m)]$. Sales within month five are then

$$(0.5)^3 mw - (0.5)(5/12)^2 mw$$

In this case, a line chart was the true and honest way to represent the sales pattern.

Confusing? Yes, it is, but in your reading you will encounter all kinds of charts, some of them honest and some not, some well-labeled and some not. Unfortunately, in some cases you must guess the intention of the chartist.

2.10 Questions and Problems

Problems to Solve

P2-1. (a) Here are data on the number of clients buying a new product in the months immediately following its introduction. Market research indicates there are 100,000 potential buyers. Fit the Bass model to these data by solving two equations in two unknowns. (The unknowns are p and q.) Predict how many clients will buy in the third month. These data represent new buyers in each month – not cumulative buyers up to that month.

month	buyers
1	1000
2	3000

(b) Actual new buyers in month 3 turn out to be only 6000. Fit the Bass model to the expanded data set using one of the following methods:

(i) A formal least-squares regression procedure using software of your choice, or

(ii) By eye. For the "by eye" method, show your work by noting the "residuals" (i.e., the amounts by which your predicted sales differ from the given data) for each of your guesses of p and q.

P2-2. Using spreadsheet software and "eyeball estimates" from Figure 2.23, construct a cash flow statement. Put the statement in a format that is suitable for a business plan. Use the spreadsheet's graphing capability to re-generate the graphical representation of the cycle. Does your re-created graph closely resemble Figure 2.23?

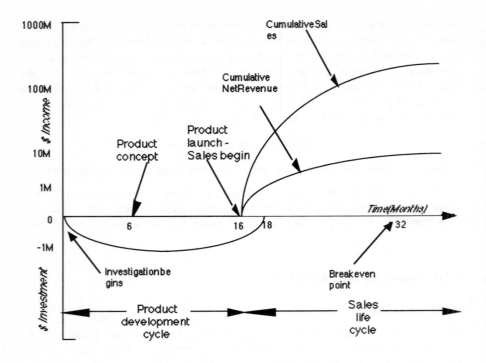

Figure 2.23 Product cycle data for the Hewlett-Packard pocket calculator, introduced in the 1970s

P2-3. Expected sales of generation 1 of a product are shown in Figure 2.24. Expected (same company) sales of the second generation are shown in Figure 2.25. Numbers on the horizontal axes represent months from the date of market introduction of each respective generation.

(a) How many months after the introduction of generation 1 must generation 2 be introduced in order to make the company's total unit sales increase in the most desirable way over time? Ignore price and external competition, and assume there is no cannibalization. Justify your answer, and draw a graph. *Hint: The "most desirable way" is probably a revenue growth path that is increasing as steeply as possible, or failing that, as close as possible to monotonically increasing.*

Figure 2.24 A simplified life cycle: Sales of generation 1

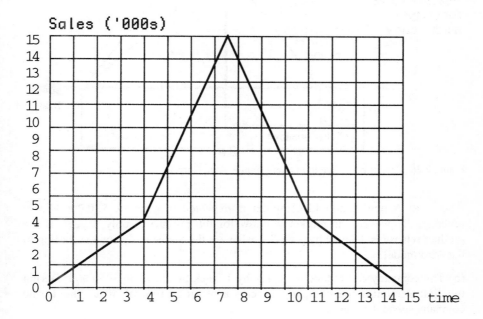

Figure 2.25 A simplified life cycle: Sales of generation 2

(b) Answer the same question, assuming that every fourth sale of a generation 2 unit cannibalizes a sale of a generation 1 unit, through month ten after gen-1's introduction. (You must presume further that the company will not withdraw gen 1 from the market before month 10.) Note that when you assume cannibalization, you must change the shape of the right tail of gen 1's life cycle.

P2-4. For the Ateq model, prove that the "lost revenue" is the "total expected revenue" times $d(3w-d)/2w^2$. This schematic for this model appears in figure 2.26. Revenue lost due to being late to market by d time units is the shaded area in this model developed by Ateq Corp.

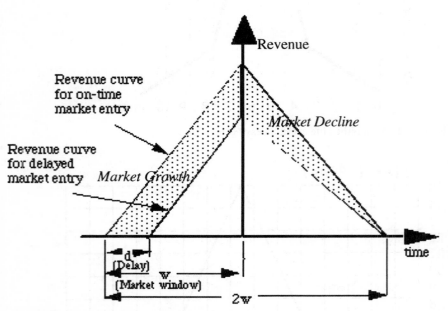

Figure 2.26 Cost of being late to market (Source: Ateq Corp.)

P2-5. (a) A new handheld computer is expected to generate $30,000,000 in revenue. However, the market window for the product is only 1 year. If the product release is delayed by 2 weeks, what are the losses in revenue according to the Ateq model?

(b) The company can reduce the delay by 1 week by spending $250,000. During the first 3 months the company expects a profit margin of 40%. Should the company spend the money?

Short Answer

S2-1. In what ways do technological products follow the "traditional" life cycle pattern of slow takeoff/rapid growth/long maturity/decline? In what way might they not?

S2-2.(a) Name a product that becomes part of its industrial customer's manufacturing process.

(b) Name a product that becomes part of its industrial customer's manufactured product.

(c) Does your answer to "b" above qualify as an "OEM" (Original Equipment Manufacturer) situation? What is the defining characteristic of an OEM? (Consideration: If I supply sugar to a candy maker, I doubt the candy maker would be called an OEM.)

S2-3. (a) A technology that starts an entire industry (and thus many product categories and a plethora of individual products) is called a basic technology. (This use of the word "basic" is different from its use in the phrase "basic research.") Name a basic technology, and indicate the industry/product growth that it has generated.

(b) Name a technological advance with less far-reaching impact, i.e., a non-basic advance.

S2-4. Name some strategies for revitalizing a mature product.

S2-5. Refer to the "Computer and Dynamo" reading summary in this chapter.

(a) Summarize the author's conception of the long technological cycle.

(b) Why is the dynamo or computer even more basic than nylon or Polyethylene?

(c) Speculate on the productivity benefits yet to be gained from today's computers

(i) if computer technology continues to advance at expected rates.

(ii) if I.T. were frozen at today's levels.

S2-6. Refer to the MRCA/Neilsen case.

(a) Will large or small advertisers ultimately benefit more from digital/interactive media? Is audience measurement on the Internet the right question, and if not, what is?

(b) What instances can you name (in other technological areas) in which a high-tech solution was initially and/or in the long run inferior to the low-tech solution it was intended to replace? Do these instances prove that "something is always lost" in quality as technology advances?

S2-7. Who is (are) the intended audience(s) of the AT&T "You Will" ads? Is there any downside risk for AT&T in this campaign?

Discussion Questions (Answers should be one or two paragraphs.)

D2-1. Several historical cases have shown that a customer or a household will eventually own two or more of a product, when in the product's early introduction stage it was inconceivable that a customer would ever own more than one. Television is a good example. Name a current technology product that is assumed to be "one to a customer." In a paragraph, outline a plausible scenario leading to customers owning more than one. Be specific about who the "customer" is.

D2-2. The first segment of the H-P pocket calculator's time axis is labeled "investigation." From your general knowledge of the era of the pocket calculator's invention, write a paragraph on the necessary differences between "investigation" phases then and now.

D2-3. Refer to the MRCA/Neilsen case.

(a) Today, MRCA is a smaller company than it was in the early 1980s. It continues to sell diary data on food consumption and non-food consumer purchases, as well as some new direct-mail related services. What more could MRCA have done to defend itself against the scanner panel companies? Why didn't the strategy described in the case work? Would a legal challenge to the bundling of store audit and purchase panel data have been feasible?

(b) Find in the MRCA case some instances of how social and family changes changed the way market research data had to be collected. What were the technological responses? How are social and family changes in the 1990s changing media viewership, purchases of household goods, and the ability of manufacturers and advertisers to learn what people are watching and buying? If you decide the latter is getting harder, do you think this will weaken the ability of our free-market economy to make and deliver what consumers really want? What will the consequences be?

(c) IRI's startup strategy was imperfect, and they did not predict the "hiccups" in their adoption curve. Should the hiccups have been foreseen? If you had been approached by IRI as an investor in the 1970s, what questions would you have asked? Would you have invested?

(d) Articulate the specific impacts of (i) technological substitution, (ii) technological convergence, and (iii) product bundling in scanner panels' supplanting of diary panels for measuring food purchases.

(e) What will individuals tolerate in terms of the capture of personal data (and advertisers' appropriation of the viewers' "personal" computers) in digital and interactive media? Is it possible to formulate and enforce standards of behavior for advertisers and website owners? Will Internet technology allow the individual surfer to foil the content providers?

Miniprojects

M2-1. Two models of technology growth are written below. Implement these models on a spreadsheet. (See this chapter's appendix for instruction on how to use PC spreadsheets for this purpose.) Experiment with varying the parameter values. You will turn in two sheets of paper for each model, for a total of four sheets. Each sheet will contain a graph of the function value (vertical axis) vs. time (horizontal axis), with the values of the parameters that generated the graph, and one or two paragraphs explaining why that particular curve is or isn't realistic for describing the diffusion of (real or imagined) products or technologies. The second sheet for each model shows the same information for a different set of parameter values.

i. The Bass diffusion model is: $X_{t+1} = X_t + (p+q\, X_t)\,(1 - X_t)$,

where X_t is the cumulative adoption rate, and parameters p and q are interpreted as the "coefficient of innovation" (external influence) and "coefficient of imitation" (internal influence), respectively.

Bass, F.M. (1969): A New Product Growth Model for Consumer Durables. Management Science, 15 (Jan.), 215-227

ii. The Fisher-Pry substitution model for successive generations of a product or technology is: $\log (s_{n-1} / s_n) = kt$, where s_{n-1} is the market share of generation n-1, and s_n is the market share of generation n. k is a parameter, and t is time. *Hints: Generations n-1 and n together comprise the entire market, i.e., their combined share is 100%. Also, your graph will look better if you let t=-20,...,20 – rather than starting t at zero.*

See Norton J A and F M Bass (1987) A Diffusion Theory Model of Adoption and Substitution for Successive Generations of High-Technology Products. Mgt. Sci. Sept. 33:1069-1087

iii. (OPTIONAL!) The Lotke-Volterra model for predator-prey populations is

$$dN_1/dt = \alpha_1\, N_1 - \lambda_1\, N_1\, N_2$$

$$dN_2/dt = -\, \alpha_2\, N_2 + \lambda_2\, N_1\, N_2$$

where the N's are population levels. You may think of this model as representing competing brands or standards stealing customers from each other; the Ns are the market size or share of each brand, and the parameters α and λ represent brand strength and marketing effort levels. Write down a competitive technologies situation that could possibly be described by these curves. How would you modify this model to represent products that are complementary (e.g., hardware and software) rather than competitive? *Hint: The parameters α and λ will be close to zero and may be negative.*

Volterra V (1936) Leçons sur la Theorie Mathematique de la Lutte pour la Vie. Gauthier-Villars Paris

PART II

ACQUISITION OF TECHNOLOGIES

3 Identifying, Nurturing and Monitoring Core Technologies

Having already defined high technology and the commercialization cycle, we must finally define "technology" and technology transfer. We then move to the criteria for choosing and identifying core technologies. (Chapter 1 established the justification for distinguishing core technologies). One criterion is the ability to master the core technology fast enough to reduce prices in a way that maintains profitability while discouraging competitive entries. Experience curves are an important consideration in timing price reductions; this chapter deals with experience curves and their computation. Again, the requisite mathematics are given in an appendix.

Outsourcing non-core technologies implies becoming a virtual organization. A focus on evolutionary and open-systems theory – a view of high-technology firms as evolutionary organizations – clarifies how even core technologies must be re-evaluated over time.

3.1 Defining Technology and Technology Transfer

Technology is knowledge used in design, products, manufacturing processes, organizations, training, software, etc. Preferably, this knowledge is *reproducible, realizable in devices,* and *transferrable.*

It must be reproducible because technology stems from science, and science rests on reproducible experiments – rather than on non-reproducible knowledge like sorcery or some kinds of artistic talent.

While the word technology implies that much of this knowledge is embedded in machines and tools, the knowledge needed to build and operate these machines must also be considered part of the technology. Below, when we explore core technologies, we will want to separate this kind of knowledge from knowledge used to turn core technologies to business advantage. This latter kind of knowledge is called *core competencies.*

Finally, technology should be transferrable. *Technology transfer* is the process of converting knowledge from on use to another – for example, from defense to civilian use, from research to application, and so on. The term technology transfer ("T^2," to cognoscenti) is especially important as it relates to transferring knowledge from its usage in one organization to another use in another organization.

3.2 Core Technologies

A company's core technologies should give it a distinctive competence. They should give the company a relatively secure distinction, at least for a period of time that is financially sensible. Core technologies should be:

1. state of the art;

2. fully tested and debugged;

3. well protected through patents, trade secrets, application know-how and a stably employed technical staff; and

4. relevant to the marketplace.

The technology should be relevant to the marketplace both in its essential appeal to customers and in its superiority to competitors' offerings. It should be well past laboratory stage and truly ready for application. These three traits have been called "market criticality," "technology competitive position," and "technological maturity," respectively.

Ideally, core technologies will have still more characteristics:

5. When core technologies are matrixed against the company's products, as in Table 3.1, the matrix should be dense. That is, core technologies should be highly relevant to the firm's products.

6. There should be few ready substitutes for the technology.

7. Preferably, the core technology is basic in the sense that it will spin off many new capabilities; even better if these new capabilities are also easily protectible.

8. The cost and availability of people to sustain the core technology should compare favorably against the ease of outsourcing the technology.

9. They should expose the firm minimally as regards regulation and legal risks.

Table 3.1 Which Core Technologies Are Used in Which Products? (Art courtesy of Cenquest, Inc.)

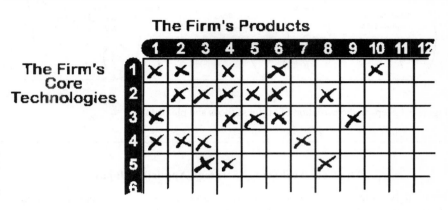

The differentiating (core) technology may be superior product engineering (Mercedes-Benz), superior product design (Sony), superior user interface (Apple, Palm Computing), or superior management of channels and marketing (Dell Computer). The firm should understand how this differentiating competence affects the bottom line, and how long the advantage may last. Dell, for example, knows the detailed impact of its fast order/delivery cycle on customer buying patterns and on inventory and component costs, and benchmarks these conscientiously against Compaq and others. But Dell still (either from fear their advantage would not last, or just as an ill-advised foray outside their core competence) botched an incursion into retail outlet selling.

The company must take care to distinguish core technologies ("differentiators") from "facilitators" (technologies which everyone must have in order to compete in the industry at all). For example, a management school dean must have a good office LAN (facilitator), but that by no means ensures success for the school; success is made likely by having a differentiator – like one of the finest technology management faculty in the world.

Core technology is a relatively new concept that replaces older business philosophies (see Figure 3.1). The Figure shows that it is a future-oriented concept. But we shall see later in the chapter that a firm may choose or be forced to review its core technologies as technological and market forces change.

THREE BUSINESS PHILOSOPHIES

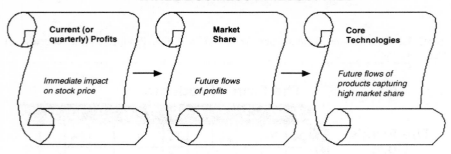

Figure 3.1 Steps in the evolution of business focus and philosophy

---

The Core Technology Strategy Applied in Japan

Previously, Japanese competitive advantage arose from inter-departmental cooperation within a company. For example, production costs were reduced because product research and development were conducted with manufacturing in mind.

"Companies that succeed in fully exploiting their own technologies for competitive advantage will be in a position to lead high-tech industries, dominate already established markets, or create new markets. Conversely, companies that fail to exploit their technologies appropriately can find themselves in serious competitive difficulty."

Many companies have started "2000 Vision" projects that will identify core business technologies, customers, and product functions.

Many Japanese companies have held on to some technologies for too long.

Even where these projects have not yet been successful they have raised the awareness that technologies must be understood "in the context of the market".

Core technologies shift over time; technologies picked now can have an influence for some time to come across several product lines.

How are core technologies managed?.

– Companies focus on a wide number of technologies. For example NEC has selected 30 core technologies, Canon 21.

– The challenge is to balance company-wide benefits with the business unit benefits.

- In the past, R&D was under a business unit which created organizational walls that inhibited sharing of technologies.

- Some companies have solved this problem by creating a company committee for each technology. This committee reports to corporate and has control over human resources company wide. It is responsible for long range planning, dissemination of information, and training.

Kokubo A (1993) Core Technology Based Management: The Next Japanese Challenge. Prism / First Quarter, Arthur D Little. Summary by Michael Funk.

Selling the Family Jewels?

IBM has developed a business model in which the company can increase its profits and sharpen its technology development by selling its core technologies.

Ira Sager's prime example is what he calls IBM's hottest growth business. "The Big Blue technology boutique" is a business direction in which IBM is selling components developed for its own systems to other companies, including its rivals for systems business. IBM is selling components such as disk drives, microprocessor chips and the little eraser-like pointer used in ThinkPad notebooks to competitors such as Hitachi, Apple Computer, and Canon. In this way, even when IBM loses a computer deal to one of these competitors, it still makes money on the components in their system. As Sager says, "Short of winning every deal, you can't beat that."

Selling components to competitors keeps IBM's factories busy and forces IBM to sharpen its technology by putting it in competition with the world's top component makers, while adding dramatically to revenues. License fees added $3.6 billion in revenue in 1994, with a growth rate one hundred times that of the entire company in the second quarter. Licensing protects IBM's intellectual property, and provides strong returns on the company's R&D budget, which was in the past folded into IBM's new products. And, if IBM can make the latest technology at the right price for competitors, it will also be creating the most competitive technology for its own computers.

As much as pursuing the component market makes sense, it can't restore or replace IBM's core business. At 10%, gross profit margins for components are only about a third of IBM's overall profit margin – better than no business at all, but not enough to turn IBM around.

Sager mentions that Hitachi's endorsement of IBM chips boosted IBM's efforts to make the PowerPC an industry standard – in which IBM ultimately failed.

Sager I (1994) IBM Knows What to do with a Good Idea: Sell It. Business Week September 19. Summarized by Cheryl Coupé.

In times of technology convergence, it is more difficult to determine a company's core technology and core competence. When a retailer builds a website, is the retailer in the software business? When a gas pipeline owner begins to carry fiber optic in the pipe, does it become a telecommunications firm? When a variety of technologies (inkjet, bubblejet, laser) can cheaply produce color copies, should a diversified firm with a laser printer operation stay in the market? Recently Tektronix decided "no" on this last question, spinning its color printer division off to Xerox. Another example (see the box below) is Texas Instruments.

Now, TI means "taking initiative."

Texas Instruments is re-examining its core businesses, assessing the real value to their customers and stakeholders. They are re-engineering how their company contributes to their customers and, more importantly, to the end-user.

TI is repositioning itself from a "demand-based" integrated circuit manufacturer/provider to a company that is out to create demand for their products by influencing the way in which IC's will be used. They are helping their customers to innovate, moving away from being a silicon provider, and upward on the value chain. TI took their patented digital mirror device (DMD) technology and contracted with Asian TV screen makers to develop surfaces that take advantage of DMD's higher resolution. Their prototypes from portables to wall-size home theaters grabbed the attention of projection manufacturers.

TI is fostering tighter connections with their customers, promoting innovation to help create billion-dollar businesses. Through their Digital Signal Processor (DSP) technology, TI developed and incorporated processes to allow for customers to hand-tailor their DSPs to include memory, power, and logic. This led to a ten-year, 600,000 DSP contract with Sony for use in the Boeing 777's audio systems.

TI is drawing a hard line when it comes to deciding in which markets to compete. Size and return on assets (ROA) are the drivers for determining the company's ongoing efforts. The manager of TI's notebook PC unit was given a mandate to increase sales to $1 billion while achieving 20% ROA, or else sell the unit.

Burrows P and L Holyoke (1995) Now TI means 'taking initiative.' Business Week May 15. Summarized by Mike Miles.

3.3 Experience Curves

In the late 1940s and early '50s, it was noticed that the cost of manufacturing airframes (aircraft bodies) decreased after several had been completed, even though labor rates and materials costs did not decrease. It was thought that employees were devising or learning better ways of assembling the craft, as they became more familiar with what was required. The trend in cost was called the "learning curve."

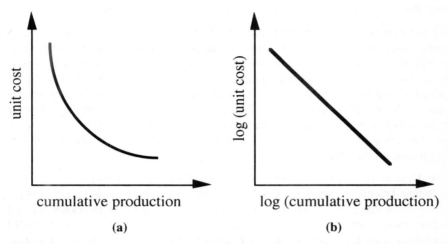

(a)   (b)

Figure 3.2 The experience curve phenomenon: (a) The cost of manufacturing one unit declines steadily as production experience increases. (b) The cost trend is becomes a straight line when graphed on log-log paper.

It was then noticed that when there was turnover in manufacturing line employees – costs decreased anyway! Because decreased costs were now evidently not a pure effect of individual learning (it seemed, rather, that employees were teaching other employees what they had learned), the graph of Figure 3.2 was renamed the "experience curve." The phenomenon was observed in many other manufacturing settings. The Boston Consulting Group and others noted its effect in petroleum catalytic crackers, in polyvinyl chloride production, in electric power production, and in the brewing of beer.

There is controversy as to whether declining unit costs are due to *the firm's collective experience* in producing the item; or to *economies of scale*. In the latter case, costs would be a function of the production level at a given point in time, rather than of cumulative production since the product's launch. In any event, it has been verified in a variety of empirical cases that unit costs drop by a fixed percentage every time cumulative production doubles.

log (Unit Cost) = -b log (Cumulative Units) => Unit Cost = (Cumulative Units)$^{-b}$

In an "80% learning curve," unit costs fall to 80% of their previous level every time production doubles. For an 80% learning curve, b = -log (0.8) / log (2).

A company's product becomes part of its customers' production processes. The customer must learn to use the product productively. If a steep learning curve (that is, fast learning, or simplicity) is *built into* the product, the take-off point in the product's demand cycle will occur sooner. So producers of industrial products must think about the "learning burden" involved in using their products.

All electronics-related companies must be thinking about when to drop prices. A price reduction – however it might be despised by stockholders – prevents competitors from taking market share, signals the imminent introduction of a newer and more capable product generation, and allows the new generation to be priced affordably yet still be seen as a big improvement in price-performance relative to the older generation. Figure 3.3 shows how the timing of price reductions is not only a fundamental strategy of the firm, but depends (inter alia) on experience-based cost reductions and on the expected actions of competitors and potential competitors.

The U.S. home appliance industry is a good example of "strategy A," i.e., reducing prices in synch with reductions in cost. This industry has always operated on thin profit margins. Add to that the fact that refrigerators are big and expensive to transport. The result is that although you (if you live in the U.S.) may drive a Japanese car and watch a Japanese TV, you do not have a Japanese refrigerator, air conditioner, or dishwasher!

In electronic markets, where prices are generally declining, the question of when to drop the price is a very fundamental one. Intel Corporation's practice is to drop prices on its higher-end microprocessors in anticipation of its competitors actions, while maintaining prices on the older microprocessors in order to (first) get customers to see upgrades as economical and (second) to squeeze competitors. This strategy is best used by a company that, like Intel, is first to market in each generation. Intel bases its price decisions on three things.

• First, its readiness to introduce a new generation of microprocessor.

• Second, its decline in manufacturing costs as reflected in the experience curve.

• And third, its anticipation of competitors actions.

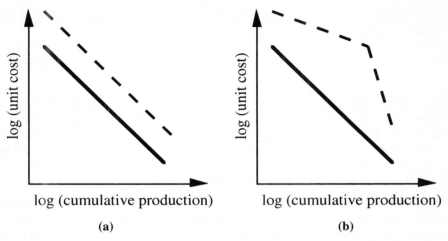

log (cumulative production)

(a)

log (cumulative production)

(b)

Figure 3.3 Pricing strategies: (a) yields small margins but prevents competitive entry; (b) is profitable until competitors enter but necessitates sharp price cuts after followers enter the market. Dotted line denotes price; solid line is cost.

The Technology Paradox

To survive and compete in today's business world, companies are following new sets of business rules: making money by giving things away; low prices and high volume; mass customization; shared technologies – and the faster the better.

Business thrives when prices are falling the fastest, because more people can afford to buy at each stage of price reduction. Also, companies can afford to give away the previous generation of hardware while making money on upgrades or service. The only thing that matters to these companies is that the exponential growth of their market is faster than the exponential decline of their prices.

Companies no longer focus on product alone; they focus on integrating many products in a common architecture. New products then don't require a new architecture, but must fit into an existing one. In integrated circuits (ICs), chipmakers' architecture strategies let them efficiently develop competencies in software and electronic components by focusing on the limited tasks in those areas that are demanded by their generation-spanning architectures.

These and other companies will mass-customize. American companies developing with multi-function chips hope to outperform Japanese firms – which tend to use single-function devices that are hard to customize – in mass customization. "Tomorrow's factories will sell customer gratification – not things."

Gross N and P Coy (with Otis Port) (1995) The Technology Paradox. Business Week March 6. Summarized by Annie Leong.

Because we've talked so much about technological innovation in this book, it is important to understand that the learning effects we saw on the experience curve do not involve innovation. The experience curve reflects declines in production costs using a fixed set of technologies. It does not reflect anything about using new, more effective technologies. But innovations do and should occur. So, how does innovation affect the experience curve?

In Figure 3.4, the experience curve looks odd, because we are not using logarithms on the axes. In view (a), you can see a normal decline in costs as manufacturing experience increases. In view (b), the graph shows one concept of how an innovation could effect the experience curve: There is a sudden drop in cost, but then the learning effect resumes at the same rate as before. In view (c), there is a second possible way that an innovation could effect the learning curve – by a sudden shift in the slope of the learning curve. In view (d), we can see how an innovation might temporarily increase costs during the period of time when everyone on the manufacturing floor is adjusting to the new procedure, but once underway, the new procedure increases productivity (decreases costs) at a faster rate than before the change.

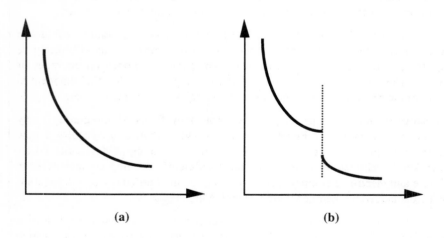

(a)   (b)

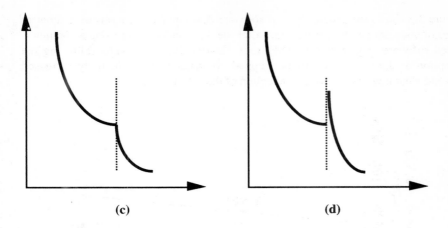

Figure 3.4 Innovation as an experience curve phenomenon: Three views
(a) Unit cost drops with cumulative production even in the absence of innovation.
(b) Innovation causes a sudden drop in cost; the same experience curve then resumes.
(c) Innovation causes a shift in the slope of the experience curve.
(d) Innovation causes a short-term productivity loss (increase in unit cost) followed by experience gains at a faster rate than before.

Now let's look at an example of how this works in the real world. The following is from a real company, and is quite typical.

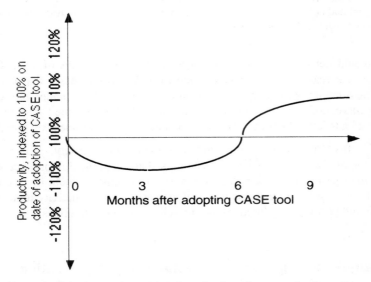

Figure 3.5 Productivity loss and recovery after adoption of a new technology

Figure 3.5 shows the productivity of software development, in terms of number of lines of code per person per day, just after the developers began using a computer-aided software engineering (CASE) tool. During the six months following the adoption of the tool, productivity dropped off. After that, productivity increased beyond what it was at the time of adoption of the CASE tool.

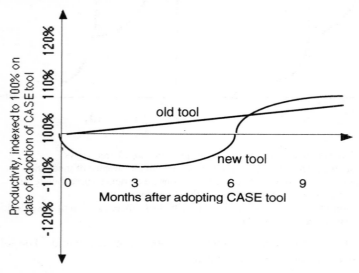

Figure 3.6 Productivity would have increased anyway (straight line), even if the new tool (curved line) had not been adopted. A buyer hopes the two lines will cross, and long-term productivity benefit will result from the new tool.

Where would net gains begin? Obviously *well after* 6 months of using CASE tools, for two reasons: First, it takes time to make up the losses that occurred during the 6 month learning period. Second, keep in mind that under an ordinary learning curve, code productivity would have increased even if the CASE tool had *not* been adopted. So the break-even point will occur well after the sixth month.

Naturally, this CASE tool was marketed as a boon to productivity. Do you think the vendor of the CASE tool told customers that adopting this product would torpedo their productivity for at least the next six months? I doubt it! We might expect the vendor to say that this tool will increase your productivity tomorrow – which should be a warning for all of us who sell productivity solutions!

3.4 High-Technology Firms as Evolutionary Organizations

A company's technological focus may have to change. The change may be forced by evolving customer tastes, or by technological change. Sometimes this technological change takes the form of technological convergence, as was the case when Microsoft almost got blindsided by the Internet. (Microsoft had been planning consumer services based on a dial-up service analogous to the pre-Internet America Online.)

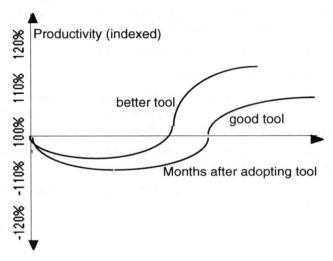

Figure 3.7 Vendors can gain advantage by designing tools that give the buyer a smaller short-term productivity hit, faster recovery of productivity, and a bigger ultimate increase in productivity.

Intellectual Capital

•Even the most modern accounting methods only track material assets and expenditures, but many business experts believe modern corporations have more intellectual capital than tangible capital – usually 3 to 4 times the tangible book value. Measuring intellectual capital has proven elusive, but most companies can tell when they have improved their use of it.

• Ernst &Young consultant Larry Prusak's definition of intellectual capital is similar to our definition of technology: "intellectual material that has been normalized, captured, and leveraged to produce a higher-valued asset."

• Dow Chemical is one of many companies to create a position "Director of Intellectual Asset Management." The current Director allows that "art and know-how" are part of intellectual assets, but has started with an easier task – organizing Dow's handling of its 29,000 patents. He found Dow and other companies have large inventories of unlicensed but potentially valuable patents. He developed a six-step program for dealing with a firm's patents:

1. Define the role of knowledge in your business.

2. Assess competitors' strategies and knowledge assets.

3. Classify your portfolio (of patents) by use, potential application, etc.

4. Evaluate. What are the costs, potentials, effort needed to realize potentials? Keep 'em, sell 'em, abandon 'em? (Keeping a patent in force over its lifetime can cost $250,000 in legal, filing fees, taxes, etc.)

5. Invest. Invest in in-house and needed external technologies.

6. Assemble your new portfolio and repeat.

• Attend to how knowledge workers interact, and provide systems to facilitate. Manage "both content and culture." Study how to keep from "losing the recipe" when workers leave the firm or work groups reorganize.

• Distinguish intellectual assets that go home (knowledge in employees' heads) from those that stay on-site (databases, networks, libraries, etc.)

Stewart T A (1994) Your Company's Most Valuable Asset: Intellectual Capital. Fortune (October 3) 68-74

A widely acclaimed Stanford University study identified the management practices that are common to very long-lived companies. These are:

• The best managers are thinkers, not necessarily charismatic, and definitely not micromanagers. They believe in procedures and policies, but procedures and policies that admit some flexibility.

• The best managers transcend trade-offs, refusing to admit, for example, that productivity and quality cannot coexist. They persist until they achieve both quality and productivity.

• They stimulate progress,but preserve the core. Their cultures may be cult-like, but they are pragmatic, and set audacious goals.

• They send consistent signals, so stakeholders feel secure in knowing what the goals are.

The "Stanford Visionary Companies," when compared to companies that did not share these characteristics, showed twelve times more share appreciation over the period 1926-1996. Apparently the research design did not include comparisons to the stock of companies that were sold, which conceivably could have netted still higher returns. But the visionary principles are appealing and the visionary companies' financial returns are impressive!

Disruptive Technologies

David Isenberg (*WIRED,* August, 1998, page 78) says, "If you're listening to your customer, it's almost preordained that you'll miss the new market. And when the new market expands to encompass the old market.... that's when companies can become obsolete."

Isenberg is echoing the thesis of Harvard professor Clayton Christensen, who argues that technology B disrupts technology A if B...

• is initially of lower cost, quality, mark-up, and complexity than A;

• appeals at first only to the market segment that demands little in the way of performance;

• potentially offers qualitatively different kinds of benefits than A;

• rapidly improves in performance; and

• thus eventually takes over the higher-margin segments.

Management, leading customers, and stockholders initially disdain technology B due to its low quality and low margins.

The personal computer, "obviously a toy," disrupted the market for mainframe computers. Christensen shows how ever-smaller hard disks ("An 8-inch drive can't possibly be as good as a 12-inch drive") disrupted their predecessors.

We know that customers are good at saying what they want with reference to a known product, and are not good at saying what they want with regard to an unknown product. What does Christensen's concept of disruptive technologies really add to that ancient wisdom? "Only two things," Christensen says, "Your current leading customers will not be your future leading customers. And your future leading customers will come from the bottom of the market."

Nonetheless, his notion has entered the common vocabulary. Cherry Murray, a research director at Bell Laboratories, is working on miniaturized television devices (*Business Week,* Aug. 31, 1998, p.83). "They'll come in at the low end, not the high end," she says, "That's the disruptive part."

Christensen C (1997) The Innovator's Dilemma. Harvard Business School Press Boston

---

3.5 Notes

"technology is..." This definition is from Mark Eaton, former head of MCC's International Liason Office.

"market criticality," "basic technologies." These terms are due to McGrath M E, Anthony M T and A R Shapiro (1992) Product Development: Success Through Product and Cycle Time Excellence. Butterworth-Heinemann Boston

"selling the family jewels..." At the Patent and Trademark Office's "Semiconductor Customer Partnership" event in Sunnyvale, CA, in 1999, one of IBM's high-level IP counsel spoke about IBM's highly successful licensing program. He said:

- Every IBM patent is available for licensing at rates from 1-5% of gross sales.

- IBM holds 30,000 patents worldwide. (Half from U.S.) - 26,000 patents pending worldwide, "mostly from U.S."

- 35% of invention disclosures published for defensive purposes.

- 20% of invention disclosures result in a patent application.

- $1,150,000,000 (yes, $1.15 billion) annual licensing revenue, all of which goes back into R&D.

(Thanks to Ed Suominen, Patent Agent, for this information)

"Recently Tektronix decided 'no'"... Wysocki B Jr (1999) Corporate America confronts the meaning of a 'core' business. Wall Street Journal November 9 (front page)

"it has been verified in a variety of empirical cases that unit costs drop by a fixed percentage every time cumulative production doubles..." Lapré M A, Mukherjee A S and L N Van Wassenhove (2000) Behind the Learning Curve: Linking Learning Activity to Waste Reduction. Management Science May 46.5:597-611. Their article summarizes the most recent research on experience curves, including limitations of and alternatives to the "power curve" formulation given by this book, and presents a new application to the steel wire industry.

"Innovation causes a sudden drop in cost; the same experience curve then resumes." This concept is due to Prof. Yuji Ijiri of Carnegie-Mellon University.

"A widely acclaimed Stanford University study..." Collins J C and J I Porras (1994) Built to Last: Successful Habits of Visionary Companies. Harper Business New York

"evolutionary organization." See:

> Phillips F (1998) Review of Laszlo E and C Laszlo (1997) The Insight Edge: An Introduction to the Theory and Practice of Evolutionary Management (Quorum Books Westport Connecticut) Technological Forecasting & Social Change 58: 321-322.

> Kamo J and F Phillips (1997) The Evolutionary Organization as a Complex Adaptive System. PICMET '97: Proceedings of the Portland International Conference on Management of Engineering and Technology July

Phillips F and D Drake (Eds) (2000) Special Section Navigating Complexity: The Future of Knowledge and Learning in Organizations. Technological Forecasting & Social Change May (special section)

"Intel Corporation's practice is to drop prices..." See Williams E (1999) Intel acts to regain low-cost market. The Oregonian (January 5) D1; (1998) Intel reduces prices for Celeron chips to bolster role in market for cheap PCs. The Oregonian (December 30) B1. (Found online at http://www.oregonlive.com)

3.6 Appendix: Mathematics Review

3.6.1 Logarithms and Exponentials

If, as is indeed the case, $8 = 2^3$, then we say that the base-2 logarithm of 8 is 3. That is, 3 is the power to which we must raise 2 to have it equal 8. This is written $\log_2(8)=3$.

When you do logarithms and exponents on a computer,

- The symbol $^\wedge$ often means "to the power of," e.g., 2^3 = 8.

- "log" may mean $\log_{10}$ or $\log_e$, depending on the particular software.

- "*ln*" always means $\log_e$.

Exponent means the same as power. "e" is the base of "natural" logarithms, and we don't have to worry here about what is "natural" about them. In this course, you should interpret both "log" and "*ln*" to mean base-e logarithms.

The basic relations in the algebra of exponents and logarithms are:

- $x^{a+b} = x^a x^b$

- $ln(ab) = ln(a) + ln(b)$

- $ln(a/b) = ln(a) - ln(b)$

- $ln(x^a) = a\,ln(x)$

- $a^0 = 1$

- $ln(e) = 1$

- $ln(e^a) = a$

These relations hold true regardless of the base of the logarithms. You will use these relations in calculating the Fisher-Pry and other models.

3.6.2 Experience Curves

Let c_0 be the unit cost of production after you have made p_0 total units. Later, after cumulative production reaches p (where $p > p_0$), your unit cost is only c. An "80% learning curve" means that each time cumulative production doubles, unit cost is reduced by 20% (20% = 100 - 80%). That is,

$$\text{if} \quad p/p_0 = 2, \quad \text{then} \quad c/c_0 = 0.8$$

To find a general formula for c, we start from the fact that the unit cost – cumulative production relationship is linear when plotted on log-log paper:

If the slope of this line is a (which is evidently a negative number), then

$$\ln c - \ln c_0 = a \, (\ln p - \ln p_0), \quad \text{or} \quad \ln (c/c_0) = -a \, \ln (p/p_0), \quad \text{whence}$$

$$c/c_0 = (p/p_0)^{-a}$$

If the learning curve is 80%, we know that

$$0.8 = 2^{-a}$$

from the equation in the first paragraph above. So,

$$a = - \ln (0.8) / \ln (2) = 0.3219$$

when natural logarithms are used throughout, and in general,

$$c = c_0 \, (p/p_0)^{-.3219}$$

when the learning curve is 80%.

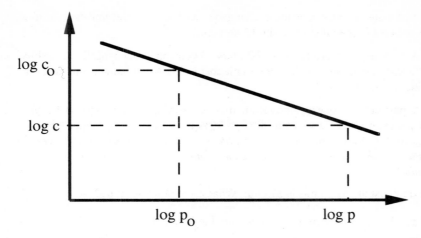

Figure 3.8 The unit cost - cumulative production relationship is linear when plotted on log-log paper.

3.7 Questions and Problems

Short Answer

S3-1. What is the difference between "market criticality" and "technology competitive position"? Between "basic" and "core" technologies?

Problems to Solve

P3-1. The cost of purchasing a machine is $10,000. The cost of maintaining it is $1000/year, and the cost of capital is 5%. What is the cost of purchasing this machine and owning it for 3 years (discounted to present value)? Ignore taxes, and presume the purchase is made in "year zero." (Depending on the business you are in, you may find it more helpful to look at this problem in terms of your customer's cost of buying and owning a machine that you are selling.)

P3-2. Referring to Appendix 3.7.2 above, prove that $c = c_0 \, (p/p_0)^{-.3219}$ for an 80% experience curve, regardless of the base of logarithms used in the computation.

P3-3. By the time TechWheels Corp. has manufactured 500 bicycles using their new titanium and carbon fiber design, their unit cost of manufacturing is $600. Assume TechWheels is experiencing an 80% manufacturing experience curve, that the 80% regime does not change through the company's 1200th bicycle, and that cost of materials also remains unchanged (salaries, too) over that span of time. What is the cost of manufacturing the 1200th unit?

P3-4. A company has produced 2,000 units. Their manufacturing line is subject to a 90% learning curve. Their current unit cost is $5. What is their unit cost after producing the 5,000th unit?

P3-5. A particular process innovation makes learning easier, and shifts your company's experience curve from 80% to 70% after you have produced p units. See Figure 3.9. *Hint: You may not be able to arrive at numerical answers for every part of this problem. In that case, your answer must be an algebraic expression.*

(a) The unit cost after p_0 units is $1,000. What is the unit cost at p_1? At p_2?

(b) Let p_0=2,500; p=10,000; p_2=15,000; and the initial cost (at start of production) is $1,060. Roughly estimate the *average* unit cost between start of production and p units as [1060 + (unit cost at p)]/2. Roughly estimate the *average* unit cost between p units and p_2 units as [unit cost at p + unit cost at p_2]/2. Show your work, and use these quantities to solve part c below.

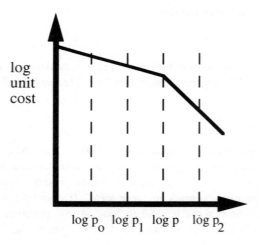

log cumulative production

Figure 3.9 Shift in learning curve from 80% to 70% after you have produced p units

(c) Referring to Figure 3.10, suppose your company is Producer #1, who enters the market at time 0. No competitors enter the market, and the life cycle is symmetric. What is the impact of the process innovation on your profits (relative to what they would have been absent the innovation)? Let m=2p and w= 1 year, with a unit sales price of $2,000. Assume units are sold instantly, i.e., as soon as they are manufactured.

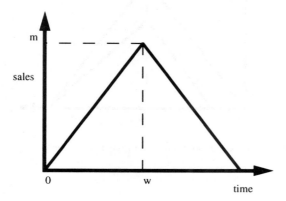

Figure 3.10 A simplified, symmetrical life cycle

(d) Suppose the process innovation in part b resulted from a $1 million research project. Would it have been worthwhile for Producer #1? If the project were undertaken instead by Producer #2 (second market entrant), and the project resulted in the same innovation with the same impact on Producer #2's experience curve, would it have been worthwhile for Producer #2? Use your answer to posit a general principle about process innovation for pioneer companies and follower companies.

P3-6. Producer #1 has decided to enter the market first (their first fundamental strategic decision.) Producer #1's second fundamental strategic decision has to do with price: Should they reduce price in parallel with dropping unit costs, or keep prices high and reap monopoly profits before dropping prices at the end of the life cycle? The first option, remember, keeps competitors out; the second maintains a price umbrella that makes follower entry attractive.

Use the data of the triangular model in Figure 3.11 to calculate Producer #1's total life cycle profits in each case, assuming the first option prevents Producer #2 from entering the market.

P3-7. Let q= the probability of a follower company entering the market, given that Producer #1 has followed the monopoly profits strategy. Using the data above, find the value of q at which Producer #1 is indifferent between the entry barrier strategy and the monopoly profits strategy.

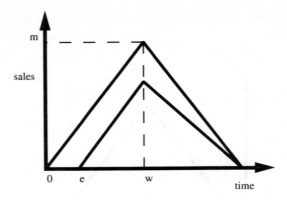

Figure 3.11 Graphical representation of the cost of late market entry

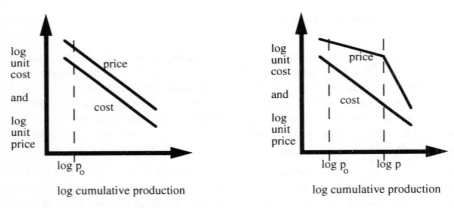

Figure 3.12 Alternative pricing strategies

Discussion Questions

D3-1. (Refer to the summary of "The Technology Paradox"):

(a) Following price cuts that may seem fatal to producers, "the economy and society will re-shape themselves to take advantage of the cheap resource." Give examples of this, other than microprocessors and memory chips.

(b) A quotation from the original article that was not in the summary was, "How do you assign prices or value in a world where quality is perfect and nothing breaks?" Try to answer this question.

D3-2. In an economics textbook of your choice, find a picture of the "marginal cost curve." Compare the labels of the x and y axes of this curve against those of the axes of the learning curve as described in this chapter. Why do the two curves give different information?   How might the information from the two curves complement each other?

D3-3. Comment on the pros and cons of IBM apparently selling its core technologies.

D3-4. Comment on the applicability of the Ateq model (see Chapter 2's Questions and Problems) to a two-firm situation. The curve on the left would show the sales trajectory for the first company to enter the market; the curve on the right shows sales of the follower company. Interpreted this way, the model implies that both companies' market shares become equal close to the end of the life cycle. Comment on the realism and relevance of this interpretation of the model.

Sannen ni moo ishi no ue.
Japanese proverb

"I exhibited the intemperance to refer to the
evening's feast as 'The Last Supper.'"
Former Martin-Marietta Chairman Norman Augustine, at a Pentagon
dinner, upon being informed by the Secretary of Defense that, "We expect defense
companies to go out of business. We will stand by and let that happen."

4 Technology Sourcing

In chapter 3 we dealt with the balance between internally developed technologies
– worthwhile only if they can become "core" technologies – and the remaining
technology that the company needs. The latter can be outsourced.

But "outsourced" is easier said than done. The company has to find the right
technologies, then form and maintain a relationship with the technologies'
suppliers. We will find occasion in this discussion to refer to Japanese practices in
scanning and sourcing, as these have been exemplary in recent years. Figure 4.1
shows the place of scanning and sourcing in the overall technology transfer
picture.

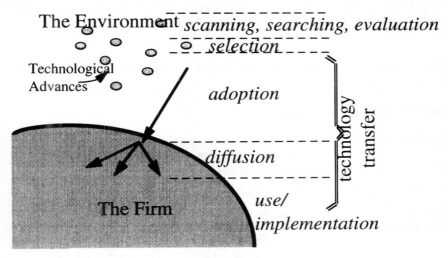

Figure 4.1 Finding, choosing, and transferring external technologies

Scientists, technology managers, and technology analysts scan the outside world for applicable technologies, screening them for cost-effectiveness, competitive advantage, and technological maturity. Ultimately, some external technologies are chosen for acquisition. This acquisition is called "adoption" in Figure 4.1. Once acquired, various labs or production lines inside the firm familiarize themselves with the technology, in a "diffusion" phase. Finally, they put it to use in a research, manufacturing, or marketing activity of the firm; this is "implementation." The adoption/diffusion/implementation phases are collectively known as *technology transfer*, although this term may be applied to the entire process depicted by Figure 4.1.

4.1 Global Scanning for Technologies

Where U.S. companies may put their best new hires in the finance department, Japanese firms put their best in external scanning. Some go so far as to send their young hires to a distant city (often in the U.S.) to rent a home, join a church and community organizations, and pick up competitors' company newsletters. There have been no publicized instances of illegality in these activities; this is "competitive intelligence," not industrial espionage. Later, these employees are promoted to the home office, where they digest information that is sent in from newer operatives.

This illustrates the great importance Japanese companies attach to global scanning. How is it done? There are many sources for information about new technologies. Different sources are more useful at different stages of the acquiring company' technology cycle; see Table 4.1.

Table 4.1 Technology Scanning: Getting Information About and/or Access To External Technology Resources

Stage of development	Sources
Scientific research	Universities and professors
	Scientific conferences
	Scientific journals
	Internet
Technology development	University-industry workshops
	Patent searches
	Engineering conferences
	Nat'l. Assoc. of Technology Transfer Executives
	National Technology Transfer Centers
	Nat'l. Assoc. of University Technology Officers

	MCC (for microelectronics-related technology)
	Foreign "grey literature" (see Note at end of this chapter)
	Mergers & Acquisitions magazine
	Asian Technology Information Program (ATIP)
	U.S. National Laboratories
Product development	Competitive intelligence methods
	Society of Competitive Intelligence Professionals (SCIP)
	National Business Incubator Assoc. (NBIA)
	Private newsletters
Process development	SEMATECH (for semiconductor manufacturing)
	Competitive intelligence methods
	Talk to colleagues, counterparts at other companies
	U.S. National Laboratories
Procurement	Vendors
	Trade shows
	Trade journals
	Business Week, Popular Science,
	Technology Review, etc.
	Talk to colleagues, counterparts at other companies

Technology obtained at "late" stages may be fed back into the next generation of your own products and processes. However, there is competitive advantage in learning about and obtaining technology as early as possible.

The CEO of a large Northwestern electronics firm offered the following as among her favorite ways of tracking technology: *Science* magazine, *Technology Review, Upside, Red Herring, Forbes ASAP* supplement, *Electronic Business, Wired,* A.T. Kearny, SRI, and PricewaterhouseCoopers white papers, the H&Q technology investment conference, AEA's public/private company investment conference, and *Upside's* technology business conference.

4.2 Ensuring Access to Outside Technologies: Managing a Supplier Relationship

An assembler of IBM-compatible personal computers maintains an office of technology assessment, to scan for and analyze pertinent emerging technologies, and an alliances office to manage the relationships with its main technology suppliers. The company's future technologies group includes, in addition to these two offices, additional functions for university relations, for internal technical career development, and for "future technologies" dialogs with strategic customers.

Of course, the most critical suppliers for this manufacturer are the companies that make the microprocessors and the operating systems for the PCs. In consequence, many (but not all) of the technologies assessed by the technology assessment office are technologies that these suppliers are working on.

The future technologies group constantly monitors the market shares, yields, quality problems, and prospective stability of several leading microprocessor and operating system suppliers − notwithstanding that Intel and Microsoft are consistently the clear leaders on most of these dimensions − and continually seek to identify potential alternative suppliers.

Because the PC company orders large quantities of processors and OS over long periods of time, it regards the supplier relationships as partnerships, that is, as strategic alliances. They are very concerned to monitor whether the supplier companies view the relationship in the same way, and whether the classic characteristics of a strategic partnership (mutual benefit, mutual commitment, and a close, dense web of inter-organizational communication) are maintained. To this end, they maintain directories of each executive's and manager's counterparts at the current supplier companies, keeping track of their rotations and promotions, and enabling new lines of communication to grow quickly when managers leave any of the companies.

They conduct technological, strategic, and financial analyses of the major supplier firms, for the purpose of understanding the approach and attitude the supplier is taking toward the alliance. At regular intervals, they produce bullet-point verbal characterizations of each relationship. Bullets include: Do they understand us? How do they treat us? What do they want from us? Are they satisfied with us as a customer? Are we valued? How do we interact? Why?

This PC maker has devised the notion of "competition for a relationship / competition within a relationship." Other PC manufacturers have other things to offer as customers of Intel, AMD, Red Hat, or Microsoft − for example, different shares of the corporate PC market and different technological capabilities. They compete to be "first in the heart" of the supplier companies, and this competition must be monitored. This is "competition for a relationship."

Within each supplier relationship is the potential for competition − "competition within a relationship." A processor or OS maker may decide to vertically integrate and compete with its former OEM customers. They may decide to buy another supplier, reducing the PC maker's range of alternatives suppliers. Finally, suppliers may use price to compete for share of margin/markup. The PC maker is limited in its ability to control this kind of competition, but needs to monitor it for its early-warning value.

Managing the technology supplier relationship is a multi-dimensional task requiring the continuous attention of many people in the firm.

4.3 Ensuring Access to Outside Technologies: Technology Incubation, Japanese Style

4.3.1 Introduction

The Japan Research Institute (JRI), established in 1989 and known from 1969-1989 as Japan Information Service Ltd., is the central research institution for the Sumitomo group of companies. With about 2600 employees, JRI is one of the three largest private research institutes in Japan. (The other two are Mitsubishi Research Institute and Nomura Research Institute.) This section deals with JRI's concepts for a Japan Technology Incubator.

JRI's ideas are distinctive. They propose to execute a three-phase incubator strategy, building a consortium incubator, industry incubators, and entrepreneurship incubators, in that order. Other distinctive features of the plan include cooperating via consortia to create and develop markets very early in the technology development cycle, and creating a modified *keiretsu* model of interfirm organization in countries other than Japan.

Although the Japan Research Institute's incubation plan is traditional in its emphasis on risk minimization, it is breaking with Japanese tradition by emphasizing non-governmental initiatives and the fostering of entrepreneurship.

JRI hopes to incubate consortia for construction industry robotics; in-situ vitrification; bio-remediation of waste sites; automobile recycling; and the energy-efficient community for the 21st century.

4.3.2 Phased Development of a Japan Technology Incubator

The Japan Technology Incubator concept has three top-level functions, introduced in a time-phased manner. The functions of the Japan Technology Incubator are:

1. *Become a Consortium Incubator.* A consortium is created for technology-business development by means of a strategic alliance of enterprises in different fields. The purpose of the alliance would be (1) to attain scientific and economic leadership, (2) to rapidly develop an emerging industry, and (3) to assist in starting small companies. Furthermore, the strategic alliance would be based on a cooperative strategy for competition. This strategy involves having enterprises in different fields (i.e., suppliers of different components, OEMs, customers and

providers of support infrastructure) create new technology products and fast-growing markets for them.

Such an alliance can minimize risk by spreading costs across the alliance members. Risk minimization would be achieved in the selection of participants for their leadership potential for developing the technology and market.

The alliance fosters market creation by maximizing synergy. The aim is that the simultaneous effect of the action of separate agencies will have a greater total impact than the sum of the individual choices.

The Japan Technology Incubator strategic alliance is different from strategic consortia or alliances that are primarily concerned with developing a technology. In the U.S., the business development of a consortium's technology is left to the individual members. In a very important sense, the market development is omitted in the U.S. The Japan Technology Incubator would use the alliance to determine how to create the global market as well as how to minimize risk by cost diversification.

According to JRI, cost diversification through a consortium involves a strategy for technology development as "chaos generation." What we think they mean is that technology development (especially development of new-to-the-world technologies) looks like a chaotic process. It is hard to see the proper order in which it should proceed. The criteria for utilizing a chaos process are to minimize cost – which means to have risk diversification by consortia participation selection – and to maximize sources, probably through an incubation strategy selection that IC^2 and JRI will research jointly.

2. *Become an Industry Incubator.* The Japan Technology Incubator would concentrate on selection of industry programs. Initially this would mean to develop an international network, possibly in the same way the consortia are incubated in the phase described above. The network would effectively use government contacts (government labs and grants), but this would stem from a non-government initiative. Finally, the industry incubator would "enforce a market network by information dispatching." In other words, once a consortium is fostered in a certain industry, efforts will be turned to incubate that industry. By controlling information about technologies and markets, the consortium will better control the market.

3. *Become an Entrepreneur Incubator.* This will require the creation of a strategy for incubation, accumulation of incubation know-how, entrepreneurship research and education, and finally incubation of incubators. In Japan, because entrepreneurs are few and because entrepreneurship education is not widespread, this step will follow the first two steps in time sequence. Ultimately, however, this step will create new kinds of business schools in Japan and a new generation of entrepreneurs.

4.3.3 Sales Points

These are the "sales points" JRI uses to solicit membership in the industry consortia. JRI translated these into English for us. In order to faithfully convey JRI's meaning, I have not heavily edited their "Japanese English." The sales points describe further distinctive aspects of the three phases of incubation envisioned by JRI.

1st - *Consortium as strategic alliance of different-field enterprises.*

"Different-field enterprises" include Japanese government ministries, private companies in Japan, small companies in Japan that are not part of the *keiretsu* groupings, federal agencies in the U.S., and United States corporations.

This JRI strategy is not new in principle, but is new in terms of scale. Japan has long had a *keiretsu* system that is tightly knit and highly cooperative, linking trading companies, manufacturers, distributors, etc. But the *keiretsu* are domestic (except for foreign subsidiaries), and exclude universities. JRI is now trying to include universities, include smaller companies, and export the cooperative strategy and structure. These three things represent the true innovativeness of sales point #1.

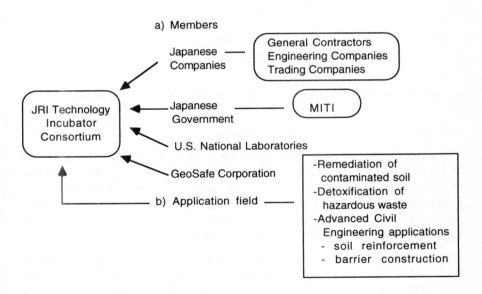

Figure 4.2 Example of sales point #1: In-situ vitrification consortium

2nd - *Risk Minimization by Consortium.*

This point includes the concepts of technical development as chaos generation, cost minimization, and maximization of success probability.

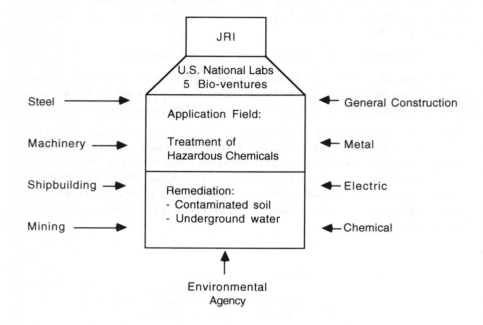

Figure 4.3 Example of sales point #2: Structure of the bio-remediation consortium

3rd - *Maximization of Synergy by Consortium.*

This point involves moving from a traditional R&D orientation to an integrated "R&D&D&M&I" orientation, that is, Research, Development, Design, Manufacturing, and Incubation. Far in advance of a product introduction – or even a product design – one must develop a strategy for market cultivation while still, as JRI puts it, "all at sea in a fog." They will achieve this by changing the way of thinking – from market prediction to market creation.

How do they propose to create markets? They will gather the big players, the firms that are likely to be the producers in such a market. Potential suppliers and customers, especially those customers disposed to being innovative "early adopters," are affected by very early marketing messages, long before a product hits the market. The companies may pool plans about target markets, and begin early-stage public relations. In this way, marketing becomes less a matter of market prediction, and more a matter of what the consortium is going to make happen.

The JRI construction robot consortium has specified thirty-two types of robots they wish to develop. Magazine articles, a book, and a television program on construction robots will be part of the market creation process – all prior to the actual development of the robots.

4th - *New Industry Creation by Non-Government Initiative.*

Traditionally, Japanese businesses wait for directions to be set by MITI and Keidanren. JRI is advancing a revolutionary idea for Japan, namely, that new industries can be created by non-governmental action. This idea further distiguishes two types of enterprises: regulation-induced businesses and deregulation-induced businesses. They state the consequence of this distinction thusly, "If regulation is changed, the market will be changed. If regulation can be changed, the market can be changed."

Two examples are JRI's Energy Community 21 Consortium, and their JCAR (Auto Recycling Consortium).

5th - *Incubation by International Network.*

Some parts of the JRI incubation work, of course, is aimed at finding U.S. and other markets for Japanese products. But others are aimed at jointly developing products and markets by and for companies inside and outside of Japan.

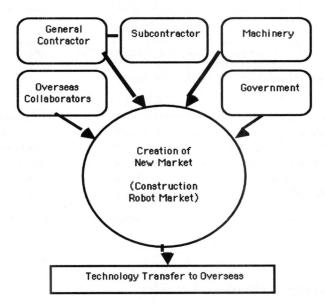

Figure 4.4 Example of sales point #3: Construction consortium

4.3.4 Conclusion

JRI contacts were frank in admitting Japanese shortcomings in entrepreneurship, and in saying that they hope their consortium incubation and industry incubation strategies, and the international networks built thereby, will result in a new generation of Japanese entrepreneurs. At present, JRI is attempting to incubate the consortia mentioned above. Each of these consortia will incubate an industry. The consortium incubator is "without walls." By the time the industry incubators come to fruition, groups of people will be working in proximity within one or more tangible facilities. This facility will begin to incubate businesses, and incorporate a training facility for entrepreneurs. These will be the new generation of Japanese businesspeople.

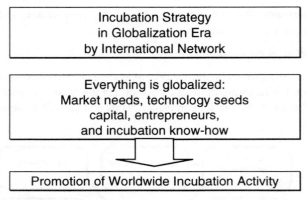

Figure 4.5 Internationalization of the incubation effort

As may be expected, this Japanese strategy shows careful long-range planning. JRI is bringing together companies and government agencies to consider "What are the technical and other obstacles to bringing this technology to full development and to market? How can we get an early start on overcoming these obstacles?" They are pioneering in needed areas – entrepreneurship and non-governmental initiatives – while respecting the need for risk minimization.

The Japan Research Institute is taking a creative yet realistic view of new technology alliances.

4.4 University Technology Transfer

Research universities are a source of new technologies for many companies. This section explains how such technologies become available.

While some university research is done under specific contracts with particular companies, by far the major portion of university research is supported by foundations and federal agencies. The latter include the National Science Foundation (NSF) and the National Institutes of Health (NIH). NSF, NIH and similar grants pay a small portion of a professor's salary during the grant period, but most goes to pay for graduate student stipends, research staff salaries, and other costs of research, including equipment, reagents and supplies, travel, and the preparation of publications. Most grants also allow some funds to partially cover the university's indirect costs (utilities, janitorial, library, the university president's salary, etc.) of supporting the research function.

Publications in peer-reviewed journals are a dominant factor in a professor's promotion and prestige; not surprisingly, then, such publications are very important to the professor. Being first to publish a significant research result can increase a faculty member's ability to attract new grants, the best graduate students, academic prizes, and job offers. Peer review and archival publication serve a crucial role in the progress of science, ensuring that the sharpest possible description of an advance, and the experiments that led to it, are recorded and disseminated for the use of other scientists. They also ensure that history attributes the advance to the scientists who first achieved it.

Business people interested in commercializing a research result must understand the powerful pull of publication – because it can work against efficient technology transfer. Here's why: Patents cannot be granted for advances that have been publicly disclosed. Academic publication puts knowledge into the public domain. Patents are granted only for technology which is not widely known, and publishing a paper is always construed as making its content widely known – even if few people read the journal!

Research done under contract to a company generally provides for the sponsoring company to retain all resulting intellectual property, and may require the professor to postpone publication until after the sponsor has had time to file directly related patents. In this case, technology transfer is straightforward (subject to the professor's ability to explain the results to the sponsor), but professors are reluctant to take on too much contract research. This is because graduate students, who are concerned about advancing their careers by making their first journal publications, cannot do so in the contract research environment.

For research work done under a federal grant, the faculty researcher must choose whether to publish an advance or "disclose" it to university technology transfer officials. Under the Bayh-Dole Act, U.S. universities may retain intellectual property (I.P.) rights to research conducted with federal funding. Universities maintain licensing offices, also known as technology transfer or technology commercialization offices, to handle this intellectual property for the benefit of the university and faculty. Upon the researcher's disclosure, the university typically decides whether to pursue a patent. If the decision is no, the I.P. generally reverts to the professor, who may decide to publish, or to file for a patent personally. If the university decides it is desirable to patent the advance, its next decision is whether to fund the patent application process internally, or to seek a licensee who will undertake the patent filing costs in return for a license to the technology. When a patent is obtained, the university often gives the original researcher first refusal on buying a license. The researcher may buy the license and start a company or sub-license the rights to the patent to a third party. If the researcher refuses the license, the university seeks an external licensee, either an established company or a start-up firm, depending on the nature of the patented research. The license cost usually includes an up-front fee plus a royalty percentage based on the licensee's sales of products using the university technology. In the case of spin-out companies headed by professors or students, universities are starting to consider taking equity in lieu of royalties.

Many universities share the proceeds from a patent according to a formula such as 1/3 to the university, 1/3 to the faculty researcher's department, and 1/3 to the researcher. The costs of filing U.S. and foreign patents on a new technology – to which often must be added the costs of defending the patent in court should it be challenged by competitors – can easily exceed $50,000 over the life of the patent and may be much more. Given that very few patents generate significant income, I.P. management is a risky proposition for a university. A few universities make money on patents; others conduct a licensing activity as a kind of perk for faculty; and others have no commercialization office at all.

How can a company interested in obtaining new technologies ensure access to university research results? Becoming a corporate member of a university research center gives a company first, but not necessarily exclusive, looks at lab results, and first option on licenses. Companies may sweeten the contract research pot by giving graduate students access to advanced equipment or other professional perks that offset the prohibition against publishing. Or they may shorten the publication delay period, counting on speed to market, rather than on a patent, as a barrier to competitive action. Other corporate university liaison officers identify promising researchers early in their careers, staying in touch with them and with their primary colleagues, hoping to influence the professors' choice to disclose vs. publish for selected research results.

It's important to realize that, from the social perspective, academic publication can also *increase* the efficiency of technology transfer: If a professor at University C, or a researcher at company D, is best prepared to build on University A's original advance, then an exclusive license from University A to company B may not allow University C or company D to learn about it, and no widely useful product will result. This is all the more true given the uncertainty of payoffs from commercialization; one commentator maintains that just 7% of original ideas (Intellectual Property or IP) are "worth pursuing." Only 27% of pursued IPs, he says, get patented and licensed. Only 32% of the licensed IPs get into manufactured products, and only 50% of productized and marketed IPs survive in the market for two or more years.   This, he notes, is an effective commercialization rate of 0.3%!

4.5 Inter-Sectoral Technology Transfer: The University-Industry Case

In this section, we will explore recent changes in the ways research is transferred from schools of management and business in the U.S. We choose this example because management research tends to be "invisible," making its transfer a more difficult process than the transfer of engineering research, and because much has been written elsewhere on transfer of "hard technologies." We will catalog the older and the newer mechanisms for transferring research; note some of the problems connected with each; mention the role of new institutions; and describe an "ideal" mechanism that is applicable to many kinds of inter-organization knowledge transfer.

Although quantitative methods had long been applied to production processes (time/motion study since the early part of the century, and linear programming since the 1950s), it was only in the 1960s that the "mathematical revolution" came to other business areas like finance and marketing. These very powerful advances in methodology put universities in the position of bestowing knowledge on industry. See Table 4.2.

Table 4.2 A History of Research in U.S. Business Schools

Era	Universities' role in research	Research Focus
pre-1960s	No research or little research	Institutional studies
1960s-80s	Universities lead industry	Mathematical modelling in operations, marketing
1990s	Industry leads universities	Team-building, leadership, entrepreneurship

Prior to this mathematical revolution, business research consisted of histories and case studies. During the heyday of the revolution, professors showed managers and policymakers how to improve operations, using new techniques invented at the universities. Examples include linear programming (for production planning, petroleum blending, etc.), conjoint analysis (for predicting customer response to product characteristics), the Markowitz model (for diversifying securities portfolios), and Data Envelopment Analysis (for evaluating the efficiency of public programs).

Now, however, the major benefits from these quantitative methods have been realized. New business problems have arisen – globalization of markets, the rise of high-tech companies, newly multicultural workforces, etc. – that are not amenable to solution by these methods. In the 1990s, most professors describe industry best-practices, and show "follower" companies how to apply the best practices. Also in the 90s, some advanced universities are participating in new, cross-sector institutions to, e.g., encourage entrepreneurship and new business formation. Examples include new business incubators, regional software industry associations, and the Japan Technology Management program (a consortium of U.S. universities that cooperates with JETRO and MITI's AIST to increase Americans' knowledge of Japanese technology industry practices.)

It must be said that not all professors accept this transition. Having spent their working careers at business schools in the decades of the 60s and 70s, some mistakenly believe that universities should always lead companies in knowledge and innovation. Actually, of course, new theory, new methodology, new data, and new problems are always playing leap-frog in every scientific discipline. Eventually the time will come when theory and methodology, the strong points of the university, will rise in importance again.

Table 4.3 shows eight ways of moving new knowledge in management from the university to industry. Of course, as I have just noted, this knowledge might arise in industry, then be analyzed, written about, and disseminated throughout industry by university professors.

Table 4.3 Ways of Transferring Technology Management Knowledge from University to Industry

1. Description	Publishing the histories of various companies' actions, experiences.
2. Application	Professor particularizes well-known theoretical knowledge to solve companies' common problems.
3. Industry Associates Programs	Industry members get first chance to own university innovations. Technology transfer is left to the member company.
4. Contract research	Company specifies a project. University carries out the project without company involvement; delivers result to company.

5. Problem-Driven Research	Jointly and simultaneously solving a company's problem and generating new theoretical/methodological knowledge. See Table 4.4.
6. University Spin-Off Companies	Start a new company to exploit professor's or student's invention.
7. University Incubators	Professors' experience and knowledge benefit new companies being nurtured within university buildings.
8. University/Industry/ Government Workshops	One- or two-day event with academic and industrial participants sharing latest research for open publication.

The first of these, historical exposition, is a powerful tool. But it takes a wise business person to apply the lessons of history. In any case, this kind of business study is often done by journalists, so university professors are not necessarily needed for it. "Application" is also a very straightforward process. Often it is done by trained agents, e.g., in the U.S. Agricultural Extension Service, who do not have or need Ph.D.s. Indeed, extension agents may be much better at implementation (for example, persuading farmers to use a new seed) than university professors.

The third and fourth methods noted in Table 4.3 are also common, but may ignore the significant "people problems" that are involved in effectively transferring research results from one organization to another. As a result, projects executed under plans 2, 3, and 4 often fail.

More and more universities are using methods 6 through 8. They engage students and the community. We have had much success with these at University of Texas at Austin and at Oregon Graduate Institute.

The fifth method, "problem-driven research," is the one that I was taught and that has (I believe) the best chance of successful implementation. It is detailed in Table 4.4.

Problem-driven research brings the academic mind to bear on new industrial problems, involves company employees at the analysis stage rather than imposing solutions on them, and can result in solutions that are generalizeable and lead to new theory. Its limitation is that it is time consuming for the skilled researcher, and cannot be standardized for widespread use by less skilled researchers. In particular, this approach is not appealing to those Ph.D.s who feel "married" to a particular methodology, such as simultaneous equation modelling. Problem-driven research requires a more general devotion to science, rather than to specific methodologies, and the researcher must be flexible about choosing the most appropriate methodologies or even inventing new ones. This often goes against the university's reward structure for the young, creative assistant professor.

Table 4.4   Problem-Driven Research: Ideal Scientific Method for University-Industry Cooperative Management Research

1. Determine that a problem is new or has significant subproblems that are *new*.

2. *In collaboration with an involved manager,* distinguish the pertinent elements of the problem.   Identify the goal that will be served by solving the problem. Examine available data and hypothesize about variate relationships.

3. Devise a *prototype* model for the practicing manager's approval.   If he or she sees that the prototype captures the major problem features and may significantly serve the specified goal, then cooperate with the manager to plan further data collection.   Begin to abstract the significant *mathematical innovations* in the model.

4. Collect data.   Use the data collection activity to give more of the enterprise's *workers a chance to "sign on"* and have a stake in the scientific solution of the problem.   Recognize that the means of organizing to collect the data will have parallels with the ways to organize for implementation of the solution.

5. Refine the second-stage model and its solution algorithm as needed. *Implement* the production version; *hand off* to the cooperating manager, who is now its "champion."

6. Refine the mathematical solution for journal publication – *with the manager as co-author.*   Look for parallel problems in other management areas that will lead to further refinement and abstraction of the new method, and constructive solution of additional management problems.   Confer with scientists who are competent to effect these refinements.

These days, more and more U.S. government grant programs require that the university perform its project in partnership with industry.   Table 4.5 provides just a few examples of programs that do this.   The programs do not, however, specify which of the mechanisms of Table 4.3 are to be used.

Unfortunately, the total available government grant money is shrinking.   An obvious solution is greater direct funding of university research by industry. Some exceptional universities have been successful in attracting direct industry contracts while maintaining earlier levels of federal research support.   But elsewhere, major culture changes will be required both on the university side and on the company side before direct industry support for university research grows.

Table 4.5   U.S. Government Support for University-Industry Cooperative Management Research

NSF-MOTI	Management of Technological Innovation Program.
NSF-GOALI	Grant Opportunities for Academic Liaison with Industry.
SBIR	Small Business Innovation Research (all agencies).
STTR	Small Business Technology Transfer Research (many agencies).
NSF	Engineering Centers. NSF Engineering Centers.

In the future, other new kinds of institutions will be needed to foster transfer of university research to society. I have mentioned incubators. University-industry-government workshops on newly emerging and interdisciplinary topics are not attractive activities from the point of view of traditional academic departments. New organizations must arise, probably within the university, to hold these workshops.

Only the largest universities can afford to patent their faculties' inventions aggressively. (The management advances discussed here lead to software patents with increasing frequency.) And even then, the activity is risky and rarely profitable, with large returns on only a few patents. New institutions are needed that will share risks and returns across many universities. It is not yet clear whether this is best achieved by academe, government, or private industry. However, consortia of universities are starting to emerge (I have mentioned the Japan Technology Management program as an example) to make more effective and attractive use of dwindling federal research funds.

4.6 Chapter Conclusion

After selecting an external technology, the company must decide on the most efficient way to transfer the technology. Sometimes this is dictated by the industry structure or by the incubation strategy, but in any case, the catalog of mechanisms for inward transfer is as shown in Table 4.6:

Table 4.6: Ways to Acquire an External Technology

- License
- Acquisition
- Spin-in
- Joint venture
- Hire
- Purchase
- Reverse-engineer

Which of these mechanisms "work better"? It depends on the organization's commitment to effective technology transfer. One research study characterizes four models of T^2:

- The "People-mover" model
- The "Communication" model
- The "On-the-shelf" model, and
- The "Vendor" model.

The researcher proposed increasing understanding of these models so that a technology transfer team, working alongside the project team, could refer to the models and increase the success of the project. While project cancellations led to inconclusive research results, a good bit was still learned about "what works."

People-mover T^2 relies on the people who own the knowledge. That is, in order to transfer the technology successfully, you must transfer the people. This model assumes the technology is not yet mature enough to be easily transferred to other individuals. Implementing the people-mover model implies all the difficulties of hiring or transferring individuals.

The communication model suggests T^2 is possible through effective documentation and presentations. It assumes that design engineers will use the documentation and presentation forums.

The on-the-shelf model suggests a technology should be packaged so that it is easy for an OEM designer to incorporate.

The vendor model implies merely that the T^2 is effected by purchasing. What is purchased is a component, not knowledge; and the quality of communication and packaging is not considered.

The four models have been listed and described above in the order of the level of technological maturity that is appropriate for each. For the vendor model to work well, the technology must be so mature that we would hardly call it T^2 at all! The research study concluded that none of the models are likely to work well alone, and that regardless of the maturity of the technology, a mix of the first three models is most likely to ensure success.

4.6.1 Technology Plan of an Academic Unit

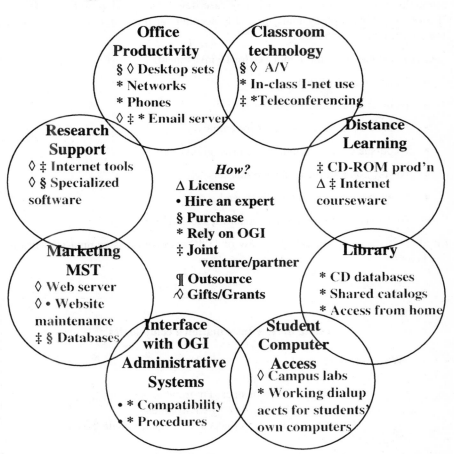

Figure 4.6   Technology plan of the Management in Science and Technology (MST) Department at Oregon Graduate Institute (OGI), 1998

The technology plan of the Management in Science and Technology department of Oregon Graduate Institute of Science and Technology utilizes several of the mechanisms of Table 4.6. See Figure 4.6. This plan is not a roadmap, because it does not include a time axis or explicitly represent the introduction of new technologies. It does, however, map each of the department's technology imperatives against one of the acquisition altnernatives of Table 4.6.

4.6.2 Acquiring Information Technology

To say that information technology (IT) is a "facilitator" is an oversimplification, as we shall see below; companies acquire IT for more reasons than simply to stay competitive in their markets. But in any case, few firms develop all their own IT. IT and IS (information systems) probably account for the majority of purchasing and outsourcing expense today.

Lucas (1999) identifies these motives for IT investment:

- To acquire infrastructure
- To meet legal requirements
- To provide productivity tools for employees
- To pursue direct and indirect returns on investment
- To stay competitive in current markets
- To enable strategic applications, and
- To develop tranformational capabilities for the firm.

An alternative and more detailed view of IT/IS acquisition is summarized in the box below.

Investing in New Information Technologies

- Why firms invest in IT
  - Expand market share
  - Prepare for future business environments
  - Improve quality of working environment
  - Improve quality of products and interaction with customers
- Types of IT applications
  - Communications and data handling
  - Mandated (government) requirements
  - Cost Reduction
  - New products
  - Improvements in quality of service
  - Strategic repositioning

- Role of IT in the evolution of work activities
  - As a tool
  - For linking services and activities
  - For outsourcing
  - Impacts on employees
- Common problem areas
  - Lack of competition (small number of vendors)
  - Inadequate planning
  - Internal resistance
  - Excessive scope of IT projects
  - Technology-driven vs. needs-driven projects
  - Software development and maintenance
- Critical issues
  - Using IT for business competitive advantage
  - Cross-functional reorganization
  - Continual user and customer involvement
  - Seeking payback
  - Auditing for performance
  - Benchmarking

Source: (1994) Information Technology in the Service Society. National Research Council Washington DC. Summary by Scott Hay.

4.7 Notes

Websites for cited technology scanning sources:

Asian Technology Information Program (ATIP)	www.atip.org; www.atip.or.jp
Business Week	www.businessweek.com
MCC	www.mcc.com
Mergers & Acquisitions magazine	www.wiso.gwdg.de/ifbg/mergers.html
National Assoc. of Technology Transfer Executives	www.nttc.edu/aft2d.html
National Assoc. of University Technology Managers	www.autm.net
National Business Incubator Assoc. (NBIA)	www.nbia.org
National Technology Transfer Centers	www.nttc.edu
Patent searches	www.patentcafe.com
Planet Science Daily Science News	www.newscientist.com
Popular Science	www.popsci.com
Science Daily	www.sciencedaily.com/index.htm

Science Friday (Public Radio)	www.sciencefriday.com
Science Now (subscription database)	www.sciencenow.org/
SEMATECH	www.sematech.org
Society of Competitive Intelligence Professionals	www.scip.org
Stanford University's Japan Technology Center	http://fuji.stanford.edu
Technology Review	www.techreview.com/

U.S. National Laboratories: Search for sites for Oak Ridge, Brookhaven, Pacific Northwest, Lawrence Berkeley, Lawrence Livermore, Los Alamos, Sandia, and other individual national labs.

By definition, Grey Literature is the information and resources that do not categorically fall into what is available via standard traditional or commercial publishing channels.... "The International Journal on Grey Literature (IJGL) is a forum for discussion of, and dissemination of knowledge about, the theory, practice, distribution channels, unique attributes, access and control of grey literature in a global context." The journal is edited by Julia Gelfand, Applied Sciences Librarian, University of California, Irvine Science Library 228 Irvine, CA 92623-9556 USA phone: 949-824-4971 fax: 949-824-3114 email: jgelfand@uci.edu.

University technology commercialization offices... The details of this pursuit are discussed on the techno-l listserv, http://www.uventures.com. University technology commercialization officers belong to the Association of University Technology Managers (AUTM).

One commentator maintains that just 7% of original ideas... Email from Chantzas C B (1999) on the techno-l listserv. November 19 (See http://cyberwar.com/~technobz/npd-success-rate.html.)

International Satellite Symposium and Conference (1997) Inter-Sectoral Technology Transfer: The University-Industry Case. Tsukuba University Tsukuba Japan March (based on the author's talk at conference)

Playing leapfrog.... Learner D B and F Y Phillips (1993) Method and Progress in Management Science. Socio-Economic Planning Sciences 27(1): 9-24 (Table 4.3 of this section is adapted from the same article.)

Start a new company to exploit professor's or student's invention.... Brett A, Gibson D and R W Smilor (Eds) (1991) University Spin-Off Companies. Rowman and Littlefield Savage MD

Professors' experience and knowledge benefit new companies being nurtured within university buildings... Smilor R W and M D Gill (Eds) (1986) The New Business Incubator. Lexington Books Lexington MA

One- or two-day event with academic and industrial participants sharing latest research for open publication.... Curtis H, Lopez W and F Phillips (Eds) (1996) The Asian Flat Panel Display: Technology and Strategy - Proceedings of a University-Industry Workshop, Portland, 1996. Oregon Graduate Institute of Science and Technology

One research study characterizes four models of T^2... Berniker E (1991) Models of Technology Transfer: A Dialiectical Case Study. PICMET'91 Portland State University Portland Oregon

Lucas (1999) identifies these motives for IT investment... Lucas H C Jr (1999) Information Technology and the Productivity Paradox: Assessing the Value of Investing in IT. Oxford University Press New York

4.8 Questions and Problems

Short Answer

S4-1. What do you imagine would be some facilitators and some inhibitors of government-to-civilian technology transfer?

Mini-projects

M4-1. Chart the decision diagram for the university patent/license process.

M4-2. Visit the websites of several major university technology commercialization offices, and inspect the descriptions of technologies available for license. Can you draw any broad characterizations of these technologies and their attractiveness to potential licensees?

M4-3. Analyze patent or "available technologies" data from Oak Ridge National Laboratory or from Pacific Northwest National Laboratories (Both labs are currently run by Battelle Memorial Institute under contract with the U.S. Department of Energy). The data for this exercise may be found on the Battelle website or the sites for the individual labs.

The exercise is quite free-form. You are to choose and "analyze" a subset of data from these sources. Analyze the data set to answer a question that is of interest to you or your company, or simply quantify and interpret a trend that you observe in the patent data or some clustering you may observe in the Battelle data. (Longitudinal analyses are preferred; you should go beyond a simple histogram of patents by tech area.) What industries will be most affected by these trends or clusters?

Other sites, including IBM or the U.S. Patent Office itself, may be used if you don't like the Oak Ridge or PNL sites.

M4-4. Here are data on patents won and patent income earned by 122 U.S. universities in a recent year. Analyze the profitability of the university commercialization activity described by these data. Is it worthwhile for all universities to attempt this? For some? Why? List your assumptions.

Table 4.7 University Commercialization Activity in a 12-Month Period in the Mid-1990s

University #	Royalties received	Licenses generating royalties	Patents issued	University #	Royalties received	Licenses generating royalties	Patents issued
1	$57,268,466	544	121	63	$9,851,391	21	6
2	$4,830,649	182	96	64	$1,916,778	14	8
3	$38,877,775	207	70	65	$385,691	30	8
4	$12,337,025	102	57	66	$253,822	34	8
5	$2,752,305	29	40	67	$122,607	15	7
6	$683,120	88	34	68	$32,806	0	7
7	$1,235,009	64	32	69	$378,569	1	5
8	$1,814,050	15	31	70	$292,461	8	7
9	$630,550	78	31	71	$88,657	10	7
10	$1,922,801	95	31	72	$88,286	10	7
11	$884,140	51	29	73	$47,274	6	7
12	$1,370,635	37	29	74	$4,923,137	10	4
13	$1,864,477	34	27	75	$1,472,771	7	4
14	$34,164,952	162	25	76	$1,150,390	1	6
15	$1,861,292	91	25	77	$327,672	20	6
16	$5,618,076	57	24	78	$212,975	13	4
17	$3,621,534	79	23	79	$209,599	26	4
18	$139,085	9	22	80	$139,192	3	6
19	$2,986,417	66	22	81	$51,887	4	4
20	$1,937,892	47	22	82	$4,102,593	2	3
21	$1,721,786	69	22	83	$2,604,173	47	3
22	$2,626,327	66	20	84	$803,021	49	5
23	$910,481	41	21	85	$480,308	19	5
24	$10,091,897	129	20	86	$453,385	22	5
25	$3,175,821	44	18	87	$108,674	13	3
26	$1,959,237	17	20	88	$100,783	19	3
27	$698,878	42	20	89	$159,307	9	3
28	$6,832,015	161	17	90	$64,712	20	3
29	$3,113,201	107	17	91	$24,334	18	4
30	$1,009,718	40	17	92	$10,598	19	5
31	$1,688,090	36	16	93	$3,078,001	8	3
32	$977,811	62	17	94	$163,281	24	4
33	$929,972	30	16	95	$135,117	3	2
34	$3,302,431	26	17	96	$125,781	6	3
35	$761,911	48	16	97	$49,777	2	3
36	$15,311,597	30	16	98	$56,393	9	2
37	$1,179,252	134	15	99	$47,348	1	2

Table 4.7 (Continued)

University #	Royalties received	Licenses generating royalties	Patents issued	University #	Royalties received	Licenses generating royalties	Patents issued
38	$2,345,877	10	14	100	$638,304	19	3
39	$1,251,939	32	12	101	$360,445	44	2
40	$374,019	18	13	102	$353,727	15	3
41	$566,087	30	13	103	$361,151	19	1
42	$473,129	3	13	104	$298,354	14	1
43	$420,465	23	12	105	$128,412	23	2
44	$4,379,997	17	11	106	$143,581	2	1
45	$1,101,987	0	11	107	$38,345	1	2
46	$721,417	18	12	108	$34,267	3	3
47	$304,099	42	12	109	$28,201	2	1
48	$869,073	36	11	110	$56,536	2	0
49	$698,665	12	10	111	$1,487	1	1
50	$692,597	16	10	112	$55,441	2	1
51	0	0	9	113	$2,067	1	0
52	$2,202,912	65	9	114	$2,933	1	2
53	$1,519,655	25	10	115	$222,824	n/a	1
54	$1,008,147	37	10	116	$148,972	7	1
55	$951,337	68	8	117	$26,124	3	1
56	$422,328	9	8	118	$24,897	1	0
57	$452,748	14	9	119	$17,461	8	0
58	$5,562,652	28	7	120	$43,888	1	1
59	$739,591	32	9	121	$530	1	0
60	$310,807	2	8	122	$822,358	15	n/a
61	307,621	5	8				
62	211,368	12	8	Total	$299,303,048	4,381	1,556

PART III

MANAGING TECHNOLOGICAL RISK

The policy of being too cautious is the greatest risk of all.
Jawaharlal Nehru

If it weren't for my bad luck, I'd have no luck at all.
Old blues lyric

5 Managing Technological Risk

5.1 About Risk

When wooing John Sculley to Apple from Pepsico, Steve Jobs asked, "John, do you want to change the world, or do you want to peddle soda water for the rest of your life?" Jobs did not ask whether Sculley wanted to forecast the future, but whether he wanted to change it. This is how today's "entreprenerds" deal with the risky future: First, identify a new, basic technology that will spawn whole new classes of products. Second, know more about that technology than anyone else. Third, inflict the technology on the world in the form of an audaciously marketed product. Fourth, watch for follow-on, spin-off, and tie-in opportunities – and take them.

Risk involves the future; the past can be "researched," but the future must be "forecasted," and business losses are attached to faulty forecasts. Risk is also bound up with ethics and social norms and values. Some high-tech development projects, as we will see below, must deal with these issues over a surprisingly long planning horizon.

The economist Frank Knight made the distinction between risk – the unknownness of future outcomes of a random vector whose distribution function we know – and uncertainty, which refers to unknown outcomes of a random vector with unknown probability law. It stands to reason that in the short run, much of the unknown future can be characterized as risk; in the long run, it's almost all uncertainty. This is illustrated in Figure 5.1, which conforms to the convention that our ignorance of the farther future is always greater than our ignorance of the nearer future. The Figure's suggestion that uncertainty completely dominates risk after 50-75 years will be explored later in this chapter.

Even in the short run, we attend too closely to risk and not enough to uncertainty. Sumitomo Bank traded commodities, and carefully studied the risk inherent in the variation of their prices. The bank's loss of $2.6 billion, due to a renegade copper trader who did not report his losses, was a consequence of uncertainty, not risk. It is tempting to think that Japan's famous "patient [investment] capital" is patient because technology corporations there are conscientious about the very long future. But I will argue later that this is not true. (The Japanese speculative bubble of the 1980s showed little evidence of long range planning, but let us lay that at the feet of the financial firms and the government.)

The magnitude of risk depends not only on the size of potential markets and the absolute size of the investment needed to exploit them, but on the fraction of the firm's resources that must be devoted to the development project. Boeing "bet the ranch" on the 767, while IBM exposed itself to negligible risk in its initial development of its PC.

Risk and uncertainty cannot always be quantified as the variance of a financial cost or revenue. Perceived (psychological or social) uncertainty is real and can affect the outcome or success rate of a business venture. A good current example is fears concerning security and privacy on the Internet.

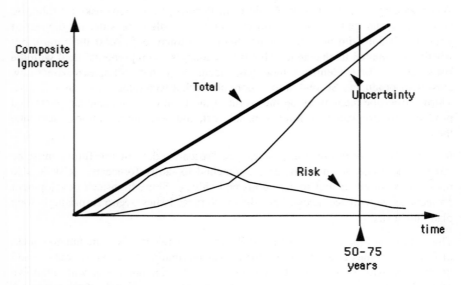

Figure 5.1 Risk, uncertainty, and the unknownness of the future

5.1.1 Dealing with Risk

Risk is not a bad thing *per se*. The business person wants to avoid or minimize the downside consequences of unfavorable realizations of risky situations. This is done in five ways: Better information for decision making; Optimal decisions; Reducing the cost of erroneous decisions; Reducing risk directly; and Pooling risk.

5.1.1.1 Better Information for Decision Making

Risk is reduced when management has accurate and timely information about the preferences and activities of customers, competitors, and sources of new technologies. This is why companies engage in market research, competitive intelligence, and technology scanning.

5.1.1.2 Optimal Decisions

Good internal and environmental information is leveraged for risk reduction by using the quantitative techniques of management science and decision analysis to make provably better decisions. An example is the use of conjoint analysis to predict consumer response to the attributes of a new product.

5.1.1.3 Reducing the Cost of Erroneous Decisions

Too large a production run of electronic components? Be prepared to "dump" them in foreign markets. A competitor has beaten you to market with the first offering in a new consumer entertainment device? Your ad agency must position your offering in a way that differentiates it from your competitor's.

5.1.1.4 Reducing Risk Directly

If left to their own efforts, the failure rate of entrepreneurs will continue at its high historical level. Companies interested in cultivating a strong base of small suppliers should support incubators and other new institutions for reducing the failure rate of startup firms in relevant industries. Companies can participate in standards-setting bodies in order to reduce the risk of offering non-compatible products, and in futures markets to smooth the prices paid for commodities.[1]

5.1.1.5 Pooling Risk

We hedge against extraordinary downside outcomes by devising insurance pools, building and loan societies, federal emergency services, and so on.

---

[1] This "structural" elimination of risk in no way implies cheating. But its meaning is well illustrated in one old movie:
Mae West (righteously, upon entering a room where poker is being played): Gentlemen! Is this a game of chance?
W.C. Fields: Not the way I play it...

5.1.2 Risk, Values, and Long-Range Planning

5.1.2.1 Long-Range Planning

History shows that liquid capital aggregations disperse after a few generations. Fortunes in cash, stocks, and bank deposits are diluted by war, division among heirs, inflexible corporate cultures, and so on. "Real" wealth, like the British crown jewels and the Vatican treasures, seem to endure longer, but the owners regard it as unthinkable to pawn these for investment purposes. Some cute fiction has been written about secret, non-corporate money managers still running the deMedici fortune after all these centuries. But 1997's Swiss banking scandals (concerning Nazi deposits of stolen funds) show this to be fantasy. If the secret bankers really existed, they would be used for these kinds of transactions.

This implies there is a top limit to planning horizons using ROI as a criterion. I think it's fifty to seventy-five years. After that, capital stops accumulating and starts decumulating. So beyond the 50-75 year point, what drives forecasting and planning? The drivers are ideas about the business environment and sociopolitical change. To the business person, new values and new social organizations mean new markets.

Anticipating the existence of a market two hundred years hence is no guarantee of being able to connect with that market's customers in the context of contemporaneous social values and so on. Nor does it invalidate the 50-75-year rule and guarantee a steady increase in your stock price between now and then. To return to the point about far-seeing Japanese firms: The existence of a 100-year plan is not necessarily an inspiration to investors.

5.1.2.2 Values, Risk, and Forecasting

The question of values cuts to the heart of forecasting, for three reasons:

- Technology and ideology are the dual drivers of socioeconomic change. This is the view that Dean George Kozmetsky has advanced at Texas for the past thirty years, and to which more and more communities and governments are being converted.

- One person's favorable outcome may be very unfavorable to another. If an outcome threatens a person's deeply held values or beliefs (abortion or irradiated food are possible examples), the most adamant resistance ensues.

- One man's unshakeable faith in an outcome (at the craps table, in the hereafter, or wherever) may be a source of great uncertainty or anxiety to the next man.

We may agree that the American business planning horizon is too short, and that net present value is not always a sensible planning criterion. Yet in the social context from which we will draw many of our next generation of employees, there is not even this rudimentary apprehension of the future. Some years back, a radio interviewer asked a teen-aged Los Angeles gang member what he expected to be doing in ten years. The young man replied, "I expect to be dead in ten years." How can you blame such a person for not having a sense of the difference between capital and yield, between spending and investing? What reason does he have to invest in the future? Do those of us with greater education and privilege share basic values and transmit them to posterity? Surely those values do not include allowing this young man's life to end in the way he expects and which reflects so poorly on our society. How have we allowed this to come to pass, and what does it mean for our future?

Let's look at some very small and very large examples of how social values and institutions interact with technology. Suppose, first, that there was a micromachine that could be cheaply introduced into rivers in great quantities. The machine would swim to the ocean, collect proteins, and convert them to a form that is not only edible but delicious when gently heated. Not only that, but the machine would, after a year, swim back upriver to its point of origin! Wouldn't investors be beating down the door to have a piece of such an enterprise? There is such a machine, of course, but there is currently no way to invest in salmon runs except through the tax-supported hatcheries and so on. Other business interests, shipping and cattle grazing for example, better fit the model of private enterprise and private property, and so have been encroaching on salmon habitat with alarming success.

People want salmon, and pay high prices for it. Here is a case of antiquated institutional structures interfering with nature's hugely efficient production of a valued food resource. My prediction is that institutions will change, but not without a lot of shouting and litigation, and in ways not currently foreseeable. For example, if I were to patent the world's first blue-eyed salmon and build a private hatchery for them, then any blue-eyed salmon caught in public waterways would be mine – or would it? The question would certainly be tested in the courts.

At the other end of the size scale, there is a clear relationship between long-range planning and macroengineering. The pyramids, stonehenge, the Suez and Panama canals, the space programs, and even some of today's large buildings and oil tankers take a long time to build and require faith in, or good forecasts of, political stability, commercial demand, availability of materials and qualified labor, and cost of capital.

The issue of values arises when we consider how some of these macro-projects were paid for. Many were motivated and financed directly or indirectly from war efforts. Others are straightforward monuments to despotic regimes. Still others exploited the religious faith of populations or exploited laborers who had no alternatives. The American industrial barons of the 19th century built a splendid network of railroads, on the backs of underpaid workers, then used the profits to donate libraries and other civic works to the American public. Hindsight tells us that these libraries were instrumental to the success of the American experiment, and to our individual lives. There is debate about whether the donors' purpose was to elevate the level of public discourse or to "tame" the immigrant workforce (Tisdale, 1997). So it is hard to come to an ethical judgment about the origin of these public works.

5.2 Technological Risk

Each step in the technology cycle involves uncertainties which may be seen as technological risk. The quintessential technological risks are: (i) a technology is chosen, from a number of alternatives, to form the basis of a new product or process. Large amounts of funds are committed to support the choice of this technology. Will a newer, better, cheaper technology, capable of fulfilling the same function, become available shortly after this choice is made? (ii) A production run is planned for a technology product. Will it overfill or underfill (like the Apple Powerbook) customer demand? More examples of technological risk arise...

- In the procurement of external technologies: The supplier may not perform to spec or schedule.

- In the development of internal innovations: The promised payoff of R&D projects is uncertain.

- The projected quality, cost, and schedule of every development project is uncertain.

- Customer demand and the shape of the product life cycle are likewise uncertain, thus affecting the timing and likelihood of recovery of investment.

- In the productive use of innovative technologies: There is risk of productivity loss as employees learn about and adjust to the new technology.

- In the combination of product and process technologies: If I use existing processes to make a new product, what is the extent of my cost disadvantage relative to a competitor who introduces a similar product made on new process technology?

- In changes and improvements in production methods: Are statistical methods for measuring the impact of a production procedure change adequate?

Still more questions about risk have arisen in our projects with the electronics industry:

- While it is known that risk-adjusted Net Present Value methods overstate the impact of risk, how do NPV methods compare to alternative methods of project valuation?

- How fast should management commit new funds to a risky project?

- How can market research information be combined with process development information at different stages of the product cycle to produce a measure of total-project risk?

- How can market size estimates be verified in time to commit to a fast-cycle project?

- How will a faulty product release (like the Pentium chip) affect the company's stock price?

Having a framework for thinking about risk, we now turn to the specifics of the new product development process, with special focus on its risks.

5.3 Product Innovation

New consumer product introductions of 1997 fit predicted patterns of innovation.

A record number of new packaged products for consumer households were introduced in 1997 – more than 25,000 of them. Companies introduced a total of 25,261 new food, beverage, health & beauty aids, household, miscellaneous and pet products in 1997, even more than 1996's old record of 24,496. Food and drink items were not introduced in numbers matching 1996, but the other categories carried the new record, with the leader being household items (up 49.9% over the number introduced in 1996, for a total of 1,177 entries).

"Packaged" items exclude clothing, some big-ticket and more durable consumer goods, and also bulk foods and chemicals.

According to Quirk's Marketing Research Review (February, 1998, page 6), the percentage of new products offering significant new or added benefits slipped since 1996 to only 5.8%. That is, 1997's new offerings were less innovative than 1996's, and significantly below the peak innovation year of 1986, when 18.6% of new products offered something really new in technology, formulation, market creation, or packaging, positioning and merchandising. Remember from chapter 1 that innovative products of Types II, III and IV are expected to comprise about 18% of the market.

Useful and appealing innovations can show up in new consumer nondurables, as shown by some of 1997's 5.8% of innovative new products. In 1997 we could for the first time buy Breyer's ice cream that stays soft enough to scoop, even just out of the freezer. New Huggies Little Swimmables disposable diapers somehow let babies swim without getting bogged down by absorbed pool water. Wrigley's Airwaves chewing gum time-releases vapors that ease nasal congestion. In Japan, Inaba tuna comes in a resealable standing pouch that is easier and safer to open than a can. And customers in the U.K. who are interested in hairless legs may buy a combination moisturizing cream and biological depilatory.

5.4 New Product Development in the U.S. and Japan

In consumer packaged goods, only two of a hundred new product projects reach the market, and typically only about 35% of these result in long term marketplace success. While this failure rate seems severe, in fact it lies in the middle of the rates for different types of technology products. In electronics, almost 8 out of 10 projects survive at least to market introduction, even if the feature set evolves considerably during the development cycle. But in pharmaceuticals, only one drug out of 5,000 initial ideas may be successful in the marketplace.

Like consumer packaged goods, software projects' failure rates also lie between electronics and pharmaceuticals. Thirty-one percent of software projects are cancelled due to uncontrolled cost, schedule, or quality. As for software functionality, in small companies about 75% of software development projects are deployed into the marketplace with 75% or more of the originally planned features. Large companies fare far worse: Their deployed software projects typically include only 42% of the originally specified features.

It's interesting to compare a new product development process typically used in Japan with the product development process used in corporate America. Since the mid-1980s, in a process of convergence[2], American companies have adopted the best practices of Japanese companies, and Japanese companies have done the same with the best practices of U. S. companies. The result of this decade of exchange is manufacturing and product development processes that are no longer terribly different. But examining the U.S.-Japanese extremes of the 1980s helps us to bracket and benchmark our own product development processes, and helps us understand why some things work and some things don't.

Part of the reason for a lower risk level in Japanese projects has to do with the government-industry relationship in the two countries. Japan engages in industrial policy much more than the United States, that is, in identifying strategic industries, channeling government resources, government protection, and loan guarantees into selected companies in strategic industries. Another reason is the

---

[2]   This is not technological convergence in the sense of Chapters 1 and 2 of this book; it is another use of the word convergence.

cross-holding of stock among companies in the same *keiretsu*. Companies that own each other's stocks do not pressure each other for short term share appreciation, nor do they threaten to sell the stock as stockholders in the United States may do.

Now, Japan having spent most of the 1990s in a deep recession, we understand that just because an industry is strategic doesn't necessarily make it competitive. Now we can see that companies and the Japanese government were simply shifting risk from one pair of hands to another.

When we look at the differences between U.S. and Japanese new product development, we see that Japanese firms freeze their designs much earlier in the product development process, or at least allow very few changes right before the product introduction. In the U.S., engineering change orders (ECOs) peak just prior to product introduction. And then, perhaps because the U.S. firms were not paying as much attention to quality as the Japanese, there is another peak in ECOs after the product introduction, which may be interpreted as fixing product defects.

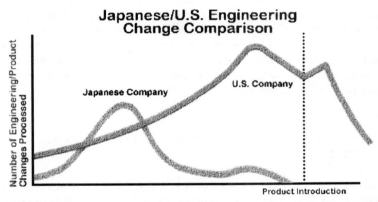

Figure 5.2 U.S.-Japanese comparison of patterns in engineering change orders (Data from American Supplier Institute. Artwork courtesy of Cenquest, Inc.)

U.S. companies tend to emphasize product design early in the development phase and worry about manufacturability and quality rather late in the development process. In great contrast, Japanese companies put a great deal of effort into quality and producibility early in the cycle.

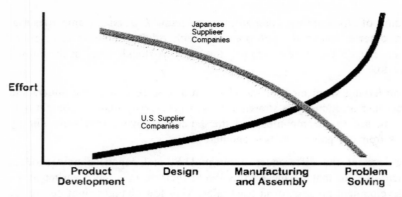

Effort

| Product Development | Design | Manufacturing and Assembly | Problem Solving |

Figure 5.3 U.S.-Japanese comparison of quality and producibility efforts across the development cycle (Data from American Supplier Institute. Artwork courtesy of Cenquest, Inc.)

We mentioned better communication between departments and functions as one way of increasing the hit rate of new products. Everyone has heard that Japanese companies are particularly good at this kind of communication, even among Japanese development teams for high technology products. Let's look at the way developers spend their time. Product developers spend time learning what they need to know in order to build the product. They communicate technical, market and process information with each other. And, finally, they spend their time *doing* – actually building the products:

$$\text{Learning} + \text{Communicating} + \text{Doing} \ \leq \ 24 \text{ hours each day.}$$

Naturally, as there are only so many hours in a day, the time spent communicating and learning has to be allocated among the available hours. What makes "high tech" high tech, is the vast amount of learning about the advanced technologies that go into a new product – as well as learning the unusual manufacturing capabilities that are needed to make the product.

$$\text{Increase in learning time} \ \Rightarrow \ \text{Decrease in communication time.}$$

Since there is an irreducible amount of time in the day that must be used for actually making the product, any increase in the learning required reduces the amout of time spent communicating. So, we must conclude that in high tech product development projects, the Japanese do not spend as much time as they would like to on communication. This is the reason for the recent Japanese emphasis on automated knowledge management.

As we look at a comparison of the classic differences between U.S. and Japanese new product development practices, remember that these are extremes and that much convergence has occurred over the last decade. In the U.S., the goal was the home run, the breakthrough product. In Japan, the goal was the base hit, or incremental improvement. We saw less R&D investment on the Japanese side and more use of licensing, more emphasis in Japan on process improvement, and more emphasis in the U.S. on product improvement.

Heavy reliance on pre-market testing	"Throw it against the wall and see if it sticks"
Much top management intrusion; reliance on frequent major reviews	Rare Reviews; frequent minor reviews conducted by engineering team
Reviews can drastically alter project funding level	Funding level more stable
Majority of ECOs late in development cycle	Majority of ECOs early in development Cycle
Excessive departmentalism	Cross-training & job rotation common
Product manager system	Group responsibility
Failures are buried	Learn from failures

Figure 5.4 Summary: Pre-convergence U.S.-Japanese comparison of new product development practices, I

The Japanese were leaders in managing new product development as a portfolio of projects, whereas products were more often developed on a one-by-one basis in the United States. Recently we've seen the U.S. move toward a platform management based new product development strategy.

On the U.S. side, a lot of market research data was collected directly from consumers. In Japan, market research consists primarily of the largely anecdotal observations of engineers and executives about consumer response, and the attention paid to purchase and shipment numbers. Where market research is a staff function in American companies and subject to the problems of communicating from one department to another, engineers and designers in Japan are directly involved with market research. Where American companies relied on surveys of potential customers before the product was introduced to market, which we call pre-market testing, Japanese market research was characterized by putting the product in retail stores, seeing if it sells and, if it does not, withdrawing it from the store. The latter strategy has been called "throw it against the wall and see if it sticks," a reference to the method used by some chefs to determine whether pasta is completely cooked.

Japanese companies led the way in eliminating major management reviews and giving more review and decision responsibility to the development team. American companies tended to be very reactive, wildly fluctuating the funding level of projects. In Japanese projects, the funding level remained more stable throughout. We've seen that U.S. companies tend to require a large number of engineering change orders or ECOs late in the development cycle – possibly as a result of changes in project funding – and that the majority of ECOs in Japanese projects came early in the project. Japanese companies led the way in cross-training of employees and rotating employees throughout many jobs and many departments. This had the result of encouraging trust in cross-departmental product development teams. In the United States, product development teams still suffer from departmental loyalties and mixed loyalties in matrixed project organizations.

In Japan, the development team is responsible jointly for the success or failure of the product. In the United States, we have a product manager system in which the success or failure of the product is identified with that particular product manager. As result, when a project fails, it's in the interest of the product manager to ensure that the project is forgotten as quickly as possible. This reduces and actually minimizes the incentive to examine the project very closely to determine why it failed. In Japan, because responsibility is more diffuse, it is much easier to dissect a failed project to see what can be learned for the future.

As an integrative exercise for this section, let's go back to the idea of Japanese market research. Knowing what we now know about the Japanese business environment and new product development process, how can we explain why the "throw it against the wall and see if it sticks" strategy is likely to work better in Japan than it would in the United States?

One explanation would cite flexible manufacturing as one of the reasons why the stick to the wall strategy works better in Japan. The Japanese led the United States in flexible manufacturing. This means that when demand for a product is not there, it's quicker and cheaper to change over the manufacturing line to a different product than it would be in the United States.

Another reason is that Japan led the United States in compressing development cycles for new products. The result is that a compressed cycle results in lower development costs, and therefore costs can be recovered more quickly. New products can be introduced more frequently, increasing the chances that one of them will be a hit with consumers. In a shorter development cycle, there is really no time to do pre-market surveys of customers. You might imagine that the throw it against the wall and see if it sticks strategy would work best for less pricey items, and certainly in consumer electronics, Japanese companies tend to produce small, inexpensive, single-function devices for the Japanese household, whereas U.S. companies tend to design bulkier, multifunction products for American households. So, of course, the transportation costs and the financial risk for the

American style home electronic products are greater, and that makes it less likely that a throw it against the wall and see if it sticks strategy would be cost effective in the U.S. Because Japanese market research has traditionally focused on the close monitoring of retail sell-through, mechanisms are in place to rapidly collect and utilize this information for decisions about whether to continue or discontinue a product.

U.S.	Japan
Goal is breakthrough product	Goal is incremental improvement (smaller, cheaper, more portable, friendlier, more convenient, more available)
Heavy product R&D investment	Borrowed and licensed product technology
Little investment in process improvement	Emphasis on process improvement
Strict distinction between pure & applied research	All research is applied
NPD managed project-by-project	NPD managed as portfolio of projects
Little attention to human design	Emphasis on ergonomics, design
Market studies are statistical analysis of opinion and intention data	Reliance on in-context observation plus purchase & shipment numbers
Market research is a staff function	Engineers, designers directly involved with market research

Figure 5.5 Summary: Pre-convergence U.S.-Japanese comparison of new product development practices, II

While we said that the throw it against the wall strategy is best for less pricey items, the fact remains that consumer prices in Japan are quite high and the Japanese retailer can charge more, relatively speaking, for a small device than an American retailer could charge. Sales and discounting are relatively unknown in Japan, and many consumer electronic items are fad-driven. All these things contribute to the efficacy of the throw it against the wall strategy.

In the distribution keiretsu system (more will be explained about keiretsu in the next chapter), the manufacturer owns the retail outlets. Consider the implications of this: It means that the manufacturer can tell the retailer what to put on the shelf. It means that the manufacturer agrees to buy back the unsold goods, thus reducing the retailer's risk. And finally, since most retailers in Japan are small shops, the store managers are able to get a first-hand impression of the customer's reactions to each product, and pass that back to the manufacturer very rapidly.

5.5 Concurrent Engineering

In the context of new product development (NPD), we can define *concurrent engineering* as the practice of turning consecutive tasks into parallel tasks, thus increasing information flow between participants in the tasks, and speeding the total project completion time. The subject belongs in this chapter because, as we have seen, risk inheres in insufficient or bad information, and in taking too long to bring a product to market. Concurrent engineering (CE) is a way of bringing more successful products to market in a quicker time, by reducing the risks of product development.

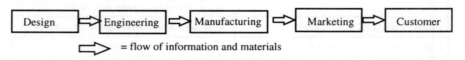

= flow of information and materials

Figure 5.6 Schematic of "sequential engineering" process in new product development, now seen as a high risk method

CE is made possible by new information technologies that enable better communication between task teams. It has been recognized by a number of government initiatives, including the conference that is summarized later in the chapter. However, definitions still differ, and some interpret CE extremely broadly, to encompass everything from Total Quality Management (TQM) to the kitchen sink. IDA (see the box below) believes that CE represents, and should continue to represent, a "systems approach" to development and acquisition, and so perhaps this diversity of views is unavoidable.

Table 5.1 Concurrent Engineering in Weapon System Acquisition: Principal Findings of the Institute for Defense Analysis

1. Concurrent engineering has been shown effective in deploying a variety of products faster, better, and cheaper.

2. This has been true in the Department of Defense (DoD) acquisition process as well as with other kinds of systems and products.

3. "There are systemic and individual inhibitors to the use of concurrent engineering in weapons systems acquisitions."

4. DoD should encourage the application of concurrent engineering in weapons systems acquisitions.

5. Concurrent engineering must remain a system-wide approach applied across disciplines; it will be "counterproductive" if it devolves into a new academic/professional specialty with the usual walls between disciplines.

6. The methods and technologies of concurrent engineering need further refinement.

7. Several companies reported concurrent engineering did not show results early enough to sustain funding within the companies.

8. "Implementation of concurrent engineering requires top-down commitment across different company functions." Benefits may not appear for several years, so successful small demonstration projects may be needed to sustain enthusiasm within the firm.

9. Such pilot projects have been shown useful and successful for this purpose.

10. Concurrent engineering was found to be useful in a variety of systems, products, technology types and technology maturity levels, both in product and production process development projects.

Source: Institute for Defense Analysis (1988) The Role of Concurrent Engineering in Weapons System Acquisition. Alexandria VA December; Department of Defense (1983) Life Cycle Costs in Navy Acquisitions. MIL-HDBK-259 April 1(NAVY)

Recent advances in computers, databases and electronic communication, and the generalization of CAD and NC tools, have rekindled industrial and academic interest in concurrent engineering design (CED). 1991 saw the first international gathering of CED experts about research priorities in CED. The International Workshop on Concurrent Engineering Design at the University of Texas at Austin enjoyed outstanding presentations by three plenary speakers, and small group discussions structured along industry lines and, later, by functional areas.

According to plenary speaker Dr. Hans Mark, government can help domestic industry speed new products to world markets by creating a supportive policy environment. The success of the U.S. aeronautics industry attests to this principle.

A representative of that industry, Patrick Kelly of McDonnell Douglas, emphasized the contributions of concurrent engineering (CE) to the goals of competitiveness and return on investment. McDonnell Douglas has realized gains toward both goals through such innovations as design for assembly/maintainability. The company has realized significant reductions in product development time and recurring cost, unplanned design changes and sustaining engineering activity, as well as greater flexibility. At McDonnell Douglas, CE is called "Integrated Product Development" to emphasize that CE integrates "from the outset, all elements of the product life cycle (conception through disposal), including quality, cost, schedule and user requirements." Mr. Kelly emphasized the importance of early and continual communication with suppliers and customers. Obstacles to further gains include the unfamiliarity of e.g. maintenance engineers with the design process, and poorly understood group dynamics in product development teams. Mr. Kelly identified education and training, improved electronic product definition, international standards, and tools for simulation and efficient communication of data among heterogeneous computer systems as the critical needs of industry for implementing CE.

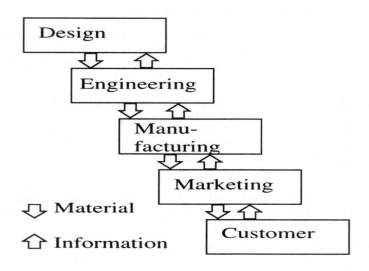

Figure 5.7 The "waterfall" model of concurrent engineering, featuring 2-way information flows

Dr. Nam Suh of MIT continued the discussion, giving technological innovation and management practices equal importance in improved industrial competitiveness. Concurrent engineering is among the management practices that are critical for success. Attention to CE is needed within firms and in the educational institutions, because CE is critical to competitiveness and because advances in computers and communications make CE ever more practicable. Although teamwork and interdepartmental communication can be achieved, to some extent, by preaching and cheerleading, these methods have no place in academia. There, basic principles and fundamental knowledge about engineering design must be generated and transferred. Dr. Suh concluded his talk by offering a list of areas in which such basic knowledge must be developed. They are CE theory, design decision-making methodologies, design aids and tools, and simulation methods.

Breakout sessions on the first day of the workshop were defined by functional area. The first group addressed research priorities in manufacturability and producibility.   Although Dr. Suh had stressed the importance of independence of design components, the members of this first group discussion saw the role of concurrent engineering as identifying the interdependencies and constraints that exist over the design/build/maintain/dispose cycle, and ensuring that the design team is aware of them. They expressed the concern that the set of constraints should not overwhelm the design team or excessively stifle its creativity. A truly creative design that satisfies customer requirements in a superior manner may justify the expense of relaxing some process constraints.

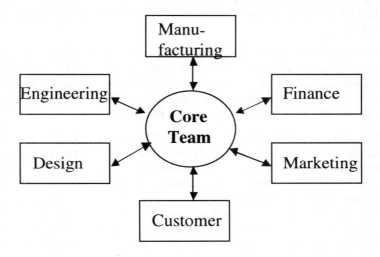

Figure 5.8 Due to the consulting firm Pittiglio, Rabin, Todd and McGrath: Core team model of concurrent engineering

The group's first research recommendation aimed at maintaining flexibility in design. How, they asked, can ambiguous objects, incomplete designs and design alternatives by represented electronically? Existing CAD tools are rigid and time consuming. How can a creative design idea be sketched electronically? Manufacturability and producibility are usually referred to as yes/no propositions, but really imply a spectrum. How can manufacturability and producibility be measured? Life cycle cost models are an important part of the answer.

A second discussion group looked at CE in assembly. The group's first insight was that input to the design process from all life cycle stages is a fine ideal that is not intellectually rigorous. At *exactly* what point in the CE process should discussion of assembly sequences, tolerances, and subassembly definitions be introduced? Other recommendations addressed safety and environmental concerns. Will disassembly be safe, and will it support easy separation of materials for recycling? Once again, tradeoffs were identified. Consolidation of parts is desirable in Design for Assembly, yet too much consolidation implies costly and inefficient procurement and inventorying. CAD and feature based design tools were described as inadequate for capturing the full range of information that determines an assembly procedure.

Statistical tools and improved sampling plans are needed to monitor quality. These tools should be multivariate and time-based. More than monitoring tools is needed, however. CAD systems should include facilities for estimating reliability of a machine, based on its digital design, and the mechanisms for integrating CE and TQM should be specified. It is not now clear how the latter two tasks can be approached.

Inspection and testing reveal information about the performance of the product and its production process, and can be seen as fitting naturally into the CE scheme. Session participants identified "lack of a body of codified knowledge on design form inspectability and testability" as a research problem. Of course, applicable databases and electronic design tools are even farther from view.

The fifth group discussed ways to optimize maintenance and reduce operations and support costs for deployed systems. They emphasized that design for supportability has a great impact on support costs. In addition to technology needs, they identified cultural changes, educational changes, and the restrictions of government specifications and standards as support issues. Design engineer's difficulty in obtaining product performance data was identified as an obstacle. The group's recommendations include ensuring that supportability issues are included in the up-front design process, developing standardized software packages for design and logistic functions, implementing a system for feedback, and developing technologies for self-healing systems to reduce repair and maintenance burdens.

The sixth group discussed CE's relationship to the design and layout of facilities and equipment. In the past, this area has been neglected; earlier work has concentrated on the use of tools of CAD, CAM, CAPP, etc. to support the concurrent design philosophy. Several recommendations were forwarded. The group suggested the creation of a publication on this topic. Alliances between universities and industries are necessary to further progress in areas such as the development of software and tools for the design and layout of facilities and equipment.

The seventh group covered economics and cost management issues. Sometimes, cost must be traded off against meeting the customers requirements; in other cases customer requirements can be met at a cost that is lower than that of current products. Therefore, cost information should be available to the design teams – not just to the accounting department – at the time a design is being developed. The group suggested a literature review of cost management as it applies to CE. Also, the group recommended identifying successful cost management practices to serve as case studies for design teams, and conducting surveys of CE companies to gain further insight. Ultimately, comprehensive cost models must be built, integrated and automated.

The eighth group concentrated on the reward system and the organizational culture for CE. The most pressing problems include the lack of understanding about the reward structure, the lack of creativity in the process design, lack of commitment to quality, and the existence of the myth of a service economy. Implicit beliefs must be made explicit, and included, where appropriate, in the product specification. Increased efforts in communication and training can ameliorate these problems, but in many ways culture is the dominating barrier to CE.

Following a plenary discussion of the above results, workshop participants dispersed into six breakout sessions organized by industry group. The unique characteristics of the computer industry were identified by the first such group, along with the critical issues affecting that industry. Research recommendations included the creation of a unified product data model from legacy databases, the development of tools to permit the concurrent development of hardware and software, the developing of software systems to support group communication, and the documentation of benchmarks for computer-supported CE implementation.

The Japanese have demonstrated the benefits of implementing solid-state electronics product design and advanced manufacturing processes in the consumer electronics industry. In order to compete in this industry, new consumer products must exhibit increasing levels of performance, quality, reliability, and maintainability, and be developed within a shorter time cycle. This situation indicates a need for improved understanding of CE tools, methodology, and the systems that support it. Some of the areas that need further research include 3-D graphic design tools, team formation, cross-cultural information systems, partial information/risk sharing system, and hands-off analysis tools.

The aerospace industry group viewed CE as having the potential to increase productivity and shorten design cycles. In order to achieve these goals, the industry needs an increase in government financial support, improved procurement practices, a well-rounded and well-educated workforce, a working template design without too many specifications, a compatible reward system, commitment by top management, measures of CE success, a team building atmosphere among engineers, and increased data exchange. Research issues specific to this industry include CE models for large, one-of-a-kind products; data sharing under government and military rules; and introduction of CE awareness throughout very large organizations.

The machinery and equipment industry discussion group chose to concentrate on batch manufacturers that produce products in low to medium volumes. The group recommended the following: formulate a life cycle concept and CE models for small to medium industries; discuss the potential cultural and organizational changes that facilitate CE application; and standardize product strategies that support CE. Standardization and modularization of parts and assemblies was a special issue for this industry.

The "CE in Education" breakout group noted an apparent lack of interest in CE within educational institutions. The design engineer of the future should be exposed to a more extensive curriculum in CE, organizational dynamics, probability and statistics, and the production/manufacturing process. Barriers to such an approach in education include the traditional disciplinary division of scholarship, and universities' emphasis on analysis over design. Team teaching and computerized instruction can help breach these barriers. With the help of educators, CE can reach its maximum potential.

Table 5.2 What is Concurrent Life Cycle Planning?

- Simultaneous consideration of all "downstream" costs (e.g. costs of manufacturing, logistics, maintenance and disposal) as early as possible in the product development process
- Simultaneous planning for successive generations of products
- Planning for the compatibility of the life cycles of all products in the firm's product line
- Simultaneous consideration of the life cycles of the product's main assembly and that of all subassemblies and components

The aim of CLC planning is optimal quality, cost, schedule, and profit.

It is clear that CE entails a more global vision of the manufacturing enterprise than earlier management concepts, and the drive to avoid suboptimization is apparent. CE is more than a "concept," because many tools and methods are now available for parts of the CE puzzle. The overarching life cycle cost models and electronic communications tools are not available, however, and specific designs for them do not appear imminent.

The four recurrent themes emerging from the workshop were models, tools, training, and culture. Participants identified measurement issues and tradeoffs that will inform future models of new product development. Tool needs boil down to expanded CAD/CAM/CAPP capabilities that intercommunicate. Design, production, assembly, and maintenance involve so many alternatives, rules and degrees of freedom that database technologies will be strained. Yet it is implicit in these discussions that these items must be automated; that is, handling them optimally will be beyond the capacities of available human expertise. Regardless of the level of automation, training is needed on multiple job stations, in the impact of design on "downstream" tasks, and on teamwork and individual responsibility. The issue of training, however, appears to be subordinate to culture. Corporate culture, and how to change it, must be better understood. What are the individual, team, and department incentives that will improve performance? What are the implicitly accepted myths that inhibit an organization's progress? Must change be top-down, or can it originate in the ranks and have an impact? The identified need was for tools for "concurrent life cycle planning," which is a more general notion than concurrent engineering or CED.

5.6 Balancing Committed Costs and Expenditures

5.6.1 Introduction

It is often observed that the greatest portion of total product development costs are committed in the design phase. That is, new product design, engineering, production and marketing costs are committed, or "frozen," much faster than they are actually expended. More specifically, several sources report a pattern of committed vs. expended monies over the development cycle similar to that of Figure 5.9.

At each phase of this cycle, the firm attempts to reduce project uncertainty by pursuing increasingly accurate estimates of life cycle cost and of the likelihood of success in the marketplace. This is achieved by means of consumer focus groups, production pilot tests, destructive testing, test markets, and so on (see Chapter 7).

It has been informally observed that a firm should not commit funds at a rate that exceeds the rate of uncertainty reduction. This is apparent from the following argument: A given return on investment (ROI) is projected for a project, and a certain investment has been committed. Given that subsequent information confirms the expected ROI, what incentive would justify committing additional funds? For a risk-averse manager, the answer is, "only a reduction in the uncertainty surrounding the ROI estimate."

5.6.2 Committed Costs

Committed costs are fixed future streams of expenditures. Within this definition are leases and other contracts (with suppliers, customers and employees), warranties, mortgages, and capital amortizations. Such commitments can reduce uncertainties of supply, eliminate some search costs, increase demand, deny opportunities to competitors, and lock in lower costs over the life of the commitment.

Designs, tooling, and other hardware are also frozen at the time a decision is made to commit funds. Another relevant category of costs, then, are those that flow consequentially from a design decision. If, for example, a design calls for aluminum rather than plastic, aluminum must be purchased and machines for forming aluminum must be acquired. According to leading researchers at another NSF design/manufacturing conference, "These detail design decisions... in large measure determine how big the factory will be, how many assembly operations will be required, and what the component purchase costs will be." Some of these costs, e.g. the incremental cost of acquiring aluminum processing equipment rather than plastic molding machines, are known. Others, such as the future relative prices of aluminum and its alternatives, are subject to uncertainty.

The first category, *committed* costs (including leases etc.), are nonrecoverable if the project fails. The second category (materials etc.), although possibly recoverable, are irreducible if the project proceeds.[3] The collection of both categories might be called *determined* costs. In the next section, we examine the foundations of Figure 5.9 and find that this distinction is meaningful.

Freezing designs and costs makes "downstream" tasks more efficient, because engineers and managers know what it is that must be manufactured, advertised and distributed, and under what conditions these tasks must be executed. Contingency plans dealing with other design alternatives can be discarded. On the other hand, of course, committed costs reduce flexibility. They make less cash available for alternative investments that may come to light. If market estimates or consumer requirements change, costs sunk into the wrong design are not recoverable. Managers ask, "What would it cost to shut down this project?" The answer to this question is, "the present value of all nonrecoverable costs." This quantity, together with speed of response, are components of the firm's flexibility.

Under sequential engineering, a prototype could be built and tested before beginning the search for the best production method. The costs of prototyping included only the cost of building and testing the prototype. Under concurrent engineering – and pressure toward rapid time-to-market – the prototyping activity

---

[3]   Naturally, negotiating may in some cases reduce these costs. This fact does not affect the present argument.

includes looking ahead to manufacturability. The cost of investigating and deciding the mode of production is now part of the cost of prototyping. This cost is now committed – and possibly expended – earlier. And, if the project is killed, these costs are not recovered. At any stage of the development process, then, bailout costs are *increased* under concurrent engineering. These increases must be offset by extra revenues from early market launch of successful alternative products, if concurrent engineering is to add flexibility to the enterprise.

Clearly, maximum flexibility would result from holding design alternatives open as long as possible, and operating on a cash-and-carry basis, making the two curves of Figure 5.9 converge. But this strategy would eliminate the benefits of cost commitment that were noted above, and is certainly unrealistic in almost all cases of interest to industry.

5.6.3 Postponing Cost Commitments

What is the relation of cost commitment to uncertainty reduction? I would argue that the prospect of reduced uncertainty is the only justification for postponing cost commitments. In other words, if no further uncertainty reduction is expected, all project costs may be committed at the earliest possible moment.

At many stages in the new product development process, data are collected and estimates made regarding product performance, manufacturing process costs and alternatives, project schedule, alternative investments, and market response. We will refer to these collectively as product, process, schedule, and market (PPSM) intelligence. These intelligence gathering activities ("tests") are scheduled, and precede decision points (also scheduled). At each decision point, we assume, management may specify a value for "fraction of remaining project costs to be committed at this point."[4] Four additional choices are made at each decision point: (i) "Go," i.e. proceed directly to roll-out of the product with no further testing; (ii) "On" to the next test; (iii) "Skip" the next test and proceed directly to the next test but one[5]; or (iv) "No Go," i.e., kill the project.

---

[4]   Committed costs as defined above are a different quantity from the expenditures management *authorizes* at each decision juncture. In the event of a subsequent project shutdown, some authorized monies that have not been committed or expended may be recoverable.

[5]   For example in Urban and Katz, "...some conditions are identified under which a test may be bypassed." Urban G L and G M Katz (1983)   Pre-Test-Market Models: Validation and Managerial Implications. Journal of Marketing Research August XX:221-234

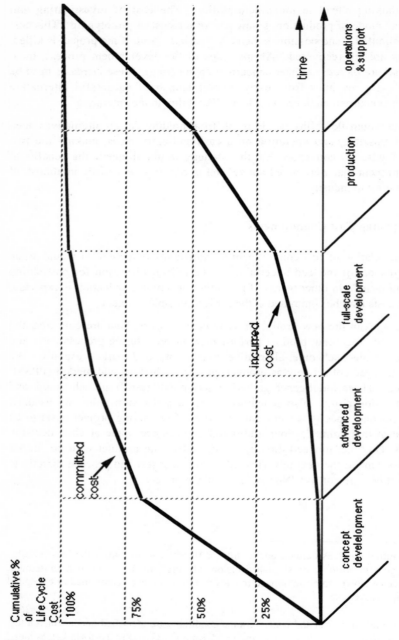

Figure 5.9 Committed costs vs. incurred costs

The first, second and third alternatives imply continuing with all scheduled production, advertising, and other "non-test" activities. Of course, for the fourth alternative, no further costs should be committed. For purposes of analysis, we assume no unscheduled tests or decisions are made.

These conditions, plausibly reflecting management practice, justify the assertion that all remaining project costs may be committed immediately if no further PPSM intelligence is anticipated. Moreover, if the aforementioned benefits of cost commitment are material, all remaining costs *should* be committed immediately under such circumstances. The following example, though unrealistic, will make the point: Suppose no further intelligence is to be gathered. The project schedule does not call for a factory to be leased until the product prototype is complete, and that activity is just beginning. Why should a lease be signed now rather than at the later, scheduled time? If a lease decision is to be made after a search for the best location and lease terms, then the search and decision should be in the project schedule, and the lease signed at the scheduled time. (We hold open the possibility that available lease terms may be so prohibitive as to merit a "NoGo" decision at that time.) If such a search-and-decision is *not* scheduled, then the lease *may* be signed immediately; there is not, nor will there be, any information to indicate the contrary course of action. There is no incremental cost attached to making this commitment now rather than later; the cost of capital is relevant only to the time of expenditure, not the time of commitment.

Figure 5.10 summarizes the possible trajectories of committed costs from a decision point "i." In the case of the "On" decision, an additional decision on committed costs must be made at stage i+1. Other decision options determine the path of committed costs for longer time spans.

5.6.4 Product Development Risk: Empirical Results

5.6.4.1 Company A

In this section we present and compare results drawn from two individual firms, one on the basis of secondary data and one on the basis of primary (original) data.

McGrath, Anthony and Shapiro offer the needed cost and risk data drawn from a single firm ("Company A") and from "other companies considered to be the best" product developers in Company A's industry. (The consulting firm with which these authors are affiliated works largely with firms in electronics-related industries.) McGrath *et al.* refer to the committed costs associated with cancelled projects as "lost investment." The attrition and cost data for Company A appear in Table 5.3.

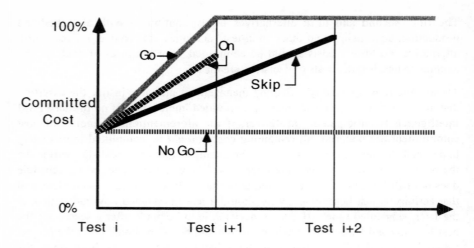

Figure 5.10 Possible management decisions at each project review

Table 5.3. Attrition and Committed Cost Data for Company A's Failed New Product Development Projects

Development stage	% of cancelled projects failing in this stage	Lost investment due to cancelled projects*
concept evaluation	19%	3%
planning & specification	26%	9%
development	37%	55%
test and evaluation	14%	24%
product release	5%	9%

* expressed as percent of all lost investment on cancelled projects

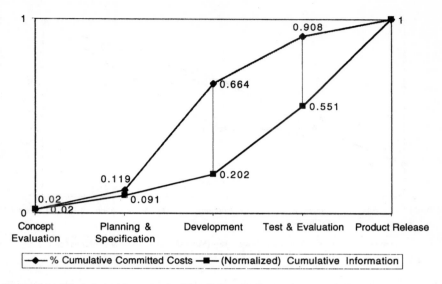

Figure 5.11a Company A's rates of uncertainty reduction and cost commitment

Figure 5.11a displays the cumulative information gain (the mathematical derivation of this entropy-type information gain measure is not included here) and cumulative committed costs for this firm, normalized to the same scale. The names used on the x-axis for the development stages are those used by the consulting firm. Note that Figure 5.11a differs from Figure 5.9 in that the lower line is information gain, *not* cash expenditures.

This firm's committed cost (the upper curve) shows the same concave shape as the corresponding line in Figure 5.9. The fact that this curve lies consistently above the risk-reduction curve shows that the firm is, intentionally or not, risk-inclined.

Figure 5.11a's comparison of rates of cost commitment and uncertainty reduction lends itself to a very simple index of risk behavior. To construct the index, simply sum the differences between the latter two rates at each development stage, then divide by the number of free points of comparison. For the company represented by the McGrath *et al.*data, the risk index is (.024+.462+.357)/3=.281. The completely risk-neutral firm would of course have an index of zero, and risk-averse firms would have a negative index.

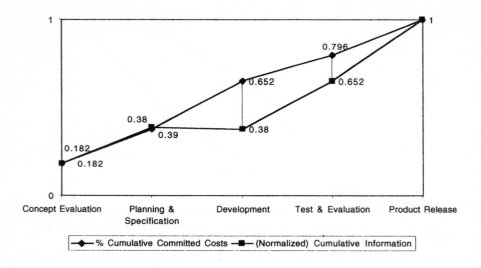

Figure 5.11b: Information gain (uncertainty reduction) and cumulative committed costs for Company A's industry benchmark firms

Figure 5.11b displays the same analysis for the McGrath *et al.* data on the industry leaders. McGrath *et al.* do not reveal how many companies are summarized in this aggregate. The closer convergence of the two curves in Figure 5.11b, and the risk index value of -0.049, show that the industry leaders in Company A's industry were slightly risk-averse but generally matched risk reduction and cost commitment more skillfully than did Company A. This analysis supports McGrath *et al.*'s contention that the leaders in Company A's industry are superior managers of the product development process, at least relative to Company A itself.

McGrath *et al.*note that the development stage is usually where most of the development investment is made. This is the stage where their industry leaders are most risk-averse. (The leaders become risk-inclined in the test and evaluation stage.) It is also where Company A itself is most inclined to take risk. They note further that "the best practice companies [shown in Figure 5.11b] developed 48 products at a cost of $60 million, while it cost the case-example company $75 million – 25% more [to develop the same number of products]." McGrath *et al.*fault Company A for scuttling an insufficient number of unworthy projects in their early stages. With equal justice, one might fault Company A for having too many bad product ideas in the first place, or for committing too much investment too early. The new product development battle can be fought on any of these arenas, and the risk profile developed in this paper does not unduly emphasize any of the arenas.

5.6.4.2 Company B

From a large, diversified manufacturer of industrial and office products, we obtained data on twenty new product development projects randomly chosen from a file representing all the firm's business units. For project planning and evaluation purposes, this firm (Company B) divides such projects into four stages: Concept, Development, Manufacturing Scale-up, and Field Test. In each stage, a project manager must estimate the updated internal rate of return, market risk, and technology risk. This firm uses only owned factories, and prefers line extensions to new businesses; by avoiding leases and unfamiliar new equipment, committed costs are minimized. Nonetheless, it is remarkable that the two curves of Figure 5.12 match so closely. Indeed in the manufacturing scale-up phase, the curve representing rate of cost commitment falls below the rate of information increase.

The raw data show that a few of the failed projects incurred expenditures in excess of the forecasted life cycle cost. Indeed, the overrun may have contributed to the NoGo decision. The upper curve of Figure 5.12 was derived using only the failed projects, for which the shutdown costs were known with certainty. The shutdown decision may have been based on cost overruns or on the perception of relatively low shutdown costs, so the extent and direction of bias in the placement of the upper curve is a matter for conjecture. (These comments apply also to Figures 5.11a and 5.11b.) Also, the location of the left end of the line at zero is arbitrary, as the data contained no projects that failed at the "start" stage. However, the fact that the lines nearly coincide does not depend on this arbitrary choice. The 50% rate of conversion of concepts into market successes, and the close convergence of the two curves in Figure 5.12, demonstrate that Company B's reputation as a superior developer of new products is well-deserved.

For Company B, the risk index takes a value of $(.10+.05-.09)/3=.02$. This index, although a useful summary, obscures interesting diagnostic features like the crossover in Figure 5.12. There, the firm in question becomes more risk-averse in the scale-up stage than it was in earlier stages.

A 1990 SAMI report revealed that of 6,960 new brands introduced in the two previous years, only 240 reached the $1 million annual sales mark. This is slightly less than 3.5%, quite a different number from the 50% post-introduction "success rate" actually achieved by Company B. The low 3.5% figure may be peculiar to the consumer package goods industry.

The overt problem in new product development is to increase the "hit rate," i.e. the proportion of concepts that become market successes. A collateral problem, no less important, is reducing the cost of failures. It is the latter problem to which the above analysis is most applicable. This section has discussed the nature of committed and determined costs, and quantified their relationship to the reduction of project uncertainty. We have introduced the concept of "product, process, schedule and market (PPSM) intelligence" and emphasized its use for jointly considering marketing and production factors in project evaluation.

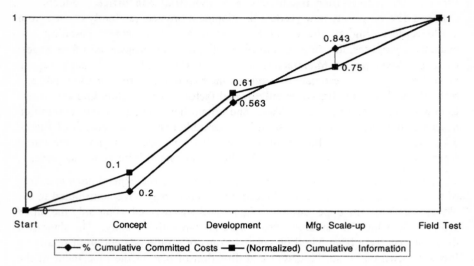

Figure 5.12 Uncertainty reduction vs. cumulative committed cost, by development stage: Data from an innovative, diversified company

5.7 Balancing the Project Portfolio

5.7.1 Capital Budgeting with Linear Programming

Suppose that we are considering n possible development projects or capital projects, and each must be funded on an "all or nothing" basis. We can write the binary decision for each project "j" as x_j, where x_j will equal one if the project is to be funded, and will equal zero otherwise. It is likely that we have a forecast of the *total* capital budget (available funds) for each of a number of future years; let's write these forecasts as b_i, where i ranges from 1 to 5 years. We have also estimated the capital requirements for each candidate project in each future year, c_{ij}. Finally, the accountants have computed a net present value NPV_j for each candidate project j. It then makes sense to allocate our capital budget to projects in a way that maximizes the total project portfolio value

$$\sum_j (NPV_j)(x_j)$$

subject to

$$\sum_j (c_{ij})(x_j) \le b_i \quad \text{for each of the five years indexed by i, and}$$

all x_j are either zero or one.

The binary requirement on x_j means this budget must be determined by integer programming. The more up-to-date versions of PC spreadsheet program optimization add-ins provide an integer programming capability. Microsoft Excel's "Solver" is an example.

If partial funding is a reasonable option for all the projects under consideration, the binary requirement can be relaxed, and it can be replaced in the optimization by the requirement that all x_j are between zero and one. The resulting, optimal x_j then represent the best funding level for each project (expressed as a fraction of the maximum funding for the project), assuming this funding level remains constant over the five years. The optimization can then be solved via linear programming; again, the optimization add-in to your spreadsheet software should be able easily to solve these capital budgets for up to a few dozen candidate projects and several years' planning horizon.

When the capital budgeting optimization can be solved via linear programming, the "dual analysis" of linear programming theory is then available to answer questions like, "How much will my project portfolio increase in value if I can obtain an extra million dollars in year three?"

This is a chapter on risk, and the capital budgeting optimization models presented in this section do not incorporate uncertainty. The mathematically advanced reader may care to look into more complicated optimization methods, like chance-constrained programming, that can explicitly deal with uncertainties in the coefficients of x_j. An alternative is to use still other spreadsheet add-ins, like @Risk™, that perturb the values of spreadsheet cells according to user-specified probability laws, essentially resulting in a simulation of the capital budget.

5.7.2 NPV vs. Decision Analysis: "Real Options"

5.7.2.1 Introduction

Discounted cash flow methods of project valuation do not allow the consideration of future decisions which would have an impact on the project's ultimate payoff.

Decision tree methods, which would allow these considerations, are easily available. Why are they not used more widely for project selection?

It is readily evident that they are used, at least implicitly, for other kinds of corporate decisions. Financial "options" (an option is a special case of a decision tree, and an options approach to making an operational decision is called a "real option") are created and traded daily around the world. To take another example, the acquisition of a fleet of buses has one value to the acquiring company if later decisions about maintenance are made in a certain way, e.g., if the oil is changed at certain intervals, and a different lifetime value if the oil is changed less frequently or not at all. To state the question simplistically: If the return on fleet acquisition is allowed to depend on reliable mechanics changing the oil regularly, why are other kinds of project selections made without depending on reliable executives to make the subsequent decisions that maximize the project's value?

One could imagine that, when a project selection today means a related decision must be made next year, the more flexible job descriptions of executives, the difficulties of tracking complex project histories, and incomplete briefing of newly hired executives may mean the later decision "falls through the cracks." Or that projects at the corporate level, involving interdepartmental coordination, are selected to minimize the need of one department to trust the other department to follow up as needed. Both of these turn out to be part of the answer.

5.7.2.2 Research Method

This issue was discussed informally with individual managers and in executive education classes, in order to compile a list of possible reasons "why decision analysis isn't used for project selection." The discussion centered around the following scenario.

> Consider an investment opportunity requiring an investment of $10,000 with an assured first-year cash flow of $6,000. The second-year cash flow is uncertain with a 50-50 chance of either a $15,000 gain or a $5,000 loss. However, the project can be abandoned at the end of the first year if new information uncovered during this period suggests that the second-year payoff will not be favorable. Dropping the project at that time involves no salvage value or penalty. The cost of capital is 10%. Should this investment be made? What is the value of the project if perfect information could be obtained at the end of the first year? Traditional discounted cash flow (DCF) analysis based on expected cash flows would have yielded a net present value (NPV) of:
>
> $$NPV = -\$10,000 + 6,000/1.10 + 5,000/(1.10)^2 = -\$413$$
>
> and would have thus signaled a rejection of the project. However, the project can be either abandoned or kept if new information is unfavorable or favorable, respectively, with a prior probability of 0.5 for each eventuality. The value of the project can be estimated by adding the value of the two options:

$$NPV(ABANDON) = -\$10,000 + 6,000/1.10$$

$$NPV(KEEP) = -\$10,000 + 6,000/1.10 + 5,000/(1.10)^2$$

$$NPV(OPTION) = 0.5*NPV(KEEP) + 0.5*NPV(ABANDON) = \$1,600.$$

Thus the value of the option given by perfect information by the end of the year is $1,600. This is approximately $2,000 more than the NPV based on traditional DCF analysis. The illustration suggests that traditional DCF analyses may understate the attractiveness of new product market ventures which typically have high levels of uncertainty associated with early time periods.

Later, twenty-five managers from (mostly larger technology) companies in the Austin, Texas and Portland, Oregon metro areas were surveyed using a written questionnaire derived from the focus discussions. The questionnaire's preface included the scenario above. Due to the small sample size, results must be viewed as exploratory. Yet they seem to suggest that the question of organizational barriers to the use of real options has wider implications for the management of the modern firm.

Decision Tree for the Option

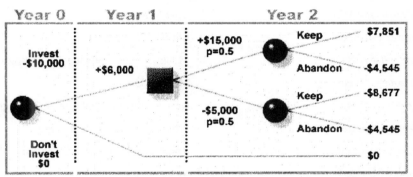

Figure 5.13 Decision tree for the investment opportunity (Artwork courtesy of Cenquest, Inc.)

5.7.2.3 Results

Respondents represented the following companies:

Emery World Wide	FEI Co.
Intel	Jantzen
Mentor Graphics	Merix

NEC America   Portland Orthopedic Clinic

Siltec Corp.   Tektronix

They represented these divisions and functional areas:

electronic design automation software   electron/focused ion beam components

health care   information systems

information technology   interconnect devices

microprocessors   OEM computers-servers

PCBs   printed wiring boards

semiconductor   silicon materials supplier

swimwear   technical software

telecommunications

and had these titles:

Business Unit Director   Cost Manager

Director Finance   Emitter Product Manager

Financial Analyst   Manager For Emerging Technologies

Manager Of Network Services   Medical Doctor

Mgr. Corp. Business Development   New Products Development Manager

Product Engineering Manager   Product Applications Engineer

Product Development Engineer   Routing Technologies Manager

Support Service Manager   System Technician

Processing Systems Development Engineer   V.P. Operations

Mgr. Product Marketing, Networking Division

They reported worldwide sales of their companies' divisions ranging from $7 million to $15 billion, with a median of $180 million.

The questionnaire addressed the companies' projects in general, but for reference each respondent was asked their firm's "most recent project" with which they were familiar. Sixteen of the 25 were "new product development" projects; others represented were "production facility expansion," "production facility development," "acquisition," and "other." The average investment required at the respondent's firm for projects of the type just named had a modal value of "$1 to $9.9 million" (13 responses), with five under $1 million and two over $10 billion. (One of the latter was a silicon chip fabrication plant; the other was not identified.) Projects of the named types ordinarily required an average of 28 months in development, though most required fewer than 24 months. The useful life of the project or investment was not asked.

In no case did finance executives alone "generally make investment decisions on a new project." Overwhelmingly (16 responses) the decisions are made by interdepartmental teams. Some are made by operations executives alone; a few by external consultants, by teams including customers or the board of directors, or "the parent company."

As for methods and criteria for project selection, fourteen answered "ROI," and eight more the closely related "Net Present Value." Three managers claimed to use decision analysis techniques; interestingly, one of these was for a product development project, one for a facilities project, and one for an unspecified "other" type of project. Ten respondents cited "other" techniques/criteria, including Economic Value Added, technology acquisition, strategic fit, payback period, shareholder value, and market necessity. Seven respondents used multiple techniques, the most common combination being Net Present Value and "other." However, no respondents using multiple techniques used decision analysis as one of the multiple techniques.

The final question focused on "reasons why your company does not use the 'options' method of valuing projects or the Decision Analysis Approach (DAA)." Each possible reason was scaled Strongly Agree=4, Agree=3, Disagree=2, and Strongly Disagree=1. The reasons appear in Table 5.4 in descending order of total score. The maximum possible score (indicating that all 25 respondents strongly agreed on one of the reasons) would be 100.

Table 5.4 Survey Scores: "Reasons Why Your Company Does Not Use the 'Options' Method of Valuing Projects or the Decision Analysis Approach (DAA)"

• Perfect information for project evaluations at future points is rarely available (or difficult to obtain).   74

• Operations executives do not like to discontinue their own projects at a future point of evaluation.   70

• All possible 'options' cannot be anticipated.   69

• More convenient to obtain complete project funding now, rather than compete for partial funding with other projects in the future.   65

• Conservative decision making avoids choosing 'options' or alternatives that involve large downside risks.   59

• Existence of entry and exit barriers will not permit project expansion or discontinuance based on DAA.   58

•Employee turnover/re-adjustment problems makes future project expansion/discontinuance difficult.   54

• Process of evaluating a project at each future decision point may incur higher costs.

• Project valuation is generally performed by financial rather than operations executives.

• DAA is more complex than ROI/NPV.   49

5.7.2.4 Discussion

The top reason for not using decision analysis was the difficulty of obtaining (or perhaps difficulty of structuring) the needed information. This seems closely related to the third strongest reason, namely that all possible future decision points cannot be foreseen in projects of the types the respondents are involved in (in contrast to our oil change example). It is unclear, though, to what extent respondents felt a true lack of information gathering, processing and forecasting in their own firms, and to what extent they were simply reacting to the oversimplified nature of the short NPV-vs.-DAA scenario that was presented as part of the questionnaire. No respondent "strongly disagreed" that it is difficult to anticipate all possible future option points.

The second most widely agreed upon reason (no one "strongly disagreed" with this one either) was that executives have latitude of action and are not constrained by procedure books – and hence cannot be relied upon to discontinue projects to which their careers are tied, even if the objective data show a low prospective payoff. Certain executive personalities are prone to believe that they can "pull projects out of the fire."

In the past, it was common to invest some funds for a project, then in interim management reviews increase the funding if the partial results indicated the project would be successful. In fast-cycle technology markets, this is no longer practicable because preparing for the management reviews is time-consuming and may cause the team to miss the market window. This trend speaks against making partial funding decisions now and modifying them later – that is, against the decision analysis approach. Survey respondents felt DAA is not more widely used because it is more convenient to obtain complete project funding or complete rejection of the project now, rather than compete for additional funding with other projects in the future. A related reason was that respondents felt there were often other obstacles within the organization (perhaps these are difficulties of re-assigning people to projects, but they are distinct from the reluctance of executives to spike their own projects) to expanding or abandoning projects in midstream.

Decision trees do allow planners to deal with project risk, though not if the "expected value" decision criterion with which most people are most familiar (Anderson *et al.*, 1991) is the sole means of interpreting the tree. Rather than utilize sophisticated interpretations ("minimum regret," etc.) of the decision tree, however, our respondents' companies find it easier just to eliminate, *a priori*, project ideas that have significant downside risk. Of course this presumes that the highly risky projects are not, as a class, vital to the organization's future.

Another relevant response had to do with personnel turnover. If a decision maker cannot be sure that an executive who is supposed to make a future decision critical to the project valuation will still be working at the firm when the decision is supposed to be made – and if the company's record keeping systems are not up to alerting the executive's successor to the full situation – then it is tempting to use a more conservative investment criteria, e.g., NPV, that does not explicitly depend on future decisions being made.

The questionnaire item "Project valuation is generally performed by financial rather than operations executives" scored 52, but contradicts responses earlier in the questionnaire that showed none of the respondents' projects were decided by financial executives alone. In addition, there was only one "strongly agree" response in all answers to the final three combined. Thus, we must dismiss the last three reasons in the list above. Significantly, this implies that the respondents believe the decision analysis approach is not inherently more complex or harder to understand than discounted cash flow methods.

Finally, it was also mentioned that choosing a portfolio of capital projects under a fixed budget can easily leave room only for projects that have positive net present values. Although planners acknowledge that a decision analysis can make additional projects look attractive, there may be no funds available to take advantage of them.

It has been suggested that projects exhibiting short time frames, limited complexity and high uncertainty are best addressed with project-team organizations, that is, matrix organizations that pull managers out of their usual functional job description. In a project organization, managers sign onto a more structured calendar and responsibility definition than in their functional job – and thus, we might suppose, are more likely to make a necessary future decision at the scheduled time. If true, this would mean decision analysis would be a preferred method for projects fitting the above three criteria. Yet, new product development projects are of short duration, high risk, and varying complexity levels, and sixteen of our 25 survey responses concerned new product projects. Within the sharp limits imposed by this survey's sample size, the proportion of preferred selection methods did not differ in any marked way between new product projects and other kinds of projects.[6] And in any case, the quite reasonable assertion that decision analysis is more necessary for high-risk projects does not negate the benefits of decision analysis for lower-risk projects. (One leading planner recommends ROI/NPV for "commercial development" projects, though he conceives of these as comprising only the later stages of the product development process, by which time "technical uncertainty and target [market] uncertainty should both have been resolved.")

5.7.2.5 Conclusions

A survey of technology company managers supports previous reports that 70% of companies use DCF methods rather than decision analysis methods to select projects. The survey revealed eight organizational reasons (the first eight in Table 5.4, plus the budgetary reason discussed in the penultimate paragraph of the previous section) why decision analysis methods are not used for project selection. (Discounted cash flow methods were not considered "decision analysis" for the purposes of this research.)

These results should reinforce what has been called "a growing recognition that project-selection models should be used to ask questions of the entire organization." The survey results bring to light issues of knowledge management, personnel management, empowerment, trust, and shrinking product cycles that affect the decision criteria for project selection.

In a sense, they call into question the very role of the executive. An executive is supposed to make decisions. But the survey shows that when an actual operational planning process requires that a decision be scheduled for a particular future time (as decision analysis would require), companies lack confidence that the decision in fact will be made on a timely and well-informed basis. As a result, sub-optimal selection methods remain popular.

These results also reinforce the ideas that management scientists have been more skillful at theory than at implementation, and that quantitative management techniques will be used more and more for pointing to fruitful areas of inquiry, rather than for arriving at optimal solutions.

---

[6]   Of the 16 new product projects, 6% (or one project) used decision analysis, 63% used ROI/NPV, and 31% used "other." Four of the five "other" responses noted multiple methods were used. Of the 9 non-product projects, 2 projects (22%) used decision analysis, 44% used ROI/NPV, and 33% used "other."

5.8 The Bigger Picture: Total Corporate Risk, and Public Perception of Risk

So, to summarize our discussions about risk in new product development: Product development proceeds in stages. Although different companies may give different names to the stages, broadly speaking, they are concept development, prototype development, market feasibility testing, manufacturing ramp up, product launch and then post-introduction operation and support.

The chances of a product idea translating into market success are quite low. Most attrition of product ideas occurs early in the new product development process. The development team's ability to implement engineering changes and design changes is high in the early stages of the process and low in later stages.

Most successful new products are, at most, moderately innovative. Most costs are committed early in the product cycle but are actually paid later in the product cycle and, indeed, even after the launch of the product. A company wishing to be rational should commit costs only as fast as it reduces uncertainty about the product's success. And, finally, companies can commit costs at a rational pace, as demonstrated by the successful diversified company we looked at earlier.

Table 5.5   Risk in New Product Development: Summary and Conclusions

• NPD proceeds in *stages.*[7]

• The stages are: Concept development; Prototype development; Market feasibility testing; Manufacturing ramp-up; Product launch; and Operation/Support. Different companies may give different names to these stages.

• Chances of a product idea translating into market success are fifty to one, or worse.

• Most attrition of product ideas occurs in the early NPD stages.

• Your ability to influence design, costs, etc. is high in early stages and low in later stages.

• Most successful new products are at most moderately innovative.

• Most costs are *committed* very early in the NPD cycle.

---

[7]   In one shop, I saw a poster characterizing these stages as 1. Enthusiasm, 2. Disillusionment, 3. Panic, 4. Search for the guilty, 5. Punishment of the innocent, 6. Awards and plaudits for the non-participants.

• Most costs are *incurred* (paid) in the production and O/S stages.

• A company should only commit project costs as fast as it reduces uncertainty about the product's success.

• Companies *can* commit costs at a rational pace.

There's Risk, and Then There's Disaster

"Disaster" is the word we use for the large, negative consequence of a low-probability event. Recently, sociologists and political scientists have added their insights to engineering analyses of notable disasters like Bhopal, the *Challenger,* Chernobyl, the *Exxon Valdez,* and the ValuJet crash. Their observations add to our understanding of risk management.

Yale sociologist Charles Perrow believes we can never ensure that complex systems will be safe. "Interactive complexity" means components interact in ways that are too numerous for designers and operators to deal with. Small errors can have disproportionate effects on the whole, and can do so too quickly for operators to react ("tight coupling"). Trying to make the system safer by adding more safety interlocks then increases the number of interactions, and may make the system less safe. There is rarely one cause of a catastrophic failure, he says, and "Failure is built into the system."

In the nuclear accident at Three Mile Island, Perrow notes, "Operators didn't know what was going on." Interactive complexity and tight coupling had combined to cause two unusual meter readings. The operators' judgment led them to believe one of the dials and dismiss the other as erroneous. But the unforeseen meltdown conditions meant that both dials were correct – a condition operators and supervisors were not prepared to accept.

Boston College professor Diane Vaughan has written a book on NASA's *Challenger* disaster. She maintains that in addition to Perrow's technical complexity considerations, we must account for the attitudes of operators and designers. In NASA's "culture of conformity," she says, engineers did not feel free to admit they did not understand how the booster rockets would react to extreme cold.

Professor Lee Clarke of Rutgers University has looked at corporate and government disaster preparedness plans, and concluded that they are "fantasy documents," assuming that disasters can occur only under ideal conditions for remediation. Exxon's planners, for example, had assumed a tanker could spill a maximum of 4,000 barrels (the *Valdez* lost 200,000 barrels), and only in good weather! Clarke says these documents are crippled by the requirement that they reassure the public about the organization's ability to control risk and remediate consequences; frankness and scientific integrity are thus nearly impossible. Thus, he goes on, official disaster response plans breed complacency and "diminish the capacity for organizational learning."

U.C.-Berkeley's Todd LaPorte has identified a small number of organizations that exhibit high performance on the safety dimension: These include San Francisco and Oakland air traffic control, California's Diablo Canyon nuclear plants, and the aircraft carriers U.S.S. *Enterprise* and U.S.S. *Carl Vinson*. These entities learn from mistakes, delegate responsibility to the most knowledgeable levels, and constantly train, resulting in "cultures of safety." But LaPorte admits there is no guarantee these practices will prevent disasters. Moreover, some high-safety cultures have reverted to ordinary organizations with average failure levels, indicating it is difficult to maintain a culture of safety. The best chances for a sustained culture of safety, advises Berkeley prof Robert G. Bea, is frequent retraining and constant concern at the top.

Stanford University's Scott Sagan has studied the U.S. nuclear weapons programs, notable for being nearly accident-free. He writes that there have been a number of *near* disasters in these programs. Such near-misses can be hidden from the public, and their aftermath might well involve more hurrying to assign blame than scurrying to prevent similar mishaps in the future.

Miller D W (1999) Sociology, not engineering, may explain our vulnerability to technological disaster. Chronicle of Higher Education (October 15) A19-A20

Technological risk fits into a broad spectrum of total-business risks that include product "shrinkage" and the fluctuations of the stock market. Of interest to technology managers is Arthur Anderson's characterization of information processing/I.T. risk factors as "access, integrity, relevance, and availability," and that risk involves not just the physical failure of these factors, but the risk of bad decision making when information systems provide managers with inaccurate, outdated, incomplete, or irrelevant data.

Marsh, Inc., a total risk management consulting firm, categorizes business risk as follows:

Table 5.6 Categorization of Business Risk

Hazard Risk	Business Risk
Financial/Market	Operational
Securities	Personnel
Commodities	Physical Damage
Interest Rate	Consequential
Currency	Criminal
Credit	Data
Political	Legal
War	Contracts
Social	Tort Liability
Terrorism	Statutory Liability
State Action	Product Liability
Regulatory	

Source: Marsh, Inc.

Marsh divides projects and situations into five risk categories: Unacceptable; Unacceptable without risk reduction via risk pooling; Acceptable with continual review and pooling; Acceptable with review and without pooling; and Acceptable. Intel Corporation also uses five levels of risk description: 1. Show stopper; 2. Low confidence/high risk; 3. Medium confidence/medium risk; 4. High confidence/low risk; and 5. Certifiable. Intel monitors the progress of projects and expects, like the companies in our new product development risk examples in this chapter, for the assessed risk score to drop from a typical starting point of 2 to better than 4 over the course of a 200-week technology development.

This is achieved by communication, integration (maintaining confidence not only that your part works, but that components being worked on by others also work), synchronization (time-based project management), and enabling (assisting all suppliers and customers to be ready). Intel builds incentives into projects (giving each department a share of the margin, and punishing teams that pass their problems off to other teams) in order to achieve communication, integration, synchronization, and enabling, and ultimately to reduce project risk. Intel would agree with Marsh, Inc., that "The problem is not simply about methodology and measures. It involves people, perceptions, and communication."

5.9 Notes

About Risk. This section is adapted from Phillips F (1998) A Technology Manager's Perspective on Forecasting. Presented to World Future Society November

When wooing John Sculley... Sculley J (1987) Odyssey. Harper & Row New York

The economist Frank Knight... Knight F H (1951) The Economic Organization. Harper Torchbooks New York

Sumitomo Bank lost $2.6 billion... Mining Journal (1996) London Metals Exchange On The Offensive. 327-8399 (October 11) http://www.mining-journal.com/MJ/11oct96.htm

Boeing "bet the ranch"... McGrath M, Anthony M and A Shapiro (1992) Product Development: Success through Product and Cycle-Time Excellence. Butterworth-Heinemann Stoneham Massachusettes

Perceived (psychological or social) uncertainty is real... Schiffman L G and L L Kanuk (1987) Consumer Behavior (3rd ed) Prentice Hall Englewood Cliffs NJ

A good current example is fears concerning security and privacy on the Internet... Wilson T (1997) Trust Trails E-Commerce Technology. Communications Week (March 24) 1

If an outcome threatens a person's deeply held values or beliefs (abortion or irradiated food are possible examples), the most adamant resistance ensues... Rogers E (1983) Diffusion of Innovation. The Free Press New York (3rd edition)

...software projects' failure rates also lie between electronics and pharmaceuticals... The Standish Group

International Workshop on Concurrent Engineering Design at the University of Texas at Austin... The conference summary is adapted from Hsu J P, Gervais J and F Phillips (1991) International Workshop on Concurrent Engineering. National Science Foundation and IC2 Institute of University of Texas at Austin. The work was presented as Hsu J P, Gervais J and F Phillips (1993) Summary Report on the International Workshop on Concurrent Engineering Design. 2nd International Conference on Manufacturing Technology in Hong Kong Hong Kong December

several sources report a pattern of committed vs. expended monies over the development cycle... Corbett J (1986) Design for Economic Manufacture. Annals of C.I.R.P. 35(1); Rasmussen A (1990) (Ed) Producibility and Product Optimization (in Phillips F Y Concurrent Life Cycle Management: Manufacturing, MIS and Marketing Perspectives. IC^2 Institute of the University of Texas at Austin); Port O, Schiller Z and R W King (1990) A Smarter Way to Manufacture. Business Week (April 30) 110-117

It has been informally observed that a firm should not commit funds at a rate that exceeds the rate of uncertainty reduction... *inter alia,* in remarks by Fabrycky W (1990) National Science Foundation International Workshop on Concurrent Engineering Design. IC^2 Institute Austin Texas October; See also Blanchard, B S and W J Fabrycky (1990) Systems Engineering and Analysis. Prentice Hall Englewood Cliffs NJ (2nd edition). (The author is grateful to Prof. Fabrycky for additional helpful comments during the research reported in this section)

These detail design decisions... in large measure determine how big the factory will be... Chryssolouris G, Graves S and K Ulrich (1991) Decision Making in Manufacturing Systems: Product Design, Production Planning, and Process Control. Proceedings of the 1991 NSF Design and Manufacturing Systems Conference. Society of Manufacturing Engineers Dearborn Michigan

"No Go," i.e., kill the project... See Charnes A, Cooper W W, DeVoe J K and D B Learner (1966) DEMON: Decision Mapping Via Optimum Go/NoGo Networks - A Model for Marketing New Products. Management Science July XII(11); Charnes A, Cooper W W, DeVoe J K and D B Learner (1968) DEMON: A Management Model for Marketing New Products. California Management Review Fall

McGrath, Anthony and Shapiro offer the needed cost and risk data... McGrath E, Michael, Anthony M T and A R Shapiro (1992) Product Development: Success through Product and Cycle-Time Excellence. Butterworth-Heinemann Boston 49

A 1990 SAMI report... Wall Street Journal (1990) Odds and Ends. April 4 (Second front page)

Balancing the project portfolio: NPV vs. Decision analysis... This section is excerpted from Phillips F (1997) Why Isn't Decision Analysis Used for Project Selection? Presented at INFORMS Dallas September

The discussion centered around the following scenario... adapted from Kensinger J (1987) Adding the Value of Active Management into the Capital Budgeting Equation. Midland Corporate Finance Journal Spring 31-42 and its variation in Srivastava R Valuation of Market Growth Opportunities in Phillips F (Ed) (1990) Concurrent Life Cycle Management: Manufacturing, MIS and Marketing Perspectives. IC^2 Institute of the University of Texas at Austin 115-130

It has been suggested that projects exhibiting short time frames, limited complexity and high uncertainty... Afuah A (1998) Innovation Management: Strategies, Implementation and Profits. Oxford University Press New York 232-235

Yet, new product development projects are of short duration, high risk and varying complexity levels... see, e.g., Haynes K E, Phillips F Y, Qiangsheng L, Pandit N S and C R Arieira (1996) Managing Investments in Emerging Technologies: The Case of IVHS/ITS. ITS Journal 3(1):21-47

One leading planner recommends ROI/NPV for "commercial development" projects... Martino J P (1995) R&D Project Selection. Wiley Interscience New York 115

previous reports that 70% of companies use DCF methods... Srivastava R Valuation of Market Growth Opportunities in Phillips F (Ed) (1990) Concurrent Life Cycle Management: Manufacturing, MIS and Marketing Perspectives. IC^2 Institute of the University of Texas at Austin 115-130

what has been called "a growing recognition that project-selection models should be used to ask questions of the entire organization."... Souder W and T Mandakovic (1986) R&D

project-selection models. Research Mgt. 29(4):36-42. (as quoted in Bordley R F R&D Project Selection vs. R&D Project Generation. IEEE Engineering Management forthcoming)

the "dual analysis" of linear programming theory... See e.g. Anderson, Sweeney and Williams (1991) An Introduction to Management Science. West Publishers St. Paul Minnesota (6th edition)

Capital investment ROI, uncertainty, and the network effect. See Miniproject MV-1 below, and Lucas H C Jr (1999) Information Technology and the Productivity Paradox: Assessing the Value of Investing in IT. Oxford University Press New York. (Lucas argues through cases that network effects increase the effective ROI of IT acquisitions. He also argues for risk-adjustment of the ROI calculation, given the high failure rate of IT insertions.)

Marsh, Inc., a total risk management consulting firm, categorizes business risk... Pinkowski M G (1999) Enterprise Risk Management. Presented at Northwest International Business Educators' Conference Seattle March

Intel monitors the progress of projects... Class presentation by McQuhae K (1998) Intel October 31; See also Krogh L C, Prager J H, Sorensen D P and J D Tomlinson (1988) How 3M Evaluates Its R&D Programs. Research•Technology Management 31(6) (November-December) 10-14

5.10 Questions and Problems

Discussion Questions

D5-1. If you have visited Japanese companies in your industry, can you give an instance where you have seen the Japanese manufacturing organization as a role model? An instance where you feel your U.S. company has a superior manufacturing organization?

D5-2. What kinds of operational decisions in real organizations might be susceptible to the "real options" approach?

Problem to Solve

P5-1. Invent data (that is, values for the c_{ij}, b_j, and NPV_j) for a capital budgeting problem to be solved by linear programming. Let the problem involve four candidate projects and a five year planning horizon. Solve your problem using a spreadsheet optimizer. (NOTE: If your spreadsheet software does not have an optimizer, it may be that it was not "installed" when you first used the spreadsheet program. If the optimizer is available for your program, it may be on your installation disk or on the spreadsheet vendor's web site.)

P5-2. There are two kinds of development projects, "big" projects and "little" projects. Your engineers have submitted to you an unlimited supply of proposals for both kinds. Big projects cost $10 million, spent evenly over 5 years. They yield revenues of $80 million, spread evenly over years 6-10 following their inception. Little projects cost $1 million over 1 year and yield $4 million in revenue, $2 million each in years 2 and 3. In turn, you are preparing a budget proposal for your department that requests $3 million in corporate funds in years 1 and 2, and 25% of the department's projects' revenues in subsequent years. Develop a pro forma 20-year plan that strives to maximize the company's returns from the projects, and shows how many projects of each type will commence each year. (This problem does not call for formal optimization, but you should do enough trial-and-error so that you are confident you have a good plan.)

Mini-project

M5-1. In the modern networked economy, few product development (capital investment) projects can be evaluated completely independently of the other investment projects that are also under consideration. For example, developing a hand-held electronic calendar can be much more valuable to a company if the company (or one of its close partners) is also developing email software or cellular phone technology. How would you modify the linear programming model for capital budgeting to reflect the nonlinear interactions or project payoffs? Write a nonlinear programming model that does this. What features of the capital investment decision does your model capture? What features does it not capture, and what are the other limitations of your model?

"Democracy is based upon the conviction that there are
extraordinary possibilities in ordinary people."

Harry Emerson Fosdick

"[Westerners] come here, they're talking about democracy, about market
economics.... And what about the economy? Oh, the free market will take care
of that, the Westerners say.... But what the hell is a free market? Is it like
Ismailovo, where they sell shashlik, icons, and toasters, where every little stand
forks over money to every Ivan who claims he's Mafiya? Is that the free
market? All right. Okay. We'll do it. How about a loan to get us started?"

David Rosenbaum, *Sasha's Trick*

6 Influence of Government Policy on Technology Acquisition and Utilization

Through the 1980s and 1990s, or so it seemed, high technology firms doubted the relevance of the federal policy making process, and spent far less on lobbying than other, more traditional industries. Legislators and regulators, in return, seemed to register high technology as no more than a blip on the radar horizon. Whether that was perception or reality, the Microsoft antitrust decision of 2000 focused Silicon Valley's eyes on Washington, and lawmakers' attention on high tech. Non-defense technology was an issue in 2000, for the first time, in a presidential election, and microelectronics executives are becoming proactive, rather than reactive, on legislation, court proceedings, and agency regulation.

This chapter offers an overview of current issues in these areas that most affect technology companies. The problem-solving orientation of technical managers can make the Washington political process – the trading of favors that results in the mysterious broadening and narrowing of the scope of legislation – baffling and frustrating. While none of the problems to be discussed here are insuperable, the industry-government culture gap will ensure that few will be resolved quickly.

6.1 Science, Technology, Management, and Policy

6.1.1. Philosophy of Federal R&D Policy

Post-World War II R&D philosophy emphasized funding of pure scientific research, that is, research for knowledge alone, without thought of application. Other research funds were set aside for applied and engineering research, especially defense-related R&D.   The pure research emphasis had some shortcomings: It was never popular with the public; it was vulnerable to funding cutbacks in times of tight federal budgets; and the translation of pure knowledge into use was haphazard.

Readers of this book will have noted its emphasis on *basic* research and basic technologies. Basic is not a synonym for pure. Basic research is conducted with the expectation of broad (and perhaps unforeseen) applicability, even if it is only "applied" to other areas of pure research. Thus, basic research may be pure or applied.

Table 6.1 U.S. R&D Funding Trends 2000

- Total R&D expenditures to rise 7.75% to $266 billion

- Federal R&D spending will increase 1% to $66 billion.

- Industry R&D spending will rise 10.6% to $187 billion.

- Remainder of R&D expenditures – $12.6 billion – will be supported by universities and nonprofit organizations.

- Private industry increasingly outsourcing R&D functions. An increasing share of U.S. industry's R&D is performed off-shore, primarily in facilities owned by the same industry.

- Congress and industry are increasingly skeptical about federal technology transfer programs.

- Federal share of R&D expenditure, now at 26%, expected to fall still further.

NIST is the National Institute of Standards & Technology (Dept.of Commerce). ATP is NIST's Advanced Technology Program.

Source: (Battelle – *R&D* Magazine forecast)

In the 1980s, industrial policy fell out of favor with the electorate. No longer would government identify and subsidize strategic industries (industries thought to be important for our economic growth and our standing among nations), nor "pick winners" among industries and companies (except when it came to military procurement). Federal support for industrial research had to be aimed at nurturing new findings early in the technology cycle, long before they are put into products that might bestow commercial advantage on individual firms. This kind of early-cycle research is called "pre-competitive."

Cost-effective R&D policy requires funding *application-driven basic research,* that is, pre-competitive R&D driven by national needs, yet conducted with the expectation of wide applicability. This is becoming more obvious to everyone involved in policy and research, but it has not become clear how this philosophy may consistently translate into grant and incentive programs that are fair and effective. Table 6.1 shows some federal R&D funding trends, and Table 6.2 gives one editorial view of how federal R&D may be made more efficient.

Table 6.2 *Business Week's* Recommendations for Getting More Science from the Same (Federal) Dollar

---

- Downsize the massive weapons labs.

- Slash the time researchers waste chasing grants.

- Fund riskier, higher potential payoff projects.

- Cut "big science" projects that have little to do with actual science, such as the space station.

- Foster more partnerships and consortia.

- End "boom and bust" funding cycles: Maintain stable funding, even at lower levels.

---

source: *Business Week,* 5/26/97, p.168.

6.1.2. Technology Transfer Policy

Any streamlining of domestic T^2 (commercialization of government and government-funded technologies) policy is beneficial as it encourages technology entrepreneurship and thus economic development. In chapter 4, we discussed the Bayh-Dole Act of 1980, governing university ownership of government-sponsored intellectual property. Other issues of domestic T^2 include the interaction of U.S. national laboratories with private companies; the design of military projects for "dual use" in the civilian economy; support for new business incubators and other new institutions for tech transfer; and technology transfer requirements for federal research grants. The latter stipulate that laboratory researchers must, as a condition of their grants, take positive steps to move their research results toward commercialization.

Domestic T^2 can touch the antitrust area as well, for example in the case of Microsoft's application programming interfaces (APIs). Microsoft controls the almost universally used PC operating system, and also authors and markets application programs such as MSOffice. It was alleged during Microsoft's antitrust trial that there are secret APIs known only to Microsoft application programmers, and that this puts third-party software developers at a disadvantage. The appeals court may decide to uphold the lower court decision to split Microsoft into an operating system company and an applications company (the latter including Microsoft's web browser). Part of the motive for a breakup would be to inhibit anticompetitive intra-firm technology transfer, perhaps including secret APIs.

International T^2 policy, especially export controls, affects the competitiveness of U.S. companies in international markets. The most visible case has had to do with encryption devices for computer files and for telecommunications. The government alleges that these products and technologies could be used for military offensives against the United States and should not be exported. Similar products developed by foreign companies then come to dominate world market share, without always truly enhancing U.S. national security.

6.1.3. Patent Reform

Several patent-related issues are at or near flash point:

- Technology companies are troubled (while other technology companies, perhaps contrary to the national interest, profit) by trivial and obvious 'business practice' patents. These allow companies to hoard patents without commercializing them; the patents are then just "licenses to litigate." Because software is often the delivery mechanism for an algorithm directing, e.g., material flow in a factory, the terms "software patent" and "business process patent" are frequently used interchangeably. While it seems the two terms ought to mean different things (the protection of specific software code falls under copyright law, not patent law), it is in practice hard to distinguish them. Amazon.com president Jeff Bezos' widely read letter opposing future business process patents nevertheless did not repudiate Amazon's right to profit from its existing one-click e-shopping business process patent.

- Patents are granted only for devices that are "new." I have put the word in quotation marks not to suggest flexibility of meaning, but because the word "new" appears in the statute. Patent examiners are supposed to search for "prior art" which might invalidate the newness of any device described in a patent application. But in practice, the Patent and Trademark Office (PTO) grants broad patents with little examination of prior art, and the secret application process does not allow the public to help the PTO identify prior art.

- The time lag in the patent process is too long for Internet companies. A three-year consideration of an application is not unusual, and six months can make or break a startup company. Venture investors now sometimes advise startups against applying for patents, arguing their time and money might better be used for product development and marketing. This could leave the companies at a disadvantage later in their life cycle.

- These troubles stem from too few and insufficiently trained patent examiners at PTO. Always present is the danger of Congress taxing patent fees, thus reducing PTO resources still further.

- Another issue is patent "harmonization." This refers to the globalization of technology business and the need to file patents in many countries at once in order to obtain wide protection for an idea. The differing patent laws in various countries make this difficult; for example, the U.S. "secret filing" system (see the box below) is not universal. In 1994, the U.S. and Japan agreed to a number of harmonization measures. The U.S. agreed to publish patent applications eighteen months after they are filed, whether the patent has been granted or not, and Japan agreed to process patent applications coming from U.S. inventors within 36 months (rather than the five or six years it has taken in the past). But still, U.S. patents are given to the first person to apply for a patent on a given device, and Japan grants patents to the first applicant who can prove to have invented the device. This and other differences inhibit business exchange between the two countries. They are thus sometimes regarded as barriers to trade in bilateral and multilateral (e.g., World Trade Organization) trade negotiations.

- Other patents, while not submarine patents in the exact definition used in the box, are held by people who have no intention of commercializing them, but hold them only for the purpose of collecting license fees from those who do commercialize. This is by no means uncommon, and in fact is a practice many large companies engage in. it is a touchy ethical issue. Factors in the ethical dilemma include whether the patent is reasonably specific, and how litigation-happy the patent holder is.

Submarine Patents, and What To Do about Them

Prior to the Patent Act of 1994, inventors received patent protection for seventeen years from the date of issue of their patent. Now, given that a patent is granted, protection extends twenty years from the date of application. This turns out to be important for protecting both inventors and other technology business people.

Pending patent applications in the U.S. – unlike in some other countries – cannot be examined by the public. Yet a U.S. patent allows the holder (Party A) to exclude all others from profiting from the invention, even if unknowing entrepreneurs (Parties B and C) have invested in and commercialized products based on a similar idea during the time Party A's patent is under consideration at the U.S. Patent Office. When Party A's patent is finally granted, A can demand license fees from B and C, who are technically infringing on A's patent. These "submarine patents," unexpectedly rising from the deep to attack earnest entrepreneurs, are another way in which the patent system – designed to stimulate innovation – can, instead, damp innovation.

Knowing this, wily inventors and patent attorneys would file extensions or modifications to their patent applications, keeping the application process going until their idea's "time has come," at which time several entrepreneurs are likely to have entered the market with products using an independently invented version of the idea. Before 1994, these inventors and attorneys had every incentive to delay the granting of their own patents in this way. In extreme examples, applications were caused to stay in process for as much as forty years. The original application may have been impractical for commercialization, or unrealistically broad in its claims, but that made no difference to Party A's ability to demand license fees successfully. In 1911, Henry Ford discovered a remedy still used today, that is, to "bust" the prior patent by legally proving its inapplicability or invalidity.

The post-1994 system removes incentives to delay applications, because doing so would shorten the time during which the inventor enjoys protection. This article discusses a quantitative study showing that the new rules benefit both inventors (who enjoy longer periods of patent protection) and technology business people (who experience fewer submarine patent attacks).

Lemley M A (1996) Submarine Patents, and What To Do about Them. Initiate II(1) (www.initiate.com)

- Life patents. Like the Internet, molecular biology has advanced spectacularly while pertinent patent law and practice has lagged behind. In 1980 the Supreme Court expanded patent protection to cover new life forms. This was followed in 1987 with a decision that expanded protection to include genetically altered animals. Finally in 1993 the law was expanded to include patents on cell lines. The effect of these changes was to open the door for inventors to apply for protection for basic life elements, including genes. In 1993 the U.S. government applied for a patent on the cell line of an Indian woman from Panama; the woman carried a unique virus and antibodies that might be helpful for fighting AIDS. There is tremendous concern that these types of patents provide unreasonable protection for the patent applicant/ grantee and no benefit to the researchers' subjects and patients. (And to other researchers: The National Institutes of Health do not license their gene patents without clauses allowing ongoing research by non-licensees.) Other controversies have to do with genetically engineered seeds that cannot, under their I.P. agreements, be re-used by farmers. Both issues have led some nations to threaten not to recognize life patents issued in other countries.

6.1.4. Competitiveness

Monitoring and understanding foreign technology policy and management practices gain importance as intellectual property (I.P.) and investment funds diffuse worldwide. Senator Jeff Bingaman's Japan Technology Management Programs, funded via the Air Force Office of Scientific Research, were effective in training professionals in Japanese technology practices and placing them as interns in Japanese companies. However, funding for this program has ceased. The R&D consortium MCC – one of the foremost monitoring stations for Japanese technologies – closed down in 2000. There is a need for universities to train technology analysts for placement in industry.

The phrase "industrial competitiveness" is often misunderstood by, and causes friction with, our foreign trading partners; they may think it means the U.S. wants to "beat" them in the game of trade. This is not true, because (among other reasons) a beaten economy could not continue as a trading partner. Americans use the word competitiveness to imply that there should be commonly understood and more or less fair rules, as in an athletic competition, and that the U.S should be able to stay in the game.

6.1.5. Trade

World Trade Organization (WTO) rules generally prohibit differential taxation of companies, quid pro quo requirements for corporate tax relief, and restrictions on the kind of companies that may benefit from government programs. These are precisely the tools that successful 'technopolis regions' have used to build prosperous clusters of technology companies! These regions, according to many studies, have been the major generators of new jobs in the U.S. throughout the past decade, and other regions are trying to duplicate their successes. The WTO officially aims to leverage the efficiency of free trade to accelerate world economic growth. But WTO rules, by working against technopolis development, may well retard economic growth rather than enhance it. Research is needed to understand the conflict between cluster-driven growth and WTO-style free trade. Both can point to demonstrated successes, but they seem to succeed in different arenas. Can they co-exist?

Unfortunately, researchers and practitioners in international trade and in technopolis development rarely talk with each other, and do not work from common data sets. One hopes this issue will be settled by research and by flexible legislation, rather than by special-interest political pressure.

6.1.6. Technology Workforce Issues

Foreign knowledge workers enter the U.S. on "H-1B" visas. These workers ease U.S. technology industry's shortage of educated engineers and scientists, yet their entry into this country is opposed by some labor and other organizations. H-1B visas should not be an issue. This is especially true as regards foreign nationals with advanced degrees, who are hardly likely to exacerbate hard-core unemployment in the U.S. – and may even ameliorate it, by virtue of their entrepreneurial energy and their demonstrated commitment to education. In any case, the 80,000 H-1B visas authorized in 1999 makes hardly a dent in the manpower needs of U.S. technology companies; the rest of the open positions must be filled by educated Americans, and this means more U.S. students must be funneled into an improved U.S. K-12 and higher education system. With luck, U.S. taxpayers will acknowledge some of the harsher realities of economic globalization, and realize that increasing school taxes (and making donations to private schools and colleges) is necessary sooner rather than later.

6.1.7. Adapting Government to Deal with Technology Convergence and Fusion

Government must prepare quickly to deal with today's very fluid technology environment. Does AT&T provide cable service or telecommunications services? The FCC must decide, and propose a policy for opening cable access to Internet providers. What complies and what violates antitrust laws in the new convergence of computing, telecommunications, and entertainment? Government is also involved in spectrum allocation for wireless communications, and broadband deployment to rural and underserved urban areas.

When most near-future economic growth will come from combining existing technologies, how can government ensure the U.S. maintains the capability of doing basic scientific research when, someday, the pendulum again swings the other way? (This question speaks, *inter alia,* to the future of the national laboratories.) As entrepreneurs flock to technology fusion ventures, how can government help them with R&D tax credits and regulatory relief?

6.1.8. Privacy

European governments have moved to protect the privacy of their citizens as regards the sale of names and demographic information from electronic databases, and the tracking of online clicking activity. U.S. citizens have been more tolerant of the collection and exchange of personal data for marketing purposes. But the Toysmart.com bankruptcy proceedings may change this somewhat. While an operating company, Toysmart.com had not only promised its customers never to share their personal data, but had joined a national certification organization that attested to Toysmart.com's customer-friendly privacy practices. Toysmart.com has petitioned the bankruptcy court to allow it to sell its customer database. Naturally, the certifying organization has requested that this not be allowed. It seems clearly in the interest of all companies to oppose the selling of the Toysmart.com database because of the chilling effect its sale would have on customer confidence in all companies and thus on all electronic and ordinary commerce.

6.1.9 Ethics Legislation

U.S. law prohibits U.S. companies from certain practices that may be common in other countries. For example, while wining and dining prospective customers is acceptable, a too-lavish wining and dining is seen as a bribe, and is not acceptable. The line of acceptability in what is known as "foreign corrupt business practices" can be a thin one, and public perception can muddy these issues further. Insinuations appear in the press, as illustrated in the box below.

Poor, Poor Motorola

By Scott Latham for the *Wall Street Journal*

This article examines the strategies and relationships Motorola activates to obtain preferential treatment by both the American and Japanese governments when selling their products in Japan. The author dismisses Motorola's suggestion that the Japanese have reneged on their commitments in the early 1990s to "open" the Japanese cellular phone market. He also questions the ethical implications of America's "revolving door" between government and industry, again suggesting that Motorola has developed and taken unfair advantage of its "special relationships" with officials on both sides of the Pacific.

Latham's anti-Motorola stance is fairly unusual for an American journalist. It shows us that there are always going to be factors that will influence the adoption of technology aside from its strengths in the marketplace. In the final analysis it may be who you know, rather than who you know how to market to, that will make or break you company's success abroad, especially in Japan.

While Latham does not suggest any laws have been broken in this case, he does make his readers think again about what is "fair" when it comes to trade issues with Japan.

—∞—

Motorola Denied Access

A Letter to the Editor of the Wall Street Journal written by Albert Brashear, Director of Communications for Motorola, in Response to Scott Latham (dated 3/4/94)

In his rebuttal, Brashear refutes Latham's suggestions that Motorola has relied on preferential treatment by the U.S. government to obtain access to Tokyo's cellular telephone market. Latham emphasized the roles of consumer choice in explaining Motorola's dismal performance in the Tokyo/Nagoya corridor, whereas Brashear insists the issue is simple a matter of market access.

If our jobs as present and future American marketers is to reach customers, we someday may find ourselves in a situation like Motorola's – where even basic market access is extremely difficult to obtain. None of the strategies we have studied can prepare us how to market to non-existent customers. The lesson we should draw from this Motorola exchange is that before we can start to differentiate our products using marketing, first we have to be in the market. Brashear is successful in persuading readers to consider Motorola's position and forces us to re-examine bilateral trade tensions between America and Japan.

6.1.10 Taxes

Tax issues of concern to high technology companies include capital gains taxes, which affect entrepreneurs and employees of startup firms when their stock goes public and when they exercise options. R&D tax credits are of concern also. Debate continues in 2000 about the taxation of e-commerce transactions, with strong advocates for keeping them tax-free and equally strong champions of using e-commerce to maintain the tax base of municipalities where dotcom businesses are blossoming.

6.1.11 Changing Roles of the U.S. and Russian National Laboratories

During the Cold War, the U.S. Department of Defense (DOD) and Department of Energy (DOE) laboratories – Los Alamos, Lawrence Livermore, Oak Ridge, Sandia, and others less famous – conducted research on atomic and other weaponry. So, of course, did the government laboratories of the Soviet Union. The Cold War and the Soviet Union are now of the past. What is to become of the laboratories?

Many are engaging in environmental cleanup (atomic research and production created a lot of pollution), dismantling of excess nuclear weapons, promotion of atomic safety and security, and computer simulation of nuclear detonations so as to reduce or eliminate live tests. The labs provided and continue to provide much basic research and a lot of civilian benefits other than national security. For example, the "star wars" anti-missile system research has been applied to the tracking and destruction of "space junk," the bits of debris left in Earth orbit by decades of spacecraft and satellites that are dangerous to new space traffic. At the Air Force observatory on Maui, advances in telescopy developed to track the space junk are now used for high-speed sports photography, including more detailed and exciting surfing movies.

The national labs have always worked with industry on cooperative R&D, mainly via CRADAs, cooperative research and development agreements, and industry has given the CRADA program generally high marks. The labs also allow private companies to use their advanced testing, measuring and production equipment. *Business Week* reported recently that Procter & Gamble was working with the weapons labs to improve statistical process control in the manufacturing of disposable diapers.

Table 6.3 Post-Cold War Missions for Los Alamos

Defense needs: Reduce the nuclear danger. Stewardship for nuclear weapons and technology, non-proliferation, and manage the legacy of fifty years of [nuclear] production.
Civilian national needs: Government-driven; agency and industry collaboration
- Energy
- Environment
- Infrastructure
- Affordable health care
- Basic research
- Education
- Space

Commercial technologies: Industry driven
- Cost-shared, market-driven research & development
- User facilities
- Technology assistance
- Entrepreneurial start-ups

Source: Los Alamos National Laboratories

The U.S. and Russian national laboratories are working together to reduce the danger of commercial reactor meltdowns by repairing the many other operating reactors of the Chernobyl type. They are trying to find legitimate employment for Russian nuclear scientists so that these men and women do not have to work for rogue nuclear states in order to put bread on the table. And they are devising countermeasures against possible nuclear terrorist attacks, which may not be missile-based.

Notwithstanding, the end of Cold War budget levels, and the trends toward smaller government and cheaper space missions, have decimated the operating funds of the national laboratories. When Texas' superconducting supercollider project was shut down, 159 computer scientists and programmers, 189 physical scientists, 429 engineers, and 535 technicians, machinists and technical staff found themselves unemployed, according to the SSC lab.

The labs try to ease their personnel reductions by encouraging employees to license federal technologies and start new companies. To this end, the labs run new business incubators and cooperate with university-run incubators. However, many talented scientists, engineers and technicians leave technical careers behind when they depart from the labs. This represents a loss of a national resource and of the public investment in the education of these individuals.

6.2 Pertinent Publications on U.S. Federal S&T Policy

Reports of the Carnegie Commission on Science, Technology, and Government are available free of charge from the Commission, 10 Waverly Place, 2nd Floor, New York, NY 10003. The reports deal with the federal R&D budget, K-12 education in science and mathematics, environmental and development economics, and linking science and technology to national goals.

American Association for the Advancement of Science (Intersociety Working Group), *Research and Development 19xx*. This series of reports engages representatives of AAAS, The American Chemical Society, the Institute for Industrial Research, and two dozen other organizations to assess the status and needs of a number of engineering and scientific disciplines and the federal agency budgets that support them. www.aaas.org.

National Science Foundation (Division of Science Resources Studies), *Federal R&D Funding by Budget Function, Fiscal Years 19xx-yy*. These reports usefully cross-foot funding agencies (Defense, Commerce, etc.) with technical areas funded (e.g., health, space technology, transportation). See also NSF Division of Science Resource Studies, *National Pattern of R&D Resources, 1999*. www.nsf.gov/sbe/srs.

National Institute of Standards and Technology, *Global Growth of Technology: Is America Prepared?* Gaithersburg, MD, 1995. This and similar reports from NIST address workforce, infrastructure, and educational needs for technological growth.

National Science and Technology Council (1997): *National Security Science and Technology Strategy.* The NSTC, a cabinet-level advisory to the President of the United States, summarizes the government's investment in national security and global stability.

National Aeronautics and Space Administration (1995): *SpinOff 1995.* Describes the application of NASA technologies to non-space and commercial arenas.

Office of Technology Assessment, Congress of the United States (1995): *The Lower Tiers of the Space Transportation Industrial Base.* OTA analyzed the structure and needs of the (sometimes small) companies that form the supplier base of the space program.

National Academy of Sciences, National Academy of Engineering, Institute of Medicine, National Research Council (Committee on Criteria for Federal Support of Research and Development) (1995): *Allocating Federal Funds for Science and Technology.* The four national academies, created by Congress, advise Congress on S&T matters. This report advises on the criteria that should be used for appropriating federal dollars to R&D.

Office of Technology Assessment, Congress of the United States (1991): *Federally Funded Research: Decisions for a Decade.* OTA, before its demise, analyzed the information needed for deciding how research funds should be allocated. Reports like this provided the data that the National Academies use to devise criteria for spending.

NASA, *Space Technology Innovation.* This periodical highlights all space and commercial NASA technologies.

NASA Tech Briefs. Another, more technical periodical from NASA focusing on design and engineering.

Space Research and Technology Transfer http://ares:ame.arizona.edu/~jeffb/technology/links.shtml. Links and articles having to do with the effects of space research on society.

National Institute of Standards and Technology (1997): *ATP Focused Program Competition 97-05: Technologies for the Integration of Manufacturing Applications.* Also *97-06: Component-Based Software* (and so on). NIST's ATP funds companies to do high-cost, high-risk industry consortial projects. These publications highlight the projects of past winners and advises new applicants on the proposal procedure.

Technology Transfer Business. This privately published (by TechNews in Vienna, Virginia) magazine is for companies engaging or wishing to engage in cooperative research and development agreements (CRADAs) or other public-private technology transfer arrangements.

R&D Forecast 2000.   Battelle Memorial Institute and *R&D* Magazine. www.rdmag.com; www.battelle.org.

Government Technology. A magazine, free from www.govtech.net, for companies selling or supplying information technology to state and federal governments.

Small Business High Technology Institute (Phoenix, Arizona. 2000): *Reauthorization of the SBIR Program: An All-American Priority.* Requests Congress to renew the Small business Innovation Research programs. These small business set-aside programs in ten federal agencies allow successful applicants to fund research toward their current or intended product lines, in stages of increasing grants. The report is available at http://www.sbir.dsu.edu/ reauthorization/Default.htm; see also http://www.sbir.dsu.edu/.

The Department of Defense, NSF, and three other agencies publish *Small Business Technology Transfer (STTR) Program Solicitations.* Like SBIRs, STTRs are small business set-asides that are mandated by many federal appropriations. STTRs are aimed at businesses willing to engage in cooperative research programs with nonprofit research institutions with the aim to commercialize results.

Science and Technology Agency, Prime Minister's Office (Japan), *White paper on Science and Technology.* This 1994 report assesses the status, organization and funding of Japanese S&T.

6.3 Japanese Industrial Policy

Let's examine the three kinds of *keiretsu* that exist in Japan. The *horizontal keiretsu* are groups of companies that perform different functions but are grouped together under a single banner, like the Sumitomo group or the Mitsui group. *Vertical keiretsu* are an integration of the supply chain which involves cross ownership and very close business cooperation among the companies. In the Japanese vertical keiretsu, it is common to have many layers of suppliers with each supplier in the pyramid responsible for only a few sub-suppliers below it. The American automobile industry has since moved closer to this kind of supply chain management. But, early in the '90s, General Motors was managing 1,200 suppliers directly and suffering from the administrative burden of doing so.

Horizontal keiretsu

Figure 6.1 Horizontal keiretsu, the example of Sumitomo (Art courtesy of Cenquest, Inc.)

Vertical keiretsu

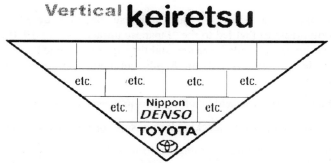

Figure 6.2 Vertical keiretsu, the example of Toyota (Art courtesy of Cenquest, Inc.)

A third type of keiretsu is the downstream, or *distribution keiretsu* in which manufacturers own their own retail outlets and distributors.

Japan's cozy government-industry relationship constitutes a very different kind of capitalism from the largely laissez-faire U.S.-style capitalism. During the Cold War, these differences appeared small and inconsequential compared to the overall opposition of capitalist versus communist systems. Now, following the Cold War, the increasingly visible differences may cause friction in the U.S.-Japan relationship. While Japan's capitalism succeeded in growing its post-World War II economy most impressively, in the 1990s it proved brittle and inflexible in the collapse of its speculative investment "bubble." The U.S., meanwhile, has experienced an unprecedented expansion in the same time frame. The U.S. and Japanese economies remain strongly linked in high technology trade and other areas; the maintenance of the binational economic relationship is very important.

Figure 6.3 Distribution keiretsu, the example of Sony (Art courtesy of Cenquest, Inc.)

Chapter 5 explained some of the Japanese product development and manufacturing practices that led to the "Japanese miracle" of the 1980s. Why has Japan's government provided greater stability of environment for long-cycle industrial projects, and less risk for highly competitive short-cycle products? Answers include:

- Most of a firm's stock is held by other firms in the same keiretsu.

- There is (accordingly) little stockholder pressure for short-term results.

- Ministries act to insulate commerce from effects of frequent changes of government.

- Ministers are political appointees but have little authority. Deputy ministers and career bureaucrats hold the power.

- The "best & brightest" university graduates are employed by the ministries.

- The collusive Keidanren, a roundtable of chairmen of largest corporations allocating government funds to strategic industries, would be illegal in the U.S. Ministries communicate the political economy agenda to keidanren. Keidanren then translates this agenda into "industrial policy" and seeds companies in competing keiretsu.

- Japanese schools and national media (NHK, Mainichi) encourage a uniform sense of national mission.

- Japan's "traditional acceptance of government intrusion in the marketplace."

Sometimes these features of the Japanese industry-government relationship can help the U.S., e.g., when MITI encouraged a "Buy American" campaign in the 1990s.

The Big Six Horizontal Keiretsu

Zaibatsu were large, family owned monopolies that existed in Japan through the end of World War II. "By the end of the war, the four largest zaibatsu with their subsidiaries accounted for 24.5% of the joint stock capital in Japan." These organizations were broken up by U.S. occupation forces at the end of the war. However, when the occupation ended in 1952 the zaibatsu gradually began to reorganize as the keiretsu.

Keiretsu contain two types of alliances: horizontal and vertical. Horizontal alliances are formed across different industries and are "characterized by cross-shareholdings and a core bank or cash-rich company." Typically, the amount of stock that each company holds in another is small (2-4%). However, overall the group will hold 17-26% of the stock. Vertical alliances are "pyramids of suppliers dominated by large manufacturers."

According to the U.S. Fair Trade Commission, the largest six keiretsu represent 32% of the capital stock in Japan, 27% of the gross assets, and slightly over 25% of sales. However, these figures are understated because several companies were not included in this study and gross assets were taken at book (not market) value.

The heads of the core companies of a particular keiretsu belong to a secretive "executive club", the function of which is to promote group solidarity. Also, these clubs determine group strategy, group actions with respect to external parties, and mediation of problems that arise between group members.

Ties between companies are also strengthened in several ways:

Intragroup financing, arranged through the core bank in each keiretsu.

Movement of directors between member companies of the keiretsu.

Reciprocal trade networks, in which member companies sell goods and services to one another.

The keiretsu are becoming increasingly xenophobic and less able to perform "healthy introspection." For example, companies employ sokaiya to disrupt stockholder meetings or manipulate proceedings in whatever way the board of directors wishes. Company auditors are typically very senior employees, are close to retirement, have no real power, and cannot be counted on to reveal problems. Members of the board of directors may hold multiple positions such as chairman of the board, president of the company, and managing director, all at the same time!

Inter-keiretsu competition is decreasing. For example, "the Ministry of Finance prohibited banks from making loans exceeding 20 percent of capital to a single enterprise. This prompted cofinancing between the banks of different corporate groups."

Yoshinari M (1992)) The Big Six Horizontal Keiretsu. Japan Quarterly (April-June) 186-199. Summary by Michael Funk.

Table 6.4 Trends in the Japanese Industry-Government Relationship as it Affects the U.S.

- U.S. trending toward continued laissez-faire "industrial non-policy," except in defense and aircraft.
- Japanese ministries are still very powerful, but...
- Finance Ministry's workload is soaring - understaffed - resources stretched.
- Finance Ministry opposes MITI's "Buy American" program – worries about lost revenues.
- Japanese firms' overseas operations and investments are beyond MITI's authority.
- Slow recovery of Japanese securities and real estate markets.
- Post-coldwar US-Japan relationship: Will US feel free to increase protectionist pressures on Japan if Japan is not needed as backbone of Pacific security?

Sources: Prestowitz, *Trading Places* ; Morita and Ishihara, *The Japan that can Say No.*

This picture of close government-industry cooperation in Japan is changing (see Table 6.4), because as firms increase their activities outside the borders of Japan, government ministries are finding it more difficult to oversee operations. Sometimes the ministries even work against each other.

Japan's attempted remedy to the crash of their country's financial industry is to deregulate. As a result, Japanese ministries are too preoccupied with financial restructuring to actually look after individual companies the way they once did. The United States and Japan are still adjusting their post Cold War relationship, and the remaining differences between Japanese style capitalism and U.S. style capitalism will either cause further convergence or further friction. We shall see.

Who Says Science Has to Pay Off Fast?

The U.S. government has persuaded Japan to stop "freeloading," that is, engaging in product development solely on the basis of research performed by other countries. The Japanese government is now funding more basic research of its own, increasing its basic research budget 16% over 1993, and Japanese semiconductor companies are now performing research that may not pay off for twenty years or more.

Meanwhile, U.S. federal and corporate research funding is on the decline [it has rebounded again in 1999-2000.   U.S. government R&D funding actually decreased almost 2% in 1996.] Corporate funding for research dropped 15% in the period 1986-1994.

Japan's mercantilist/nationalist approach to technology transfer has eased, and the Japanese have allowed "leakage" of production technology through offshore ventures and alliances. Japanese science agencies now accept foreign participants, and these participants' patent rights are protected.

Gross N, Carey J and J Weber (1994) Who Says Science Has to Pay Off Fast? Wall Street Journal March 21. Summary by Frank Goshey

6.4 Technology Industry Policy at the Local and Regional Levels

6.4.1 Technology and Regional Economic Development

County and municipal governments administer the road building, schools, zoning, property taxes, and marketing that affect a technology company's location decision. Now that it has become evident that a critical mass of technology firms in a region like Silicon Valley or Seattle can create great wealth, more regions are bending every effort to building their own "technopolis," hoping to enjoy the clean industry, growing job rolls, and increased tax base that will result.

It is beyond this book's scope to go into great detail about a firm's relationship with local governments. In short, however, companies want to be good citizens, and municipalities want to give companies incentives to be good citizens. Anytown, USA, fosters its economic development by persuading existing companies to relocate or start new divisions there; by helping existing local companies to succeed (perhaps by helping them find new export markets); by helping new companies start up in Anytown, and by participating in new institutions and alliances that further the latter three objectives in such new ways as are dictated by the changing technological and economic landscape. Public/private incubators and investment capital networks are examples of this kind of institutions and alliances.

Economic development objectives are met in two ways: by direct incentives, and by self-investment. Direct incentives include tax breaks for companies that are expected to offer a large number of new local jobs. This tactic is favored by Washington County, the center of Oregon's Silicon Forest. Rent-sharing on a building, temporary waivers of regulations, and other kinds of relocation assistance are also examples of direct incentives.

Austin, Texas attracted the consortium MCC in the 1980s with a dramatic self-investment. The University of Texas at Austin collected philanthropic pledges for chaired professorships and new laboratories. It told MCC that if MCC chose Austin, the university and the community would invest in itself in this way, increasing the number of science and engineering graduates needed by MCC's projects and members, and attracting scientists of top reputation to Austin. (There were some direct incentives involved in the MCC effort as well.) Other cities successfully use the lure of new airports, highways, rail systems, and communications networks to convince companies that it will be easy to do business there.

The next section illustrates these principles and some others, in the context of a region that is trying to attract a certain kind of technology industry. The Balearic Islands are a province of Spain in the Mediterranean Sea. The provincial capital, Palma de Mallorca, used a grant from the European Community to pursue a self-investment strategy, namely an attractive facility for net businesses and teleworkers.

6.4.2 ParcBIT and Balearic Economic Development – A Consulting Report

ParcBIT is intended to help trigger a diversification of Mallorca's economy away from its dominant reliance on agriculture and traditional tourism. Its name, Park for Balearic Innovation and Technology, implies the islands' diversification will rely on new, different, and perhaps unforeseen directions, though probably these directions will involve new technologies.

I felt honored by Sr. Font's invitation to join the expert panel for the ParcBIT design competition. Perhaps the obvious place to start this chapter is to explain why, with no training in architecture, I was invited to judge an architectural competition.

The IC^2 Institute is one of the world's premier institutions for the study of technology as a driver of economic development. The Institute was a key factor in the transformation of the Austin, Texas region into what *Business Week* magazine calls a "technology hot spot," that is, a fast-growing regional economy that clearly owes its growth to both start-up and established technology companies. IC^2's contributions included a role in attracting the research consortium MCC and the manufacturing consortium SEMATECH, founding the renowned Austin Technology Incubator and the Austin Software Council, and researching and pioneering graduate curricula in technology management and entrepreneurship.

The result has been the "IC^2 Model" of technology based economic development, and the "Austin Experiment" that demonstrates and proves this model.

Also, due to earlier employment in the private sector, I have a background in tourism market research. Thus I am aware of impending threats to traditional tourism, such as shorter vacations and an increase in "weekend getaways" – especially for Americans and most especially for two-career families in which both spouses cannot take two weeks' holiday at the same time. Other trends involve the consequences of the increased divorce rate – single-parent holidays and the complex logistics of gathering for vacation the children from multiple marriages. Still further trends include, most significantly for Mallorca, the work holiday and the special-interest or segmented market tourism strategy. ParcBIT will involve attracting teleworkers, entrepreneurs, and business conferences, and possibly the families of the persons making the vacation destination decision. ParcBIT aims to attract short-term visitors, vacation home owners (continual medium-term visitors), and permanent residents. All of these will choose Mallorca because of the island's traditional attractions as well as the amenities provided by the innovation park. I emphasize tourism because I see ParcBIT not as a revolutionary new direction, but as an enhancement and modernization of tourism in the Balearics.

In the remainder of this section, I will summarize what seem to be the imperatives for a regional technology development strategy, and offer initial observations about their relevance to the Balearics. In that context, I will then offer some reactions to the architectural competition, and ideas on how information technology will affect the design and use of space at ParcBIT.

6.4.2.1 Technology Based Economic Development

Achieving technology based economic development depends on attention to these things in the early stages:

Globalization and the Transformation of Competitive Advantage. No nation has a natural transportation cost advantage with regard to information. Rather, nations build competitive information advantage via investment in education, telecommunications infrastructure and software. Corporations can, and do, make daily use of information from sources around the planet. One hundred years ago, information traveled around the globe by sailing ship, arriving months after transmission, if at all. Today, it arrives instantly, regardless of its geographical origin. However, closing a sale on a high-value-added item, or maintaining customer satisfaction with an expensive, high-service-content product, requires face to face contact. Regions hoping to profit from these kinds of products must have fast air transportation to customer sites.

Political Organization. A local community must organize to gain the support of governors, presidents, and representatives to national and supranational legislatures, so as to benefit from all possible grants and appropriations for technology development.

Mental Set. All elements of the community must have an accepting, or at least tolerant, attitude toward change and innovation, toward entrepreneurship, and toward other sectors, agencies and community elements.

Strategic Positioning: Choosing a pony to ride. With limited resources, a community must leverage its natural advantages to produce a profitable fit with a very focused investment in technology. Examples include Austin (software and semiconductors), North Carolina (biotechnology), Tsukuba, Japan (government laboratories), and Kansai City, Japan (private companies' research laboratories). An examination of other regions' success factors and failures aids the positioning exercise.

Making it Work: Riding that pony. The positioning choice is followed by an effective balance of attracting relocating companies in the chosen industry; retaining and nurturing indigenous companies; starting new companies; and building innovative institutions to support the latter three activities (for example, the Austin Software Council).

The Changing Role of Cities. To the extent that people have lived in or traveled to cities for proximity to information, cities are obsolete. Future cities will exist to emphasize the creative spark and the human satisfaction that comes from proximity to other people.

The Role of Universities. Universities should produce not just graduates, but a steady flow of new ideas, devices and spin-out enterprises that can be commercialized for the region's economic benefit. Universities in technology growth-pole regions are partners in economic development with governments and companies.

Tourism, Music and the Arts. Technology entrepreneurs and engineers are of course educated people who appreciate a fine quality of life in their communities – preferably coupled with a low cost of living. They are attracted by many of the same amenities that attract tourists. A thriving artistic community and a healthy tourist trade can quickly find common interest with the promoters of environmentalism and clean, modern technology industry.

Marketing the Region. An art/tourism/high-tech alliance is an attractive product. Nonetheless it takes sophisticated public relations, advertising and media executions to ensure a steady flow of attention, visitors and funds.

Role of Innovation Parks. Innovation parks spur the region's growth by providing an attractive setting where an atmosphere of creativity can flourish, and where a critical mass of technologists, scientists and thinkers can build on one another's ideas. The park serves as a showcase for industry and a magnet for students.

Building the Community: The role of the visionary. In every case I've studied, one influential individual has translated all the above elements into meaningful objectives for his or her region. This person has formulated a vision, evangelized it, and marshaled the political, popular, and financial support needed to make it real. This visionary individual has inspired others in the region to translate the objectives into programs and execute the programs. These people may become absorbed in the particulars of the programs they are implementing. But they are always able to rely on the visionary to maintain the integrity of the vision. I cannot overemphasize that this visionary must be a very extraordinary individual, combining technological knowledge with planning acumen, personal charisma, and political savvy.

Figure 6.4 The IC^2 model of value-added economic development

Orchestrating the Technopolis – Overture. The first step in formulating a regional strategy, as I mentioned, is to gather global examples. Then the question is, can these examples be translated into local success? Answering this question is an opportunity to kill three birds with a single stone. An initial regional conference or workshop gathers expert opinion on the region's potential strategies and outlook for success. It also draws local people who can implement programs into the orbit of the visionary. As a public relations event, it draws early, perhaps nationwide, attention to the region and tests the waters regarding public opinion and potential support for moving forward.

Orchestrating the Technopolis – First Movement. The support of professional and community organizations must be obtained, and new organizations started where necessary. A comprehensive plan should be drawn. The plan should be ambitious but flexible, and is not necessarily written down. Fast success of the entire plan should not be expected. But seed money should be obtained while early momentum exists, and must be used to achieve and display small early successes.

The picture of the IC^2 Model, Figure 6.4, implies foremost that the initiative for economic development comes from the local community. In an age of rapid technological change, only the local region can identify its own best opportunities in a timely way. The picture shows that success depends on the cooperation of the government, academic and business sectors. These sectors collaborate, perhaps in new ways, to bring together entrepreneurial talent, capital, technology, and business management expertise that will identify and satisfy new market needs.

6.4.2.2 The Balearics: Combining Tourism and Technology Development

There are only three ways to increase tourism revenue: Bring more people; get them to stay longer; and get them to spend more money per day. ParcBIT will attract people who will stay longer than charter-flight tourists, and who will spend more money in the Balearics. These will be individual teleworkers, sales teams, telework task groups (called "skunk works" in the U.S.), and conventions, meetings, and management retreats. Eventually, ParcBIT will attract not just company delegations, but companies – startups, spin-offs, new and relocating divisions, relocating small and mid-size companies, and their vendors.

I believe Mallorca has the potential for attracting the desired traffic. Its advantages include a perfect climate with clean air, good air transport service, and some strong university departments. The university, IUB, now has new school of hospitality and hotel administration that can be an effective partner in technology based tourism.

There is a proud heritage of artistic and literary achievement, and celebrity artists and performers vacation in the islands. English, Spanish and German are widely spoken. The clustering of British, German, and Scandinavian expatriates makes an interesting diversity that holds the visitor's interest. The Ministry of Finance and Economy seems skilled and energetic in dealing with the challenges I observed in Palma, which include a water shortage, a government tradition of centralized decision making, and pockets of conservative sentiment in the business community.

6.4.2.3 The ParcBIT Masterplan International Ideas Competition

From the Ministry documents circulated prior to the competition, one could abstract the following judging criteria:

1. Sensitivity to ParcBIT goals and requirements (i.e., meets requirements).
2. Innovative contributions to concept and objectives (i.e., goes beyond requirements).
3. Conceptualization of vision.
4. Balance and interspersion of use.
5. Ecological design solutions.
6. Interest of building typologies.
7. Technical interest.
8. Integration with landscape.
9. Use of resources.
10. Infrastructure design.
11. Architecture of buildings.
12. Technical feasibility.
13. Financial viability.
14. Understandability to the general public.

Naturally, different competing entries addressed these criteria with different emphases. I must say, though, that none of them seemed to attempt seriously to address all the published criteria. In particular, I was surprised that there was no state of the art discussion of intelligent buildings. Nor was there discussion of the many European cultures that will be represented at ParcBIT, and how to accommodate the various lifestyle preferences they will bring. Strikingly for an innovation park competition, some teams seemed biased against technological change. I sympathize with the desire to preserve the best of one's culture, but I believe that preservation of the past can coexist with embracing the future.

This was my first opportunity to work closely with a group of architects, and I was impressed by their wide range of interests and cultured learning. It is rare in academia and in business to find people who can grasp equally the "soft" or cultural considerations and the "hard" (mathematical, engineering) side. Many of the architects I met in Mallorca can do this, and I hope ParcBIT can continue to benefit from their very creative input.

Finally, I learned something important from the architects that I will use in my own work. That is that certain types of human behavior, e.g., innovativeness, can be "engineered in" in an enterprise not just by organizational or motivational means, but also by the physical design of working space.

6.4.2.4 Technology, Entrepreneurship, and Use of Space

How do new telecommunications technologies affect the use of space? Without pretending to know all the answers, I would like to offer these preliminary thoughts on a subject that was under-addressed by the contestants. All of them potentially affect the layout of ParcBIT.

- Teleworkers keep strange hours in order to communicate worldwide. Local services, e.g., cafes, snack bars, laundromats, must be open 24 hours a day. Services to support fast-cycle, competitive innovation must include printing, photographic developing, etc. – that is, all the 24-hour services needed to help the entrepreneur produce the physical materials needed when it finally becomes necessary to stop teleworking and visit the customer face to face.

- High-tech industries compete by driving down fixed costs, including real estate costs. It is only a small exaggeration to say that such companies can be content to work out of tents. Expensive real estate will not be a source of business prestige, especially if the clients never see the home office.

- Several of the most famous Silicon Valley companies got started as ideas scribbled on napkins at bars. Picking up on this phenomenon, some U.S. bars in high-techh areas now have their own Internet nodes as well as multimedia facilities that allow customers to bring and play their favorite sports, music or computer graphics videotapes.

- Entrepreneurs may feel isolated from their families, due to their long working hours and frequent travels. Office buildings and living spaces must be arranged so as to facilitate access from workspace to child care space, and a balance of company-oriented family activities with private family time.

- Entrepreneurs, even teleworker entrepreneurs, need face-to-face interaction with other entrepreneurs, to share technical tips and social reinforcement. Space should be arranged to encourage encounters. This means paying attention to traffic patterns in hallways, for standing encounters; small corner conversation alcoves, small and large conference rooms, and auditoria.

A life of entrepreneurship can involve extraordinary stresses. Remedies include partying and meditating, and these too require different kinds of spaces.

6.5 Notes

Rosenbaum D (1996) Sasha's Trick. Warner Books New York 204

microelectronics executives are becoming proactive... O'Connor R J (2000) Mr. Tech Goes to Washington. Upside February 113-122

Cost-effective R&D policy requires funding *application-driven basic research*... This generalization of Vannevar Bush's post-World War II R&D philosophy has been championed by me and more currently by Donald E. Stokes of Princeton's Woodrow Wilson School.

> Learner D B and F Y Phillips (1993) Method and Progress in Management Science. Socio-Economic Planning Sciences 27(1):9-24

Stokes D E (1997) Pasteurs Quadrant : Basic Science and Technological Innovation. Brookings Institute Washington DC

Amazon.com president Jeff Bezos' widely read letter... See Lessig L (2000) Online Patents: Leave Them Pending. Wall Street Journal (March 23) 23; Martinez M J (2000) Amazon Executive Calls for Stricter Patent Guidelines. The Oregonian (March 10) B7. (For examples of business process patents, see www.public-domain.org/patent/business.)

Ashton W B and R K Sen (1988) Using Patent Information in Technology Business Planning – I. Research•Technology Management (November-December) 42-46. This article discusses using patent trends to assess a firm's competitive position.

The differing patent laws in various countries... El-Badry S and H Lopez-Cepero (Eds) (1995) Tricks of the Trade: Intellectual Property in the United States and Japan. IC^2 Institute, University of Texas at Austin

PatentCafé is a resource for those interested in patent issues: www.patentcafe.com. Their magazine, Cafezine, is at www.cafezine.com, and they offer a new new digital rights management resource, www.lockmydoc.com.

"The corresponding articles of the European Patent Convention are 60 (right to a European Patent, see text at the bottom), 62 (right of the inventor to be mentioned), 81(designation of the inventor) as well as Art. 4 of the EPC Protocol on Recognition. They can be found electronically for example under http://www.epo.co.at/legal/epc/e/ar60.html and following http://www.epo.co.at/legal/epc/e/arti4.html... The European Patent Convention makes a reference to the national patent laws of the member states which contain different regulations." Source: Thomas Moser INNOVATIONSAGENTUR 1020 Vienna / Austria. See also http://www.ipr-helpdesk.org/ipwire.

Venture investors now sometimes advise startups against applying for patents... Roberts B (2000) The Truth About Patents. Internet World (April 15) 72-84

Life patents. This paragraph adapted from Mark Rehley's summary of Marsa L (1996) Patent Wars. OMNI Winter 17(9):36-41 (Life patents can inhibit research.... Statement of Maria C. Freire, Ph.D., Director, National Institutes of Health Office of Technology Transfer before the Senate Appropriations Subcommittee on Labor, Health and Human Services, Education and Related Agencies, January 12, 1999); For both sides of the patented seed issue, see Ehrenfeld D (1997) A Techno-Pox upon the Land. Harpers (October) 13-17; and Shapiro R B (1998) Trade, Feeding the World's People and Sustainability: A Cause for Concern. Center for the Study of American Business Washington University in St. Louis (April)

The Russian national laboratories. See Hecker S S (2000) Russia's Nuclear Cities: A Business Opportunity and a National Security Imperative. IC^2 Institute, University of Texas at Austin April

The U.S. House of Representatives Science Committee lists currently pending bills at http://www.house.gov/science/106th_bills.htm.

industry has given the CRADA program generally high marks... Gibson D V, Jarrett J and G Kozmetsky (1994) Customer Assessment of CRADA Processes, Objectives, and Outcomes at Martin Marietta Energy Systems, Inc. Oak Ridge National Laboratory and IC^2 Institute, University of Texas at Austin. See also Beardsley T (1997) The Big Shrink: Federal labs are developing new chipmaking equipment. Who will reap the benefits? Scientific American December 15-16

6.6 Appendix: Terminology of the Japanese Government-Industry Relationship and Allied Areas

Dango	Collusion in bidding for public construction projects.
Gyoseishido	MITI's "adminsitrative guidelines," used to direct industrial policy. Although technically these are informal, practically speaking they carry the weight of law.
JETRO	Japan External Trade Organization. Responsible for facilitating foreign joint ventures, etc., and for gathering industrial intelligence.
JFTC	Japan Fair Trade Commission.
Kaisha	corporation.
Keidanren	Federation of Economic Organizations. The elite organization of Japanese business leaders. Exerts considerable influence on government policy.
Keiretsu (= kigyo shudan)	Trading group. A collection of companies tied by mutual stock ownership and strong connections with a single large bank. Members of a keiretsu are expected to do business almost exclusively with other members.
Keizai doryukai	Japanese version of the American Committee for Economic Development. One of the Big Four business organizations. Concerned with developing a Japanese ideology of business.

Mainichi Shinbun	with Asahi Shinbun, Yomiuri Shinbun, and Japan Times, the major national daily newspapers.
Marunouchi, Otemachi	Tokyo's financial districts.
Meishi	Business card.
MITI	Ministry of International Trade and Industry.
Nihon Keizai Shinbun	Japan Economic Journal (newspaper). It is called the "Wall Street Journal of Japan," though not connected with WSJ.
NIKKEI	Nickname for the Tokyo stock exchange.
Nikkeiren	Federation of Employers' Associations. After Keidanren, the most important of the Big Four business organizations. Focuses on labor/management issues, productivity, and modernization.
Nissho	Japan Chamber of Commerce and Industry. Oldest of the Big Four. Lobbies for the interests of Japan's small and medium sized enterprises.
Ringi(sho)	The ringi system of decision making is widely used in large Japanese corporations and government agencies. It enables lower management to obtain policy guidance from superiors. The request for decisions or guidance is a document known as a ringisho. The document is circulated to all concerned, upward, downward and laterally. These other executives signify approval by affixing their seal to the ringisho.
Sanken	Industrial Problems Study Council. Committees of top executives charged with studying and advising the ministries on specific business problem areas. Membership overlaps with zaikai, which is focused on more general issues of policy directions.
Sogo Sosha	Corporation.
Sokaiya	"Ringers" hired to control stockholder meetings.

| *Zaibatsu* | A trading/holding company. |
| *Zaikai* | "Business circle" of the leaders of the Big Four organizations. The activist power structure of the Japanese business community. |

(Sources: These definitions were gathered in part from U.S. Dept. of Commerce, Bureau of International Commerce (1972) Japan: The Government-Business Relationship, and Business Tokyo magazine.)

6.7 Questions and Problems

Short Answer

S6-1. Patents are usually said to encourage innovation by protecting its fruits. What is one way in which patents can discourage innovation? How? Can you suggest any remedies?

S6-2. NIST's Advanced Technology Program (ATP) funds high-cost, high-risk collaborative development projects. Is basic research ever "high-risk"?

S6-3. This question requires you to look up a Business Week Online article by John Carey, Why Washington is Anointing Flat Panels, May 16, 1994. The article describes one of the last gasps of industrial policy in the U.S., namely a subsidy to the flat panel display industry.

(a) Who wins and who loses under Flamm's plan for subsidies to U.S. flat panel display (FPD) producers?

(b) What role does military demand play in the growth of the U.S. FPD industry?

Discussion Questions

D6-1.   In teams of 2 or 4, *debate* one of the following issues relating to the differential progress of technology and social norms. Look for background materials on these topics, on the Web or in a library. Teams will be assigned to argue "pro" and "con" on each question your instructor selects for debate. Teams should meet privately for 20 minutes prior to the debate, to decide what arguments to present and to select a spokesperson. Allow 2-1/2 minutes each for primary presentations pro and con, and rebuttals pro and con:

(a) Limits should be placed on the number of children sired by a single human spermbank donor.

(b) International treaties should limit overfishing.

(c) Transplantation of vital organs should be limited (worldwide) until it can be ensured that all donations are voluntary.

(d) Anonymous Internet remailers should be banned.

D6-3. (a) Discuss causes of Japan's increasing emphasis on basic research. Specifically, consider the impact of reduced US government R&D expenditures; the end of the cold war; US-Japan trade pressures.

(b) What do you think is Japan's strategy for realizing industrial benefits from this R&D? Is the strategy at risk when non-Japanese researchers are permitted to participate in designated projects?

D6-4. Criticize the Post-Cold War Mission of Los Alamos National Laboratory. Speculate on how the talents of scientists and engineers laid off from the national labs (as a result of federal budget cuts) might be put to use for the benefit of society and the satisfaction of the scientists and engineers.

D6-5. Choose a US Government publication from the list in §6.2, or any other government publication dealing with S&T policy and budget. Write several paragraphs explaining:

- who published the information and why

- what kinds of information is in the publication

- why it is or isn't relevant to a technology manager.

Bring the publication to class to share and explain to classmates.

PART IV

NEW-TO-THE-WORLD PRODUCTS

"An optimist is someone who thinks the future is uncertain."
Walt Whitman

"Next week a doctor with a flashlight will explain to us where
market forecasts come from."
Scott Adams' character Dogbert

"Never underestimate the bandwidth of a Federal Express package
stuffed with diskettes."
Bruce Sterling

7 Researching Technology Markets in a Fast-Cycle World

Information gathering can and should start early. *Early,* as in long before the product is launched, and if possible before the product is conceived – ideally, at the point when a generic technology comes into existence. This generic technology may be capable of generating many products in many different, and perhaps new, industries.

In this chapter, we will see how to get and use the information that is needed to decide which industry to enter first with a new technology. We will then follow the innovation chain down to the product's post-launch period, identifying the information that is needed at each stage, and remarking on where this information can be obtained and how it can be analyzed.

Of course, some of this information is about technology (scanning for technology information was covered in §4.1), some is about competitors, and some is about customers. The chapter title thus uses the word "markets" in a very general way to cover all of these, but you may prefer to use the phrase "PPSM (product, process, schedule, and market) intelligence" that was introduced in Chapter 5.

7.1 Forecasting

In Chapter 5, we noted the difference between passively predicting the future and actively creating the future. Successful technology managers do both; we might cite Gordon Moore, who transformed many industries by creating Intel

Corporation and yet devised an accurate and famous prediction tool known as Moore's Law.

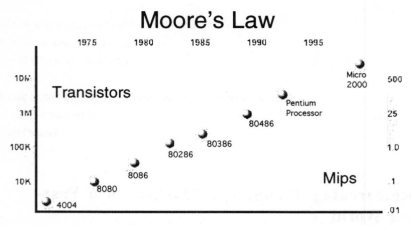

Figure 7.1 When plotted on semi-log paper, Moore's Law presents a straight-line picture of processor power increasing over time.

We noted limits to trend extrapolation, notably that time series forecasts (which depend on known probability laws) are soon overwhelmed by uncertainty – the state of the future in which even the probability laws are not known. In any case, we further noted, straight line or S-shaped forecasting models are not likely to be accurate over more than 50-75 years. We have mentioned also that in the dynamic technology world, disruptive technologies sneak into the market from "under the trend line." In this chapter we will learn that forecasts – including Moore's Law – are also limited by physical laws.

So we may conclude that most long-term forecasting is used to decide on a company's strategic positioning, whether to enter a particular industry, whether to purchase a particular company, or what one's core technologies should be. While these are immensely important decisions, the bulk of PPSM information is used for decisions about the short term future of product development projects, product feature changes, or advertising campaigns. In fact, much information collection is not about the future at all, but is an attempt to find out what's going on *now,* or even to find out what happened in the recent past.

There's not much of a trick involved in collecting information. Most readers will recognize that they collect it without trying, and even feel swamped by incoming information. The real challenges, for companies as well as for individuals, are:

- winnowing information that *can and will be used* from information that (i) can be used but won't be; (ii) is irrelevant; (iii) is misleading or wrong; or (iv) is right, but in an unusable format.

- displaying information of the first type above, in a way that is understandable for decision makers and stakeholders.

- using information of the first type to effect needed changes in the organization.

7.2 Technology Forecasting

Technology Forecasting uses logical tools and techniques. It describes and identifies the rate and direction of change and the implications of that change. The objectives of technology forecasting are to:

1. Define and communicate technical realities and environments;

2. Anticipate and distinguish between possible, probable and preferable futures;

3. Anticipate change.

Technology forecasters attempt to predict the environment and the rate of acceptance of new technologies in the future based on previously observed rates of substitution and mortality of comparable technologies. For each client and technology, they seek to distinguish the "preferable" future and determine which types of decisions should be made. Also, they examine the likelihood of probable and possible futures and the impact of current decisions if the possible or probable futures come to fruition.

7.2.1 How Did Technology Forecasting Get Started?

Scientists in the early 1950s examined the number of components in aircraft. Using trend analysis, they noted a rapid increase in the number of components per aircraft. Given the reliability of the vacuum tubes used in the earlier aircraft and the projected number of components in future aircraft, scientists were concerned about the aircraft's weight and reliability as they added vacuum tubes. They suggested examining newer technologies like transistors.

Scientists used logical analysis and tools in a repeatable experiment to analyze current and future trends and the implications of those trends. This allowed them to realize a discontinuity was approaching and the need for new types of components would require a different type of electronic part.

7.2.2 Why Do Companies Do Technology Forecasting?

In today's world there is increased complexity in all systems, processes and procedures, and a greater reliance on those systems. Fifty years ago, there was little worry about a hard disk crash, since all records were stored on paper. As systems became increasingly complex and accomplished more, companies and individuals developed a greater reliance on their power to simplify and automate transactions.

As companies become more dependent on technology, it becomes more important to have knowledge about the ways technology could change and adapt in the future, in order to innovate and adapt to a changing market place.

Technology forecasting strengthens the planning process. Many companies rely on one set of assumptions for planning, but technology forecasting encourages forecasting on several sets of assumptions. It is important to consider whether the changes in technology and the market are linear or structural. Will the future hold steady but increasing rates of change in familiar technologies, or will a discontinuity change the type or use of technology altogether?

According to John Vanston of Technology Futures, Inc., a technology forecasting exercise may be occasioned by interest in a single technology; a focus on a company's needs; interest in a technological area encompassing more than one research thread; a look at overlaps and convergences among diverse technology areas; or a mixed approach looking both at needs and technologies. He goes on to say that a forecasting exercise should have a clearly stated objective, schedule, scope, and methodological approach (tool kit).

7.2.3 What Are the Tools of Technology Forecasting?

This section is adapted from a summary (by David Blankenbeckler) of John Vanston's book *Technology Forecasting*. There are, according to Vanston, five types of tools for technology forecasting:

1. *Surveillance.* Surveillance includes scanning, monitoring and tracking. These tools enable the user to continually examine existing technologies and their modifications, and look for new technologies entering the market. Scanning, in this context, is a broader activity than the technological scanning described in Chapter 4. Here, it involves identifying social, political, economic and other environmental factors that might affect your company. Monitoring focuses on a particular development, and is a more structured activity than scanning. Tracking is even more focused, consisting of continuous measurement of many characteristics of a development known to be of importance to the firm.

2. *Projective.* These techniques are based on the theory that the future will be very much like the past. Projective tools include trend extrapolation, substitution analysis (via the Fisher-Pry, Bass and other models), and examination of historical historical precursors of a current trend. These are analytical tools designed to determine how future technologies will be accepted and at what rate, or in the case of substitution, how quickly and easily products substitute for other products. Delphi rounds, nominal group

discussions, and structured interviews may also be called projective techniques.

3. *Normative.* Normative tools include focus groups, morphological analysis, and interviews. These tools ask the user to pace themselves in the future and set goals, to plan backward to meet future goals. This is accomplished by trying to identify the perceived future needs of society and forecast technologies that will be developed to satisfy those needs. Normative tools include the impact wheel, morphological analysis, and relevance trees. All are variations on the technology mapping techniques discussed earlier in this book, although morphological analysis focuses on the customer benefits conferred by a new technology and asks what alternatives might deliver the same benefit.

4. *Expert Opinion.* Expert opinion draws on the intuition and beliefs of leaders in the field. It can be done with single individuals, surveys or in a group format.

5. *Integrative.* This tool is really a combination of the other four tools. It attempts to collapse all the information into a coherent planning process. Integrative tools include scenario planning. A number of mathematical modeling techniques may be brought to bear to aid the integration.

7.2.4 Using These Tools

We know that industries and markets adopt technologies in similar patterns. Technology forecasting attempts to use those patterns to predict how quickly other similar technologies will be adapted and when the correct time to enter the market occurs. This can be accomplished with trend analysis and substitution analysis. These two tools show that while many forms of consumer electronics have had rapid growth in sales, the number of years to the growth break has been uniformly short. A company using this information would determine how much time is left before the growth break to determine whether or not to enter this market.

Another tool is simply the knowledge that information has a life cycle like products. Where information is in the life cycle can tell us who knows about the technology and the type of customers they might be. For instance, if information on technology is still in the early stages, it will be contained mostly in science fiction and technical journals. The consumers of this literature are probably innovators, but are not opinion leaders. As information moves through the life cycle, it reaches mass media, such as *Time* or *Newsweek.* Here it is consumed by the average American, but at this point probably has buy-in from opinion leaders and early adopters. Finally, information late in life is analyzed and reviewed, as in historical analysis and retrospective doctoral theses. At this point, we have reached all potential customers and should start to look for a new product or an improvement to start on rapid growth again.

Dateline 1873: Sir John Erickson, Surgeon General to Queen Victoria of England, said, "The abdomen, the chest and the brain, will be forever shut from the intrusion of the wise and humane surgeon."

Dateline 1895: Lord Kelvin, who was president of the Royal Society and later had the temperature scale named for him, said, "Heavier-than-air flying machines are impossible."

Dateline 1920: H. M. Warner, one of the Warner Brothers, said, "Who the hell wants to hear actors talk?"

Dateline 1943: Thomas Watson, then Chairman of IBM, said, "I think there is a world market for maybe five computers."

Dateline 1977: Ken Olson, then President and Chairman of Digital Equipment Corporation, said, "There is no reason anyone would want a computer in their home."

Dateline 1981: Bill Gates of Microsoft said, "640K [internal memory] ought to be enough for anybody."

As our little tour through history just showed us, no matter what technique you use there is no substitute for examining your assumptions very closely. Those people made forecasts that make us laugh today. We have to give them credit for trying to forecast the future at all, but they should have followed up on each and every one of those forecasts by asking themselves, "But what if I are wrong?' You may want to pause for a moment and go back over each of those quotes. Ask yourself, "What was the faulty assumption underlying each one of them?" Explore what principle from this book they might have applied to avoid being led astray as badly as they were.

In great contrast to these, we may point to a forecast that has proven to be true for a far longer period of time than its originator probably ever imagined. This is Moore's Law (Figure 7.1), created by Intel co-founder Gordon Moore. In 1965, Moore predicted that the density of transistors on a microchip would double every two years. Although sometimes the doubling period has proven to be 18 months rather than exactly 24, Moore's Law still holds true, almost 35 years after its first formulation.

Like many of the forecast laws used in technology management, a linear trend appears when plotted on semi-log paper. You can see that transistor density, which is basically a proxy for the price-performance ratio of the microchip, has changed by nearly five orders of magnitude over 30 years. I would venture to say that a lot of people lost a lot of opportunities to make a lot of money by failing to believe that Moore's Law would continue to operate into the future. However, most people now see a limit to Moore's Law. That limit has to do with quantum effects as the circuits get closer and closer together.

Heisenberg's Uncertainty Principle tells us that we never know the exact location of an electron. In most circuits, precise location doesn't matter because we are pretty certain the electron is close to where it's supposed to be. But in ultra-dense circuits the electron could appear in the wrong component at the wrong time, destroying the chip's functionality. Microprocessor manufacturers are expected to hit the wall with regard to this quantum effect in about 2025. Then what will become of Moore's law? One thing we may see is the trend in this graph simply leveling off. Another possible outcome is that electronic chips will give way to photonic chips. This could allow Moore's law to continue far into the future, at its current pace or at a different slope. Perhaps ways will be devised to protect the circuits from quantum effects, or even to leverage the quantum effects in creative ways that work to the end-user's advantage. Which of these do you think will happen? Why?

Technology forecasting relies upon many other exponential growth laws that are also subject to some ultimate limit due to physical laws. Say we plotted "the maximum speed of aircraft" on a vertical axis and time on the horizontal axis. The data points representing the new aircraft introduced in each year would show exponential growth in maximum velocity if plotted on ordinary graph paper, and the data would show linear growth if plotted on semi-log paper. But can this trend continue into future production? Just what is the physical limit of the speed of a jet aircraft? Did you guess that it's the escape velocity of the earth's gravitational field? That is the outer limit – if a jet went any faster, it would be a spacecraft! And since we're designing jets for use within the planet's atmosphere, we have no use for one that operates at greater than escape velocity. So, in the future, the imaginary graph we have just drawn of aircraft velocities will show a leveling off, perhaps asymptotic to the escape velocity[1].

Actually, Moore's Law gives us the opportunity to draw together three great theoretical threads of this book:

- learning curves,
- technological substitution, and
- technological convergence.

How is it that Moore's Law could continue without exception for more than 30 years? *Could this law just represent the learning curve of the entire microprocessor industry?*

Probably not, or at least it's not that simple. The learning curves we've seen in most industries account for only one or two orders of magnitude in cost

---

[1]   The *Pearl curve* lends itself to trend extrapolation in these technology forecasting situations. The formula is $y = L / [1 + a \exp(-bt)]$, where y is the level of use or size of the application market at time t; L is the upper limit of this level; t is time; and a and b are parameters to be estimated.

improvement. We know that fresh learning is needed each time a new fabrication plant is opened to produce a new generation of microprocessor. And at least eight generations of microprocessors are involved in the current graph of Moore's Law.

Each of these generations of chips accounts for slightly more than one order of magnitude of learning improvement, but keep in mind that each generation of chip also supplanted its previous generation in a process of technology substitution.

As generations of DRAMs came and went, the total sales of DRAMs exhibited a smooth curve. We saw generations of mainframe computers come and go, and the sales curve was smooth for all mainframe computers. We saw it for different kinds of cord materials for automobile tires – as each came into fashion and another went out, the total market for corded tires generally increased.

We observe the same phenomena for generations of microprocessors, which from a consumer's point of view have exhibited ever-increasing performance for a generally stable or decreasing price per chip. This cascading effect, as processors are smoothly substituted for each other, makes for the smoothness of Moore's Law. Now, where does technological convergence fit in? Microprocessors are at the heart of a digital revolution that caused the convergence of computing, telecommunications, data connectivity and networking, and digital television and audio. Convergence increased the demand for new equipment to replace older model television sets, VCRs, stereos, personal computers and telephones. This caused a huge new market for devices that were never heard of nor dreamt of before. And it is consumer demand that finances the R&D that results in greater and greater densities of transistors on a microchip. Technological convergence like this ensures that Moore's Law continues as long as physically possible.

But interesting things happen when we start to compare *two* loglinear trends. Figure 7.2 compares the exponential decline in prices of Random Access Memory, or RAM, with the similar decline in prices for magnetic hard disk storage devices.

In the late 1980s the trend toward smaller, denser hard disks changed the picture dramatically. If it hadn't been for smaller and denser disks, the price trend of hard disk capacity and the price trend of RAM would have experienced a crossover in the late 1990s.

What would have happened had that cross-over occurred? In 1997 and 1998, many industry executives were pushing the idea of a "Net computer" that would connect only to the Internet and eliminate the need for a local hard drive. By 1999, it was apparent that Net computers were showing no sign of taking a significant share of the PC market, so we could say that the invention of very small hard drives saved the "personal" in personal computing.

At Oregon Graduate Institute, an online course called "Principles and Trends in Technology Management" is taught from this textbook, and provides another example. The course is Internet-based, but also uses a CD-ROM as an adjunct to the Web access. This is because in the 1999-2000 time frame, the Internet's bandwidth limitations do not support the high-quality video and audio transmission that is needed for a high-quality learning experience. OGI provides this content through the CD, while maintaining a uniform browser-based interface for the student.

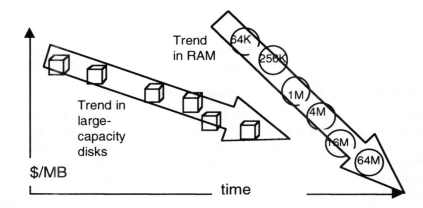

Figure 7.2 Comparison of price trends: Winchester disks and random access memory

If you are interested in these questions of trend forecasting, you might make another graph similar to the hard disk/RAM graph. Your graph should have two lines on it: one is the trend of the access speed of CD-ROM drives in a PC, and the bandwidth of the Internet available to the average home or office computer. When will the two lines intersect? What would you conclude about the date at which you will no longer need a CD-ROM to take this course? To what extent would you discount your answer due to the greater reliability of a CD drive over Internet delivery? Are there any laws of physics like the quantum limit to Moore's Law or the escape velocity of the earth that would limit Internet bandwidth or CD access times? What about wireless Internet access? Are there any physical laws that would limit that bandwidth?

7.2.5 What Impact Do the Technological Trends Have on Us?

There are many new types of technology that are just emerging from science fiction. Virtual reality, collaborative technologies, biotechnology and neural computing could all have an impact in the next few years. This tells us our perspective about using technology and its impacts should be longer. For example, virtual reality was only mentioned in science fiction in the late 1980s. Currently, everyone is aware of the concept, but it will be several more years before widely useful applications are developed. This means our perspective on technology and impacts should be about a decade.

Another impact of technology is the generational difference among the users. A generation of workers will enter the marketplace soon who don't need to be "brought up to speed" on information technology. This means that businesses will spend more time finding applications for technology rather than training employees to use technology.

Crossovers

It is worthwhile to identify and act on **crossovers** when they occur.

In 1984, MRCA Information Service's consumer purchase tracking service first published the fact that, for the first time, more U.S. households owned cats than dogs. This was a sign of greater urbanism and trends toward apartment and condominium living. It was a sign to marketers that the future lay more in miniature and under-the-cabinet kitchen appliances, and less in tractor mowers.

In 1990, Americans first made more visits to "alternative health providers" than to conventional M.D.s. The next half-decade saw explosive growth in herbs, vitamins, and natural food stores.

In 1995, for the first time, more PCs were sold than television sets. This may imply that the "set-top box" for home delivery of interactive multimedia will be a dead-end technology. It seems unwieldy to use a peripheral machine to make an analog TV interactive (although younger Nintendo enthusiasts may disagree). PCs are intrinsically interactive, and their increased penetration of the home market may indicate that they will be the natural delivery vehicle for multimedia.

7.3 Technology Assessment

Whereas technology forecasting looks at the timing, function and cost of new technology, the purpose of technology assessment is to evaluate and predict the *effect* of a new technology. The (professed) main use of technology assessment is as a guide to commercial development of a technology. It is also done in the interest of the public or as a study of history. The quality of technology assessment today varies; the average level of assessments produced can be substantially improved, in the opinion of various practitioners. Much of technology assessment now deals with the potential for social change inherent in a new technology, so objective metrics do not often emerge from a technology assessment exercise.

Also, much of the technology assessment philosophy is only tangentially related to the market potential of a technology or a product (see Table 7.1). Indeed, the language used by TA practitioners betrays a distinct lack of marketing orientation. However, we include a short discussion of technology assessment here because its varied considerations are worth thinking about when one is introducing a product that is new to the world.

Technology assessment is often performed by government agencies, notably in the United States by the U.S. Office of Technology Assessment (OTA). This is a definition "that is consistent with OTA's statute":[2]

Technology assessment is the identification and examination of the direct and indirect effects or impacts of technology on people, institutions, and society. Technology assessment includes consideration of the likely or possible economic, social, political, and other effects of technology, and analysis of the range of policy options that might address these effects as well as the technology itself.

The steps in the process of technology assessment are roughly as follows. It is paraphrased from an Internet communication from another OTA employee. You can see that the usages of the words "technology" and "product" are somewhat confounded.

1. Create an impartial description of what the technology is, and its purpose.

2. Test the design features of the technology and insure that they add value to the intended use of the product. Use testing methods that thoroughly cover all aspects of the technology's use.

3. Codify who is influenced by the technology and decide whether they are part of its target market.

4. Perform a vigorous cost analysis of the technology, including intangible costs such as quality of life and ethical costs. Include the effect the technology has on environmental issues – resources, pollution and infrastructure.

5. Analyze the benefits of the technology. what are the payoffs in jobs, knowledge, human rights, etc.?

6. Evaluate the competition. What are the other choices besides this product? Are they also appropriate choices? More appropriate?

7. Look at the effect this technology has on a number of markets. OTA distinguishes the "natural market" for which no selling is needed (perhaps an

---

[2]   kindly provided by Barbara Ketchum and Fred Wood, then of OTA. Congress decommissioned OTA in 1996. See Bimber B and D H Guston (1997) Technology Assessment: The End of OTA. Technological Forecasting & Social Change February-March 54(2&3) (special issue)

example is videoconferencing for distant consultation during surgery); the "assisted market" for which some information must be conveyed, but no hard sell is needed; and the "artificial market" whose members have no "real" need for the product but may buy it if certain appeals other than relative advantage are emphasized in the marketing.

8. Synthesize the above issues to provide support for a recommendation.

9. Make a recommendation if your research and analysis warrant it. Keep in mind that technology assessment may not always produce a clear direction.

Table 7.1 What Must Be Understood about Every New Technology

- its social and environmental impact

- its technical performance

- its safety

- its synergies with other technologies

- its market potential

- product development costs

- time windows of opportunity

Japan's New Growth Industries

Japan must take steps now to develop future growth industreis. It is critical to build the base for these industries before Japan's aging population imposes burdens on the economy. Industries must develop in response to market demands and not from government policies that protect or promote selective activities.

Kimura does not support his assertion that markets must be free to operate without government intervention. He states that government must take an active role in housing through subsidies, in energy development through tax credits, in health care through a stronger social service infrastructure and leisure through mandated vacation laws, day care and recreational facilities construction.

Kimura identifies eleven fields and predicts they will experience an above average growth if certain requirements for growth are met. These eleven are: Housing; Information/communication; Energy; Environment; Health care/social services;

New distribution methods; Culture/leisure/self-actualization; Urban environment; International exchange; Human resource training/mobility; and Business support.

The author cites two methodologies used to develop his list of industries: a needs-driven method and a technology-driven method. He concedes his projections are imprecise and must be used with caution. But, he concludes, "Japan needs to strip away the fat from its industrial structure and start building the industrial muscle of tomorrow."

Kimura Y (1995) Japan's New Growth Industries. NRI Quarterly Spring. Summary by Clayton Abel.

7.4 Technology Appraisal

The IC^2 Institute's research and implementation activities in technology commercialization have given rise to a method for analyzing opportunities related to new technologies. We call this method "technology appraisal," in order to distinguish it from what is traditionally meant by "technology assessment." The latter term, especially in the sense it is used by the U.S. Government, implies a purely technical assessment of the innovation. Technology appraisal, on the other hand, involves identifying one or more product markets in which the innovation may represent a profit opportunity; identifying complementary technologies which, in combination with the innovation in question may point to other market opportunities; choosing the industry/market that best justifies an initial push to product development; and identifying the firms that are potential licensees or partners in new consortia for product development.

Under a grant from NASA, the IC^2 Institute's Center for Commercialization and Enterprise founded and manages the NASA Technology Commercialization Centers (NTCCs) adjacent to the Johnson Space Center in Houston and the NASA Ames Center in California. As a part of their task of incubating new businesses based on NASA technologies, the NTCCs engage in precisely this type of technology appraisal on a regular basis. This section describes the procedures, databases, and networking and primary research techniques used in technology appraisal, with examples from the NTCCs' work.

7.4.1 The NTCCs

Each NTCC has a staff of two or three persons which includes the Site Director and a Technology Specialist. Some administration is done centrally in Austin, and the Technology Licensing Managers, who do most of the technology appraisal work described below, are housed in Austin.

As a result of the Technology Licensing Managers' recommendations, NASA makes technology transfer agreements with entrepreneurs and companies that will commercialize the selected NASA innovations. These agreeements are of three types. The first is a license to a NASA-patented technology. Licenses comprise about one third of the agreements signed so far under the NTCC program. The second is a CRADA, or Cooperative Research and Development Agreement, which often signifies a nonexclusive license to market the innovation. The third option includes other, usually less formal, knowledge exchanges. The latter are called "Space Acts" after the legislation that authorized them.

7.4.2 The Ames NTCC

Like Austin, Sunnyvale is an entrepreneurial community with sophisticated investors and entrepreneurs. Located in Silicon Valley, Sunnyvale enjoys one of the nation's best-developed venture capital presences, and many of the companies entering this NTCC are VC-backed. However, the business plan screening used to select entering companies is not as stringent as it is at the Austin Technology Incubator, and a 90-day trial plan is in place for entering companies. Fifteen companies were in place at the NTCC in 1995. In general, the technology transfers effected by the Ames NTCC are driven by "entrepreneur pull."

7.4.3 The Johnson NTCC

Located in Clear Lake, Texas, about midway between Houston and Galveston, this NTCC is located in an area populated mainly by existing NASA contractor companies. Some of the entrepreneurial companies are funded by federal Small Business Innovation Research (SBIR) grants. In general, the technology transfers effected by the Johnson NTCC are driven by "technology push." There are six companies in this NTCC. Although the Johnson NTCC has attracted fewer companies than Ames, 120 of the 150 NASA innovations investigated so far by the NTCCs have emerged from the Johnson Space Center.

The differences in organization, environment and technology between the two NTCCs are summarized in Table 9.2.

Table 7.2 Two Entrepreneurial Environments (1995)

• NASA Ames NTCC

    • Silicon Valley entrepreneurial atmosphere

- Sophisticated entrepreneurs and investors

- Venture capital presence

- "Entrepreneur-Pull"

- 15 companies in NTCC, not stringently screened

- Basic NASA research

- NASA Johnson NTCC

    - Little basic research – a mission-oriented space center

    - Existing NASA contractor firms, plus some SBIR firms

    - "Technology-Push"

    - Six companies in NTCC

    - 120 of the 150 NASA innovations examined by NTCCs.

7.4.4 The Appraisal Procedure

This section is a conversation with Thomas Farrell, Technology Licensing Manager at IC^2's NASA Technology Commercialization Centers. His remarks relate to Table 7.3.

Table 7.3 Technology Appraisal: The Linear View (Adapted from Kozmetsky and Kilcrease, 1994)

Step	Detailed Tasks
1. Brainstorm potential markets	a. Focus on fit between technology and a real or potential market need b. Revisit as appropriate
2. Identify product characteristics of each market	a. Determine how the technology might be embodied in a product or service b. Features/specifications/benefits should be well-defined for effective communication.

3. Research and analyze current situation in each market, independent of new technology	a. Identify the industry or industries which currently serve this market b. Determine market size, growth rate trends, channels, pricing, etc. c. Identify and analyze direct, indirect, current and future industry competitors d. Explore customer needs, perceptions, buying behavior e. Determine technological regime and levels of technical and market uncertainty
4. Determine whether there is a real or perceived need in each market for the proposed new product	a. Conduct primary research where necessary b. Revise the proposed product characteristics where appropriate c. Determine the value of the new product to potential customers
5. Identify factors necessary to bring the proposed new product to each market	a. Analyze barriers to entry b. Identify necessary complementary assets and capabilities
6. Determine the impact of the new product's entry	a. Determine the nature of the innovation (radical vs. incremental) and its impact on the technical regime, market power, etc.
7. Determine the best market entry strategy	a. Analyze relevant organizational factors (access to complementary assets, etc.) b. Analyze relevant strategic factors

Thomas Farrell: We have broken down the technology appraisal process into three phases. The three phases evolved as the NASA NTCC plan was being implemented. The essential idea is that two things must be done: 1) Find the technology. That is the premise that underlies all the work we do: NASA has technology, and we need to find it and understand what it is. Once we've found it, the next thing we need to do-- and what is not usually done by NASA -- is understand how good it is and what is really the market opportunity for it. Typically and historically, NASA has made that determination simply by advertising the technology and waiting to see whether anyone shows an interest. If there is any point where we feel we really add value in this phase, it is in phase two. In the third phase, after analyzing the technology and the market, we actually execute some kind of transfer in order to move the technology into the market.

In this way, the NTCCs function more broadly that the NASA technology databases that exist at centers throughout the country, for example at Texas A&M University. Although they are changing a little bit what they do, their modus operandi is community outreach. They make themselves known to the community, so the business community knows to go to them when they have a problem that they think can be solved by federally developed technology.

The latter centers have catalogs and databases. If someone calls them and asks for the latest thing in laser optonics, the database throws off a list of 18 references, 3 DOE, 5 NASA, the names of the government laboratories where the research is, and the names of the scientists the business person must talk to.

Our approach is technology driven, though with the confidence that markets can usually be found. This is in contrast to the previous NASA-sponsored centers, which are based on waiting for random industry contacts.

Phase 1: Technology Identification and Definition. The first phase, phase one, is technology identification and definition. This is an activity that depends heavily on interface with the scientist. Each of our NASA commercialization centers has a staff member whose job it is to find technology . Their job is to procure leads from the utilization office, patent counsel, scientists, or third parties. These are, or were at the time we designed this procedure, the sources for leads about a new technology. What we are now doing is a little bit different, because NASA's internal structure has changed since this time. The original precept was that NASA doesn't know what they have, and NASA doesn't have the people to look for promising commercializable technologies. We need people from our organization (according to this original approach) to go in there and find the technology because only they will know something that has commercial potential when they see it. That may or may not have been true, but the interesting thing is that we are more or less prevented from doing that now. There are sufficient gatekeepers into NASA now that we don't have the free rein that we had when the program started a year ago.

The technology specialist finds the technology, and when the IC^2 technology specialists have found that technology they put their arms around it and try to understand what it is that they have found. Typically, the only understanding that NASA has is captured in some kind of disclosure or patent that they have issued. They rarely have an understanding of what that means in terms of benefits to a potential commercial user. So one of the things that is really vital in this process is that the technology specialist adds value to the identification process by asking the right questions: Does the inventor know what alternative technologies are out there in the commercial marketplace? Often, they have no idea whether there is a commercial product that does the same thing or not. But they are interested in understanding where their invention stacks up against other technologies.

This whole process is, incidentally, to a large extent captured by the database that we have in Austin. So you see that the technology summary is really a report that is compiled by the specialist. Based on that report, a decision is made whether this is a technology that is worth pursuing. It is really a decision that is made without gathering any more information.

The premise behind this was that we would have more technologies being found than we know what to do with. That is why we built a capability to filter early on. In practice, that is not the case. We could look at every technology in the kind of detail we have in phase two and make a decision based on that.

Phase II: Opportunity Analysis. The next thing we look at is: Has the licensee been identified? Really what that is getting at is identifying a dual track. Where we are either looking at a technology that has no commercial partner, or we are looking at an outside company out there that is interested in licensing it. Or there is a company that wants to come in to the incubator that knows they want to license it. If there is some one there who wants to license it, we put it through a fast track. In phase two there is not a whole lot of difference between the two in terms of what we do in Austin. The main difference is, when you have a license you have already identified, we have an additional step, which is to determine: Is this best commercialized by licensing into a large company or should it be a start-up company. Whereas if there is already someone there the decision is kind of already made for us. If there is someone already there we assume that the potential licensee already has information about the market, technology, competition.

So the first activity, if there is a licensee, is to capture that information and identify where the gaps are in their current understanding. An example would be an entrepreneur comes to the site knowing about a technology in NASA that they want to commercialize. They want to get into the center, but they do not have a business plan. They already have some information; they know about the technology, they know what the market is, but they don't have enough to put together a business plan. So, what we would do in that case is fill in the gaps. Most of what we do would be in the market analysis area.

Fred Phillips: TLM is?

Thomas Farrell: Technology licensing manager.

Phillips: And the technology specialist?

Farrell: That is the person on the site.

So really, for a start up company that has already been identified, our value-added in the process is to help and to understand their market. We are doing somewhat similar work, but with a different objective, if there is no one in place. If there is no one already in place, really our objective is to find someone. The first thing we have to do is identify and research potential applications and associated markets. Basically, it is a brainstorming process where we generate ideas for what you can do with the technology. Hopefully, there will be one that floats to the top – one or two markets that are obviously the strongest applications. If there is not it means we have a lot of work to do because we have to look at each one in some detail. Out of this will come a market analysis report with objective of: do we have something that is commercially viable?

Phillips: Now the brainstorming: Does it involve searches of on-line databases?

Farrell: Yes, it ties into here, but it is somewhat separate. We have some documents to show you how we do this; we break the process out a little more in terms of doing secondary research and primary research.

Phillips: Do you call people who might know--do some personal networking as well?

Farrell: Right, maybe talk to some professors here at U.T. who work in that area.

This is an area that some people doing tech-transfer put a little more emphasis on. There is kind of an assumption about what level of expertise you need in a specific technology to make this determination. We probably erred on the side of not getting really technical specialists involved. If you compare to something like BMDO, the Ballistic Missile Defense Organization (the former Star Wars organization) they will take technology and go on the road. They will have a road show, come to Austin and get together everyone from the semi-conductor industry and say: Here is a technology that we think might be relevant to the semi-conductor industry, what do you all think of it? Looking back, I would say that is probably necessary in some cases.

So that is something that, if we had come across situations where we thought that it was really necessary, we would have done it more. But most cases, the technologies were not complex. One of the things about Johnson [Space Center, near Houston] is that they are an operational center. They have a lot of widgets. That is what they do. They build things to go into space, to help execute a space mission. They do not have basic research. There are fewer things you can do with something that is being productized.

Phillips: At Ames they do more basic research?

Farrell: In theory they do. At AMES, actually, we have very little information about what really they do. That comes back to John Gee's [Director of the IC^2/NASA NTTC near the NASA Ames Center in Silicon Valley] belief that "it is the entrepreneur that makes this thing happen. I am not interested in going and applying technology with AMES and letting the technology drive this process. I don't think that works. I think you start with an entrepreneur, and find them something at NASA."

Phillips: I have a slide that I use in class that divides these innovations into three categories. One is a minor product extension like a self-focusing single lens reflex camera and another is new product category like single lens reflex camera and the other is innovation that causes a whole new industry--like the camera. If it is something that you all judge could potentially start a new industry, then there are a lot of gaps and it takes a visionary to make that happen. If the product category was PC's then let's suppose it takes a Steve Jobs to do that. But if it is a self-focusing single lens reflex camera when SLR's already exist then--who cares? Finding a visionary isn't going to make any difference. The technology makes the difference. If it is going to create a new industry then it is both.

Farrell: What you have just been talking about kind of gets at... assuming we found something that is commercially viable, we now need to decide what to do with it. Do we start a company or do we license it? The truth is we have rarely got in to this question. Usually it is obvious that it is a license, or in a startup there is already one there.

Is the innovation...

a new technology that will start a new industry

a new product that creates a new product category?

a new model or a brand extension in a mature or existing industry?

(multimedia)

(home video)

(self-winding camera)

Figure 7.3 One of the key questions of technology appraisal

Phase III: Execution. One of the questions is, what kind of innovation is it? Is it an incremental improvement? For instance, if you have a better type of focusing system, there is no sense in starting a company--you want to license it to Canon, Nikon, or whoever. There are too many complementary assets that are controlled by existing companies.

We have to decide if it really makes sense to license it or do want to try to start a company (Table 7.4).

Table 7.4 NASA Technology Commercialization Centers: Commercialization Options

- New venture
- License to existing firm
- Hold for further internal development
- Terminate

That pretty much wraps up what we do in opportunity analysis here in Austin. The one other side of it which is done on site is, if you want to think of what we have been talking about as the demand side, the supply side is do we have something that is available for license, do we have something that is going to be developed in a short enough time frame that it makes sense to license it? That is two things we look at that we call development status (see Figure 9.3), and the availability. Availability is things like is it patented, has it even been disclosed. If it hasn't been disclosed, then it is not available. If it was developed by a small company under contract, the small company has a two-year window to claim rights to it. NASA can't do anything until those two years have elapsed. So, there are a number of reasons why it might not be available to license. Development status is just a concept – do they have a prototype, do they have a working prototype that has been certified for space flight or actually been up on the shuttle.

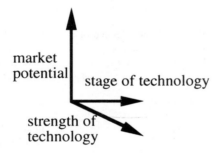

Figure 7.4 Three-dimensional classification of technologies

There is kind of a third decision here. That is if it is neither available or well developed, is it strategic enough of a technology that we should be interested in it anyway. If either of those is a yes, it means that we will continue with what we are doing in Austin. If we get a no on either of those, then we don't really do anything.

Phillips: Do you search patent databases yourself?

Farrell: We do, sometimes. Often we will assume that it has been done in getting the patent if there is one already in place. It depends a little on the technology. I spend some time in the engineering library going through technical papers to see who else has done work in a similar area. That kind of follows the technology assessment. We don't really do work that would support a patent application.

Phillips: Is there an information source on CD-ROM or that you can dial up directly that has patent information?

Farrell: There are a couple now on the Internet. They are not as comprehensive as the patent library down at UT, and the engineering library. They have all the patents on microfilm. Whereas the searches you do on the Internet, you just pull up the text. It is pretty nearly impossible to understand what is going on without looking at the pictures.

Table 7.5 Technology and Market Information Sources (Adapted from Kozmetsky and Kilcrease, 1994)

Category	Selected Titles	Uses
General Economic Indices	Statistical Abstract of the US CPI Detailed Report Survey of Current Business	Broad economic forecasts; demographic statistics; consumption
Industry Data	Manufacturing USA S&P Industry Surveys US Industrial Outlook DataPro Directories	Analysts' overview and forecast for specific industries
Corporate Directories	Ward's Business Directories Hoover's Handbook CorpTech Directory	Detailed information on specific competitors, customers or suppliers within an industry
Financials	SEC Online	Annual reports, 10-K, 10-Q's.
Analysts' Reports	Value Line InvestText	Financial Analysts' reports on specific companies. Often provide insight into industry trends and competitive positions.
CD-ROM Business Databases	ABI/Inform	Ability to search business literature by specific subject
CD-ROM Engineering Databases	Compendex Plus	Search engineering literature by specific subject
CD-ROM Computer Science Databases	Communications of the ACM Inspec	Ability to search computer science literature by specific subject

Technology Journals	Technology Review New Scientist Magazine Nikkei Weekly TKA	Periodicals monitored to identify and track emerging technologies
Patent Information	PTO database WWW "Law" section	Full Patents on microfilm & WorldWideWeb. CASSIS patent search database on CD-ROM.

Farrell (cont'd): Let's take the licensee identifying leg first. A person interested in licensing is either an entrepreneur or an established company. If it is an established company, then we will put together a licensing application, and hold their hand through the process. We really have not, in practice, got into that. We have done it, a little bit, with start-up companies.

There are some issues here that we didn't anticipate in putting this together, such as conflict of interests. We are being paid by NASA to commercialize NASA technology. Is there a conflict of interest with our helping a potential licensee get a license from the entity that is paying us to do the work? We are being paid to look after NASA's interests, and yes, we are helping a licensee that obviously has their own interests. NASA, presumably, wants to get a good deal out of the license. They want the highest royalties they can get, and so on. The licensee wants to pay as little as possible. There is information that is going to be, to some extent, proprietary on each side. It is kind of like a lawyer acting for both parties in litigation. There is a question of equal access and opportunity for all potential licensees. If that exists and is maintained, there is no conflict of interest.

Phillips: So have you ever refused to help someone fill out a license application?

Farrell: No. When it comes down to negotiation it is still the NASA patent counsel that has control over that side of it. IC^2 does not.

Phillips: Someone would have to construe really finely to make an issue out of that. The grant exists to help NASA transfer the technology, so the grant probably doesn't obligate you to maximize NASA's revenue.

Farrell: No, but it is an interesting issue. My understanding is that one of the inherent problems in licensing technology is that the licensors are usually represented by a group of attorneys. It is usually a legal activity. Their biggest concern is three years down the road when this has turned out to be a gene splicing technology, that no one can come back and say, "you mean you gave that away for ten thousand dollars and here is this company making one hundred million a year on it? You really screwed up." So typically, the licensor is careful that they get as much of the value through the license as they can get. In that sense, even though our contract is not really to maximize revenue for NASA, there is still a little bit of exposure there.

We are still on the first leg with a startup company. It pretty much goes back to the commercialization center director. If necessary they will put a team together, although, in this case the team is already in place. This is something in the ATI [Austin Technology Incubator] model that probably wouldn't happen; they already have a team in place.

The next thing that has to happen is that the business plan has to be put together, and then have capital raised. I think we have, with a year of experience, broken down this box a little bit more in terms of what exactly that means. We have certainly done that in California. You have to have a business plan, you got to have money to get into the center. We have the 90-day program, where the buyer has to convince me that he has a shot. I will let you in and we do a business plan together, do market valuation. If you don't do that by a certain time you are out.

So there is a little bit more to this than is captured in this chart. The notion here is that you put a document together, the site director or a group of reviewers determine whether you have a good business. Actually, there are two parts to the business plan: one is a plan that which would be used to get the license. If it is a startup company NASA pretty much wants a full business plan. If it is an established company, they want to see a marketing plan. So there is a document that NASA wants to see to give them a license. If it is a startup, it is a full business plan for finding capital. The site director, in theory, is involved in that.

On the established company route, as part of the market research to be done in phase II we will have identified companies that are or might want to be in this industry. The market report we put together will probably feed into some kind of document that summarizes the opportunity as we see it. Rather than sending out an article like a *NASA Tech Briefs* article that has very little on what the market opportunity is for the technology. We will put together something that says, "yes, this is a good technology, but here are the competing technologies, here is how big we think the market is." It is bridging the gap. Businesses want to know if there is money to be made in this. They usually don't have the resources internally to figure that out, they just take a patent and see where they go. We try to do some of that for them. When we contact them that is the kind of carrot we will try to organize around.

If we generate any interest, selecting licensees has never been a issue, at least not for us. When they have been found we help them get a tech-transfer agreement in place, in either a license or an agreement or whatever it takes. Then we write it up.

7.5 Market Information and the Total Life Cycle

So far we have mentioned global scanning, expert opinion surveys, and other methods for information gathering and analysis. The reader is familiar with still others, such as pre-market test models and focus groups. Table 7.6 emphasizes the data gathering techniques appropriate at each stage of the innovation chain. These are the stages that make up the innovation arrow presented in Chapter 11. A detailed discussion of each method of data acquisition and analysis is beyond what can be accomplished in this book. The message to be taken from this chapter, though, is: Use Table 7.6 to plan your data gathering activities. Start early, schedule carefully, and follow through completely.

Table 7.6 Market Research for the Total Technology Cycle

Innovation Step	Information-Gathering Activity	What You Will Find Out
Scientific breakthrough	Scientific communication	Other relevant lab work in same discipline.
Idea for a new industry	Consortium-building; Technology forecasting	Who the players will be; Best available technology.
Technology development	Interdisciplinary discourse; Technology assessment	Relevant work in engineering, other sciences; Contact with engineers; What's it good for?
Invention	Global technology scanning	New capabilities, emphases of major players.
Patent application	Patent search	Will your idea be protected?
Business plan development	Competitor intelligence; Expert opinion	Attainable share; Cost of selling.
License	Market forecast	How much to charge for license.
Product development • Concept	Concept evaluation	Prospects relative to other product ideas.
• Engineering/ prototype	Focus groups; Reliability testing	Appeal of product's "information package" and attributes.
• Manufacturing ramp-up	Development team communication;	Internal, external information; Expected growth rate.

	Pre-test market	
Product launch	Test market	Reality check.
Post-launch	Customer satisfaction surveys; Customer tracking	Needed product modifications; Usage patterns; Product quality via repeat buying rate.

IntelliQuest, Inc.

Company Background

IntelliQuest is an Austin-based market research firm that specializes in serving the needs of high technology industry. While still relatively young, the company has grown dramatically over the past decade. The company's success, which has mirrored the growth of computer-related products generally, has stemmed from its ability to understand and react to a rapidly-changing market. As the computer industry matures and grows, IntelliQuest will have to compete with more firms doing market research in technology sectors. The company's primary challenge is to maintain its competitive advantage in an increasingly crowded field.

IntelliQuest's Niche

Much as a biological species can survive by finding a protected niche, a business entity can prosper if it is able to serve a select market left unattended by its competitors. From a Darwinian perspective, finches were able to survive on the Galapagos Islands because they were ideally suited to that environment and flourished in it. Companies can achieve success in a similar fashion if they can find a protected niche, evolve to exploit it, erect barriers to entry, and then emerge to compete in the market. Until recently, Apple Computer had been a shining example of this strategy – by protecting its proprietary operating system, the company had made strong profits – while only gaining a ten percent share of the personal computer market.

Market research is not a new concept. Firms have been around for decades offering companies information related to their current or potential customers and the products that they consume. Many firms have established themselves in traditional consumer markets, including packaged goods and branded items, building solid reputations based on time-tested techniques and experience with a number of product lines. IntelliQuest, as a new-comer to market research in the mid-1980s, could not realistically compete in a densely-populated business without differentiating itself from the pack.

The company has specialized in high-technology industry, which was a relatively new and unfamiliar sector to market researchers in the mid-1980s. As an early entrant into that market, IntelliQuest was able to develop its expertise while creating important relationships with high-tech clients, many of whom had their first exposure to substantive market research through IntelliQuest. The firm's contacts and experience in a narrow subject matter gave it an initial advantage over competitors with broad client needs and little exposure to issues pertinent to high-tech companies.

IntelliQuest has since expanded its niche by performing customized research projects, by creating unique tracking products, by using advanced research designs and techniques, and by pushing the use of technology in research methods. The company does no work with scanner or volumetric data collection that are the stock and trade of many research firms. Instead, IntelliQuest has used its distinctive abilities and knowledge to successfully implement complex questionnaires and investigate brand preference at a very high level of detail. Technology markets move and change at such a comparatively high rate of speed that generalized market research firms are unable to keep up. IntelliQuest has differentiated itself from the crowd by performing fast, complicated, and accurate product research in a niche that few others have entered successfully.

IntelliQuest's Research

IntelliQuest typically does research in such areas as brand awareness, pricing strategies, market segmentation, consumer profiling, and product design and adoption. To illustrate these issues, if Dell were considering a new line of notebook computers, the company would want to know how consumers perceive Dell's brand name, what price the market will bear for notebooks, who is buying them, why they bought notebooks previously, and what product configuration is optimal given the preceding factors. IntelliQuest can analyze any of these issues individually, or in a comprehensive product analysis, depending upon the needs of the client.

One of the most important analytic tools that the firm uses to guage consumer desires is conjoint analysis, which is a technique that makes a respondent choose among competing product characteristics until the optimal mix is achieved. For a computer product, a conjoint respondent could be asked to compare systems based upon processor speed, memory, hard disk space, and price to determine the trade-off between better components and a higher price. Most products have significantly more than four characterstics, however, which can make the process very tedious for the respondent. In addition to the design complexity, conjoint studies are also challenging because the researcher must target respondents who are truly involved in the purchase decision. Due to the difficulty in administering the conjoint tool and the rapid pace of change in technology industries, IntelliQuest must provide innovative methods of delivering surveys in order to stay ahead of its competition and produce meaningful results.

Advanced Research Techniques

IntelliQuest offers its clients a number of innovative techniques of data collection. Many of these techniques are relatively new, and have some pitfalls associated with them. CATI (Computer Aided Telephone Interviewing) is used by many good research firms use telephone interviews, but the research is difficult. Interviewers are often blocked by voice mail, gatekeepers, and identification numbers. The interviewee must complete the entire client interview, which may mean keeping a reluctant participant on the phone for 45 minutes. It sometimes takes 13 or 14 phone calls to get the interviewee to take part in the survey. Furthermore, the integrity of the resulting information requires at least a 70% response rate, making completion more important.

Another way IntelliQuest has reached its survey audience has been to send a disk by mail that contains a conjoint exercise. This approach requires the respondent to turn the disk in on time, but generally results in more honest answers, a greater willingness to participate, and greater use of open-ended questions. The company has also experimented with fax surveys, although insufficient standards between fax machines can result in less than ideal results. The fax survey is quick and easy to implement, but usually has a heavy nonresponse.

A similar, but more reliable approach has been to implement a modem-to-modem response, whereby the respondent can fill out their survey and send it electronically directly into IntelliQuest's database. The is probably the best way to go, because there is no online system needed for respondants. However, the company then limits its sample size because it can only mail surveys to people who have modems. One alternative that is still developing is the use of the Internet and the World Wide Web. This is potentially a long-term prospect for information, because the current sample of internet users is not representative of the overall customer base. Not only that, but respondents' ISPs and on-line services often charge for all access, which makes it costly for the respondent to participate.

Research Challenges

High technology firms are confronted with many problems that other sectors may not have to deal with. Products such as computers are subject to very short product life cycles, segmented markets, and a highly competitive environment. These factors make it difficult to do accurate market research, because by the time the study is complete, the product may already be outdated. In addition, many technology purchasers are low-incidence respondent targets, can be difficult to locate, and are often "survey weary" decision makers, which means that survey results can be skewed or biased.

Pushing Technology in Research

In light of the fast-paced competitive technology market and the difficulty in locating reliable respondents, IntelliQuest has responded by using technology to improve its data gathering. One method has been to use customer identification services to target a new segment of buyers. In this way, IntelliQuest, uses product registration as a method for after-market database marketing, which enables them to develop an integrated data collection, analysis, and reporting system.

Another innovation has been to create the IntelliQuest Technology Panel, which is a carefully balanced and screened population of pre-recruited purchase influencers and users of technology products. This collection of respondents gives the company a consistent source of information from willing respondents who are previously identified as meaningful. The panel is attractive those involved because high technology is a fast-moving, popular industry, and people like to be involved in it. These rapid response data collection methods insure highest research validity and the most meaningful results for the firm's clients. Thus far, IntelliQuest has used its FAX/Optical Character Recognition, Disk by Mail, and Disk by Internet techniques to elicit responses from the technology panel.

In-class presentation by Intelliquest Senior Project Manager Jay Shutter, April 5, 1995. Summary by Andrew Elder.

7.6 Further Reading

For more details on the life cycle data gathering and analysis methods noted in Table 9.6, see the following. Some of these publications, of course, address more than one of the life cycle stages.

7.6.1 Scientific Communication

Compendex Plus CD-ROM database of current engineering literature; Inspec CD-ROM database of current computer science literature; *Communications of the ACM;* others.

Also relevant are the publications of the Institute for Scientific Information, in Philadelphia. Their *Current Contents* contains tables of contents from the world's leading scientific journals. Inquire to ISIORDER@ISINET.COM.

7.6.2 Consortium-Building; Technology Forecasting

Gibson D V and E M Rogers (1994) R&D Collaboration on Trial.   Harvard Business School Press Boston

Vanston J H and D C L Prestwood (1989)   Technology Forecasting, New Product Development, and Market Evolution. in R Smilor (Ed) (1989) Customer-Driven Marketing. Lexington Books Lexington MA

Phillips F Y (ed.)(1992) Thinkwork: Working, Learning and Managing in a Computer-Interactive Society. Praeger Westport Conneticut

7.6.3 Interdisciplinary Discourse; Technology Assessment

Cooper W W, Gibson D V, Phillips F Y, Thore S and A B Whinston (Eds) (1995) IMPACT: How IC^2 Research Affects Public Policy and Business Markets. Kluwer Amsterdam

The Office of Technology Assessment was closed down by Congress in the budget cuts of the 1990s. Their publications can be found in libraries.

7.6.4 Global Technology Scanning

Leading-Edge Technologies. (periodical) The IC^2 Institute, The University of Texas at Austin in Association with KPMG High Tech Practice. 2815 San Gabriel, Austin TX 78705.

(For Japanese technologies): New Technology Japan, published by JETRO; DJIT Digest of Japanese Industry & Technology; Japan Technical Affairs; others.

Many technology brokers and consultants compile industry studies. An example is *Flexible Circuits for High Density Applications* (2000) from TechSearch international in Austin, Texas. This report covers production trends, technology trends, applications and markets, and a directory of material suppliers and manufacturers.

7.6.5 Patent Search

Use the CASSIS patent search database on CD-ROM. The University of Texas at Austin has a microfilm database, PTO, that contains the full text and figures of the patents.

The Internet Patent News Service is on World Wide Web at http://sunsite.unc.edu/patents/intropat.html.

7.6.6 Competitor Intelligence

McGonagle J Jr and C Vella (1990) Outsmarting the Competition: Practical Approaches to Finding and Using Competitive Information. Sourcebooks Naperville IL

Savidge J (1992) Marketing Intelligence: Discover What Your Customers Really Want and What Your Competitors Are Up To. Business One Irwin Homewood IL

McGrath M E (1995) Product strategy for high-technology companies: how to achieve growth, competitive advantage, and increased profits. Burr Ridge IL Irwin Professional Pubications

7.6.7 Expert Opinion

Jackson J S (1985) The Medical Laser Market: 1985-2000. Professional Report Graduate School of Business The University of Texas at Austin

Stern L H, Lacobie K J and Z X Zygmont (1989) An Assessment of Potential Markets for Small Satellites. Virginia Center for Innovative Technology Herndon VA

Brink S (1989) An Evaluation of Opportunities in the Development of New Distribution Pathways for Recorded Music. Master's thesis Graduate School of Business The University of Texas at Austin

7.6.8 Market Forecasting

Chambers J C, Mullick S K and D D Smith (1971)   How to Choose the Right Forecasting Technique. Harvard Business Review July-August

7.6.9 PPSM Information

Phillips F Y and R K Srivastava (1994) Cost Commitment vs. Uncertainty Reduction in New Product Development. IC2 Institute working paper.

McGrath M E, Anthony M T and A R Shapiro (1992)   Product development: success through product and cycle-time excellence. Boston Butterworth-Heinemann

7.6.10 Concept Evaluation

Ram S and S Ram (1993) A knowledge-based approach for screening product innovations. Cambridge Massachusetts Marketing Science Institute

McGrath M E, Anthony M T and A R Shapiro (1992) Product Development: success through product and cycle-time excellence. Boston Butterworth-Heinemann

7.6.11 Focus Groups

Templeton J F (1994) The Focus Group. Probus Chicago

Greenbaum T L (1993) The Handbook for Focus Group Research. Lexington Books New York

7.6.12 NPD Team Communication

Hayes R H, Wheelwright S C and K B Clark (1988) Dynamic Manufacturing Free Press New York (chapter 11)

7.6.13 Pre-Test Market

Pringle L G, Wilson R D and E I Brody (1982)   NEWS: a decision-oriented model for new product analysis and forecasting. Marketing Science 1(1):1-29

7.6.14 Test Market

Kotler P (1988)   Marketing management : analysis, planning, implementation, and control. Englewood Cliffs NJ Prentice Hall (6th edition)

7.6.15 Customer Satisfaction Surveys

Schumann P A Jr, Prestwood D and A Tong (1994)   Innovate! : straight path to quality, customer delight, and competitive advantage. New York McGraw-Hill

Barnard W and T F Wallace (1994) The innovation edge : creating strategic breakthroughs using the voice of the customer. Essex Junction VT

Griffin A and J R Hauser (1992) The voice of the customer. Marketing Science Institute Cambridge MA

Hanan M and P Karp (1989) Customer satisfaction : how to maximize, measure, and market your company's "ultimate product." New York American Management Association

Hayes B E (1992) Measuring customer satisfaction : development and use of questionnaires. Milwaukee ASQC Quality Press

Self-assessment guide for organizational performance and customer satisfaction: based on the Presidential Award for Quality criteria. Washington, DC, Federal Quality Institute (For sale by the U.S. G.P.O., Supt. of Doc., 1993)

7.6.16 Customer Tracking

Phillips F Y (1990) Tracking the Introduction of a New Product: What Really Happens? in Phillips F Y (Ed) Concurrent Life Cycle Management: Manufacturing, MIS and Marketing Perspectives. IC^2 Institute of the University of Texas at Austin

Sudman S and R Ferber (1979) Consumer Panels.   American Marketing Association Chicago

7.7 Tracking the Introduction of a New Product: What Really Happens?

When a new product is test marketed, the tracking of its success is very carefully arranged, often using an expensive custom tracking service. When the same product is rolled out on a national basis, the situation is very different. It is only one of a great many new products introduced every month. As such it will not enjoy the focused attention of the retailers or of syndicated market research (tracking) services. Consumers, even those who buy the product, will have trouble identifying it for market research purposes, especially if the data collection vehicle is organized on a product category by product category basis.

In the early stages of national rollout, except for extremely easy-to-categorize new products, retailers will be experimenting with placement of the product on store shelves. Market research questionnaires will not have been modified either to add a new category for the product or to prompt respondents to place the new product in the desired category blank in the questionnaire. Reputable research firms will be reluctant to do the latter, since too overt a prompt will amount to an advertisement for the new product and will bias the respondent. They will prefer to tell respondents, "enter products of type X in the space for category Y on page 3." Of course for a unique and innovative new product there will be no other products of "type X" for some time to come. There isn't even a word for type X!

Research firms are not omniscient. They find out about (and set up to record purchases of) new products in only three ways: an employee of the research firm just happens to see the product advertised or for sale; a sample respondent reports a purchase of a new item; or the manufacturer informs the research firm that a new product has been introduced. The first way is obviously highly unreliable. The second can be error-prone. The third way is surely best. But, understandably, notifying research firms can be a low priority for a brand manager in the middle of a busy national rollout campaign.

What kinds of respondent errors are likely in the reporting of a new product purchase? On an unfamiliar package, the consumer may mistake one of the product's descriptor words for the brand name. If the brand name or the name of the item is a foreign word, a synthetic word or an unusual word, the repondent may misspell it on the questionnaire, making the item difficult to trace to a manufacturer. Without the manufacturer's name, even the Universal Product Code (UPC) can be difficult to trace. If the consumer discards the package before the callback from the research firm, no additional detail can be gathered. MRCA Information Services, the leading diary panel firm, maintains dictionaries of common product misspellings, but such a dictionary is useful only for established products. Special computer runs must be made to match UPCs if it is desired to reconstruct the first reported purchases of a new product.

Diary panels do not ask respondents to enter every product they buy, but only products for which there is a category listed in the diary. If the category is not listed, the consumer, very reasonably, simply will not enter the purchase. Purchases of a new product thus may be missed initially. MRCA diaries provide an "extra space" category that the respondent may use "if you cannot find a place to enter a purchase you think should be in the diary". MRCA will categorize an entry in this section if it is identifiable and in a desired category. Items not falling into a diary category are not recorded.

The huge number of new products introduced every week makes it hard for a panel service to give special attention to identifying a single new one. Yet because a new item's success hinges on repeat buying – and since panels are the instrument of choice for measuring repeat buying – panels are the preferred way of tracking new products. Furthermore, profitability and continued corporate support for a new product depend on its achievement of a certain minimum share of category volume. So a category must be identified in some manner.

In scanner panels, categorizing a product is done by the research firm rather than by the consumer, but the same problems apply for new products. The purchase will be recorded under its Universal Product Code, but will not be reported to clients as part of the appropriate category until that category is clearly identified. As in diary panels, the novel appearance of an unknown UPC may indicate either a new product or an error. In diary panels, additional consumer-provided details are available for verifying that a new UPC actually means a new product. Scanner panels do not have recourse to this backup redundancy, and must depend on the manufacturer for notification of new products.

Many mathematical models of new product introduction depend heavily on the first few purchases in order to fit parameters. The real story behind panel measurement of new product purchases shows that this is precisely the time when measurement is most problematic. Of course, surveys are also used to monitor new product growth.   But respondent recall errors in surveys introduce measurement problems too, and at least in panels respondents report the correct date of purchase.   Diary and scanner panels can be used to track new introductions, but they do so more reliably in retrospect than in real time.

To end on a positive note, 22 of the 25 most successful new brand items in the period 1970-1984 were in well-established categories (notably pet foods and frozen entrees).   When you read the three exceptions, try to think what you would have made of them as a consumer or a researcher, on seeing them for the first time.   They were: granola bars, stove top stuffing, and a diet sweetener.

Table 7.7 Top 25 New Brands, 1970-1984

Brand	**Manufacturer**	**Annual Retail $**
* Lean Cuisine	Nestle	$ 447 Million
Chunky Soup	Campbell	276
* Le Menu	Campbell	238
Minute Maid Refr. Orange Juice	Coca-Cola	224
Maxwell House Master Blend	General Foods	185
* Kool Aid Sugar Free	General Foods	171
Yoplait Refrigerated Yogurt	Yoplait	167
* Prego Spaghettii Sauce	Campbell	165
Brim Coffee	General Foods	156
* Dinner Classics	Armour	155
Tender Vittles Cat Food	Ralston Purina	130
Weight Watcher Single Dishes	Heinz	115
Tyson Frozen Fried Chicken	Tyson Foods	114
* Chewy Cereal Meal Bars	Quaker Oats	113
Folgers Flaked Coffee	Procter & Gamble	112
* Equal Diet Sweetener	Searle	110
Pringles Potato Chips	Procter & Gamble	106
Stove Top Stuffing	General Foods	106
Swanson Hungry Man Dinners	Campbell	106
Tasters Choice Decaf.	Nestle	104
Stouffer Frozen Pizza	Nestle	104
Cheerios Honey/Nut	General Mills	103
Stouffer Oriental Entrees	Nestle	101
Meow Mix	Ralston Purina	96
Mighty Dog	Carnation	94

*Represents new brands introduced since 1980.

Source: SAMI $ Volume 52 weeks ending 6/22/84.

7.8 Notes

Section 7.2.3 is a summary of Vanston J (1988) Technology Forecasting: An Aid to Effective Technology Management. Technology Futures Inc. Austin Texas. Summary by David Blankenbeckler.

...adapted from Kozmetsky and Kilcrease.... Kozmetsky G and L Kilcrease (1994) Technology Market Assessment: Fact-Based Decision Making for Successful Technology Commercialization. IC2 Institute, The University of Texas at Austin May

See the following for more on NASA technology transfer amd subjects related to technology appraisal:

Bopp G R (Ed) (1988) Federal lab technology transfer: Issues and policies. Praeger New York

Gibson D V, Jarrett J and G Kozmetsky (1995) Customer Assessment of CRADA Processes, Objectives, and Outcomes. IC2 Institute, The University of Texas at Austin

IC2 Institute (1993) NASA (Field Center Based) Technology Commercialization Centers: Value-Added Technology Transfer for U.S. Competitive Advantage. The University of Texas at Austin January

IC2 Institute (1994) Technology Knowledge Access: Leading-edge Technologies and Alliances. The University of Texas at Austin Summer 2(1)

Kodama F (1995) Emerging Patterns of Innovation: Sources of Japan's Technological Edge. Harvard Business School Press Boston

Lacobie K J, Stern L H and Z X Zygmont (1989) An Assessment of Potential Markets for Small Satellites. Virginia Center for Innovative Technology November

NASA (1993) Space Technology Innovation. NASA Office of Advanced Concepts and Technology (November-December) 1(6)

NASA (1994) NASA Commercial Technology. Agenda for Change July

NCMS (1994) Legislation will encourage NASA, DOD to develop dual-use technologies that will benefit the private sector. Focus National Center for Manufacturing Sciences November

Vanston J H and D C L Prestwood (1989) Technology Forecasting, New Product Development, and Market Evolution. in R. Smilor (Ed) (1989) Customer-Driven Marketing. Lexington Books Lexington MA

7.9 Questions and Problems

Short Answer

S7-1. Describe the differences between technology forecasting, technology assessment, and technology appraisal.

S7-2. Define the steps of technology appraisal.

S7-3. (a) What barriers prevented smooth outflow of technologies from NASA to the NTCCs?

(b) Can you suggest additional dimensions for evaluation that should be added to the three axes in Figure 7.4?

Discussion Questions

D7-1. Refer to the summary of the Kimura article. (a) The article describes market research based on social demographic data, aimed at forecasting the growth of entire new industries. What functional/organizational unit in a U.S. company should be responsible for this kind of research?

(b) Do you believe this kind of analysis can be valuable for the long term development of sales of your firm's product? If so, how?

D7-2. (a) Under what circumstances are the expert-opinion (scenario, Delphi) forecasting methods more appropriate than the quantitative (Fisher-Pry, Pearl curve) techniques?

(b) Discuss a past application of one of these methods (or a variation) in your company.

D7-3. Prominent companies do not just predict social change; they create it and form it. Discuss the role or potential role of "vision" in creating future demand for products based on your company's critical technologies.

D7-4. How can technology forecasting be used to advance corporate goals in the areas of

(a) Deciding which of a number of similar technologies to use as a platform for products?

(b) Planning a research and development program?

(c) Putting a dollar value on a technological innovation?

(d) Evaluating external opportunities and threats?

Miniproject

M7-1. In teams of about six, perform a Delphi-type forecasting exercise on one or more of the following commercial/technological advances. To proceed,

1. Each group member will write down a year he/she believes to be the expected date of the named advance.

2. A moderator (who may also vote) will list the written dates and display to the group the range, mode, mean of the votes.

3. Each member will have the opportunity to explain his/her forecast. Open discussion is then permitted.

4. A second written vote is then taken. The moderator records the votes, computes the mean and reports this to the instructor.

5. Your instructor will reveal what a panel of experts* has forecasted as the date of this advance. You will be asked to share your observations about the process of arriving at your forecast. Compute the variance of your first set of votes and the variance of the second set of votes. Which is larger? Why do you think the larger variance is larger?

Choose from the following list:

A. Nuclear fusion is used commercially for electricity production.

B. One half of the waste from households in developed countries is recycled.

C. Aquaculture provides the majority of seafood consumed.

D. Hand-held microcomputers are used by the majority of people to manage their work and personal affairs.

E. More than 30% of computations are performed by neural nets using parallel processors.

F. PCS has a 10% share of the market for voice communications.

G. Half of all goods in the U.S. are sold through electronic commerce.

H. Due to automation, factory jobs decline to less than 10% of the workforce.

I. Ceramic engines are mass-produced for commercial vehicles.

J. Parents can routinely choose characteristics of their children through genetic engineering.

K. A manned mission to Mars is completed.

L. High-speed rail or maglev trains are available between most major cities in developed countries.

---

* The Washington University Forecast of Emerging Technologies used a variety of forecasting techniques to predict the dates of 85 advances. Delphi respondents included prominent futurists, forecasters and technical experts. Their average variance on the 85 items was about three years.

"A new technology is something like an undocumented alien; no matter how worthy, the forces keeping it an outsider are stronger than those that can ease it into full citizenship."
Feigenbaum, McCorduck & Nii, *The Rise of the Expert Company*

"I hear, and I forget.
I see, and I understand.
I do, and I remember."
Confucius

"The inventor can't do it all, you've got to change people.... Human inertia is the problem, not invention. Something in man makes him resist change."
Thomas Alva Edison, February 1921, in *Popular Science*

8 Adopting New-to-the-World Products

8.1 Introduction: Barriers to Adopting NTW Products

In this chapter, we will look at what can be learned from the theory of the diffusion of innovation, from cognitive learning theory, and from anthropological and organizational behavior studies of culture change in organizations. The theoretical presentation is kept to a minimum (but with lots of references) so that we can move on to the theories' practical implications for (i) why there are barriers to customers accepting innovative products, and (ii) how sellers – and buyers – can overcome these barriers.

The barriers include learning burdens, and the fear of learning burdens. They also include hard costs of switching from the present methods of getting the job done. Most of the other barriers can be seen in terms of *risks* of adopting the innovation. These are the risks associated with behavior change at the personal and organizational levels, and reluctance to accept these risks can result in powerful forces opposing the innovation. We detail these risks briefly and illustrate the problem with an example (the Dvorak keyboard) before moving on to our theoretical excursions.

8.2 Perceived Risk

There are several forms of perceived risk.

Table 8.1 Perceived Investment Risks

Financial risk	Potential loss of money if the product does not perform satisfactorily
Functional risk	Possibility that the innovation will malfunction or fail to live up to performance expectations
Physical risk	Possibility that the new product will inflict physical harm on the user
Psychological risk	Potential psychological discomfort resulting from a poor choice
Social risk	Potential loss of face or respect in the eyes of relevant others, caused by their rejection of the innovation

Source: Schiffman L G and L L Kanuk (1987)   Consumer Behavior. Prentice Hall Englewood Cliffs NJ (3rd edition)

In *Rethinking Business-to-Business Marketing* (Free Press, 1991), Paul Sherlock confirms that business buying decisions are driven by the emotional and psychological character of the buyer, who uses logic to rationalize the decision after the fact. This is a devastating conclusion, from the point of view of an engineering-oriented seller or a seller who likes to emphasize the low operating expense or easy maintainability of her product. The later life cycle segments (the late adopters and laggards) will in general put more emphasis on the rational reasons for buying. They can be reached by "life cycle cost marketing," i.e., by emphasizing reliability, easy logistical support and so on. But they are highly averse to taking the risks noted above. They are reassured not by *analysis* of cost and risk, but by the *demonstrated* low cost and risk experienced by earlier customers. Despite the appearance of rationality, this need for reassurance is part of the "psychological character" mentioned by Sherlock. It is not wise to minimize the role of emotion in the buying decision.

8.3 QWERTY vs. Dvorak[1]

The common computer and typewriter keyboard is known as the QWERTY keyboard. (I will leave it to the reader to discover why it is called QWERTY.) QWERTY is difficult and inefficient because it is:

- hard to learn
- slow to use
- and requires a lot of work (finger movement).

Because the hammers on early typewriters would jam when a typist typed quickly, QWERTY was invented in 1873 to *deliberately* slow typing. But by 1932, newer typewriters were more mechanically efficient, and wouldn't jam as much. At that time, August Dvorak used time/motion study to create a keyboard with home row

$$A O E U I D H T N S,$$

which letters account for 70% of all typing. (22% of typing is done on the upper row of a Dvorak board.) E and T, the most common letters, are hit with the strongest fingers. Vowels on the left, consonants on the right means hands usually alternate strokes on the Dvorak board.

The benefits (less typist fatigue, errors etc.) are proven for Dvorak board. It is accepted by ANSI and other standards bodies. But...

- manufacturers
- sales outlets
- typing teachers
- and the installed base of typists & machines

remain committed to the status quo, i.e., QWERTY, even fifty years after the introduction and proof of superiority of the Dvorak board.

---

[1]   Some of the details of this and the next section were adapted from Rogers' E (1983) Diffusion of Innovations. New York Free Press (3rd edition)

8.4 Innovation and Social Systems

Innovation occurs within a *social system,* i.e.[2]

- a set of *interrelated entities* (individuals, families, organizations) that
- are engaged in cooperative activity to
- accomplish a *common goal.*

Features of a social system are:

1. The entities are not identical. I.e., there is social structure within the system. This structure, especially the *communication structure* (who talks to whom under what circumstances) affects the diffusion of an innovation through the system.

2. *Norms* vary even across segments of a social system. Norms define the *tolerated range of behaviors* for members of the system.

3. *The most innovative member of a social system is often regarded as deviant, and is accorded low credibility in the social system.* More effective in bringing about change are the *opinion leaders* and the *change agents.*

4. *Opinion leaders* are not usually at the top of the social heap. They are not the chiefs or medicine men. Their leadership is of an informal nature. An O.L.'s opinion leadership stems from his/her

- technical competence
- social accessibility
- conformity to system's norms

When compared to their followers, O.L.'s are

- more exposed to external communication
- more cosmopolitan
- of somewhat higher social status
- more innovative, and are

- at the center of many communication networks.

---

[2]   The phrases in italics in this section are those of Rogers, op.cit., upon which I could not improve.

Change agents may be on the buyer or the seller side. They

- take initiative on change, and
- are often degreed professionals
- who use opinion leaders as levers.

8.5 Classification of Adoption Decisions for Innovations

Sociologists classify innovation adoptions as

- individual vs. system wide
- collective vs. authority decision

Everett Rogers (in *Diffusion of Innovations,* op.cit.) tells of a social worker who is sent to persuade village women in Peru to boil drinking water as a disease preventative. This story cuts to the heart of the problem of innovation diffusion.

The social worker, a young unmarried woman from the city, was not a credible change agent in the eyes of the married village women. (These women were the authority figures who had to be persuaded to boil their families' drinking water.) The social worker won the trust of two village women. One of these, however, was a known hypochondriac. The other was a migrant from a distant region who had never become accepted as a village insider. Neither woman was an effective opinion leader. Further, the social worker was ignorant of local superstitions regarding the "hot" and "cold" spirits inherent in all materials (including water). Her explanation of the reasons for boiling water conflicted with these traditions, confusing the village women. Finally, the impossibility of performing or explaining a controlled experiment made it impossible to demonstrate unequivocally the health benefits of boiling. The program failed, and the social worker departed without having changed the behavior or the health of the villagers.

The germ theory of disease – i.e., that very tiny animals (bacteria) cause illness – proved to be an impossible sell in the Peruvian village. Not only is the theory implausible to people who are accustomed to fearing only larger animals, but the bacteria themselves are invisible and intangible, as is the process by which they cause disease. Throughout history, it is only religion that has been successful at marketing the intangible and the invisible. Now that many of the products sold by high tech firms are based on electrons, DNA, nanomachines, photons, and even quantum effects (not to mention stock index futures and the like), religion's corner on the intangibles market has been diluted.

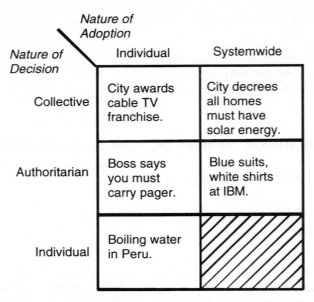

Figure 8.1 Examples of innovation adoption, by kind of adopter and kind of decision

But we're getting ahead of the chapter's agenda. Let's look at two more aspects of innovation diffusion, then see what can be learned from learning theory.

8.6 The Adoption Process

According to Rogers (1983) the adoption process consists of two main stages. In the *initiation* stage (to put it in marketing terms), the customer organization prepares to purchase an innovation from a vendor. In the *implementation* stage, the customer integrates the innovation into the buying company. Each stage has sub-phases:

Table 8.2 Stages in the Adoption of an Innovation (Rogers)

I. Initiation
        A. agenda-setting
        B. matching
II. Implementation
        A. redefining/restructuring
        B. clarifying
        C. routinizing

In I-A, a problem is perceived by the company, and a search is begun for an internal or external solution. In I-B, a possible solution is identified, and the fit between problem and solution is examined. II-A is a process of customizing the solution and adjusting the behavior of the adopting organization to accommodate the solution. In II-B and II-C, more is learned about the implications for the organization of having adopted the innovation, and finally, the innovation loses its newness and its use becomes routine.

8.7 Classification of Innovations

Rogers (1983) classifies the characteristics of innovations as follows:

Relative Advantage
> * *economic*
> * *social-prestige*
> * *convenience*
> * *satisfaction*

Compatibility

> Is the innovation consistent with the experiences, prior beliefs and norms of the customer?

Complexity

> *Perceived difficulty of understanding and use.*

Trialability

> Can the customer *experiment with the product on a limited basis?* (Elsewhere in this book, we have called this characteristic *divisibility.*)

Observability

> *Degree to which the results of product use are visible* to the prospective customer.

"Relative advantage" may be understood as the objective benefits of buying the product. Rogers lists cost, convenience and satisfaction as such benefits; we might add schedule, training and service. Selling relative advantage is akin to what we will later call selling life cycle cost. This strategy includes emphasizing the objective benefits just mentioned, and emphasizing the "ilities."

Table 8.3 The "ilities"[3]

reliability	disposability
maintainability	marketability
testability	distributability
producibility	manability
supportability	manufacturability
quality	assemble-ability
recyclability	safety

To the list of innovation (product) characteristics we can add the risk factors that were listed at the beginning of this chapter, and also a final characteristic, *recourse:* Can the customers change their minds if they find they have made a mistake? Does it cost the buyers much to change their minds?

Higher levels of any of these characteristics (except complexity and risk, which are usually seen as better when lower) generally result in faster adoption. But, Holak and Lehmann (1990) show industrial customers value compatibility far more than (objective) relative advantage characteristics. This finding reinforces that of Sherlock, cited at the beginning of this chapter: Emotional and psychological considerations dominate the buying decision, even for technical products.

"**Cranberry sauce**? What's that?" asks the British housekeeper in the movie *Shadowlands.* Not a surprising question, given that cranberries are grown only in the U.S., Chile and Ireland, and non-U.S. sales of cranberry products amount to only 5% of a $2 billion market.

---

[3]   Although many of the ilities seem to be of more interest to the maker than to the buyer, all affect buyers' costs. Life cycle costing involves more than simply emphasizing the qualitative features listed above; it can involve very detailed calculations of usage and disposal costs. For a complete guide to life cycle cost calculations, see Fabrycky W J and B S Blanchard (1991) Life-Cycle Cost and Economic Analysis. Prentice-Hall Englewood Cliffs NJ

Formed in 1930, Ocean Spray Inc. is a cooperative of 150 cranberry growers in Massachusetts, Wisconsin, New Jersey, Oregon and Washington. Some cranberry growers in Canada and grapefruit farmers in Florida are also members. Ocean Spray was the first to develop cranberry juice drinks in 1963, following an ultimately groundless 1959 health scare involving pesticides used on cranberries. Ocean Spray now controls 67% of the cranberry market. Their goal, according to spokesman Chris Phillips, is to expand their $70 million international market tenfold by the year 2004.

The failure of Ocean Spray's cranberry sauce introduction in Japan, and the current touch-and-go status of its introduction in Australia, show that while not high-tech, cranberries are an innovation in non-U.S. markets. "The fruit's bittersweet taste is unlike any other, requiring painstaking palatal education," says the *Wall Street Journal.* Ocean Spray's CEO John S. Llewellyn adds, "We're introducing an unknown fruit in an unknown brand in a foreign market that's unknown to us."

Luckily, a Harvard University study sponsored by Ocean Spray revealed that cranberry juice is effective for prevention and treatment of urinary tract infections. This added momentum to marketing tactics that included blending cranberry juice with popular local drinks (black currant juice, in Great Britain) and giving away lots of free samples. Ocean Spray now deems the entry into the U.K. a success.

Also central to marketing this innovation across borders was the question of product name. In Japan, Ocean Spray shortened the name to "Cranby" for ease of pronunciation. By a stroke of luck, "hoshien pei" translates from Chinese as "healthy refreshment," and sounds something like Ocean Spray. The company sells juice products in Taiwan under this name. The exact French translation, "airelle de myrtille," while beautiful, was thought to be too long for commercial efficacy. The company is taking a risk with French distaste for Anglicization of their language, and will call the fruit "le cranberry" in France. But there's no need to be too literal: Ocean Spray might note that the New Zealand gooseberry did not sell well in the U.S. until the name was changed to "kiwi fruit."

Pereira J (1995) Unknown Fruit takes on Unfamiliar Markets. Wall Street Journal (November 9) B1

Griffith J (1995) Appetite Ripens for Cranberries. The Oregonian (November 25) C1

8.8 Stages of Developmental Learning

Several years ago, Apple Fellow Alan Kay, in a visit to The University of Texas at Austin, discussed how people draw circles. Ask a small child how to draw a circle, Kay said, and the child will lay pencil to paper and say, "You move a little, turn a little, move a little, turn a little..."

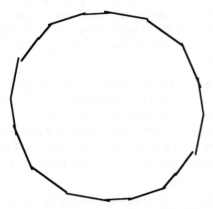

Figure 8.2 Drawing of a circle: Example from the kinesthetic stage of learning

Ask an older child the same question, and the older child will take up a thumbtack, string, and pencil, then draw all the points that are equidistant from a center point:

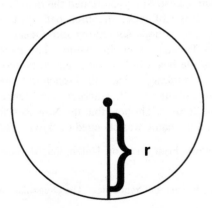

Figure 8.3 Drawing of a circle: Example from the visual stage of learning

Finally, Kay said, ask a teenager to draw a circle. The teenager will respond with "one of the most dangerous pieces of knowledge known to man":

$$x^2 + y^2 = c^2$$

Figure 8.4 Equation of a circle: Example from the abstract stage of learning

The first of many points to be drawn from this story is that researchers in developmental psychology, notably Piaget and Brunner, have identified distinct stages of cognitive growth in human children. Though Piaget and Brunner use different names for the growth stages and delimit them slightly differently, the cognitive modes that a child displays at each stage have been repeatedly confirmed.

For our purposes, referring to the circle drawing exercise, the earliest kind of learning is *kinesthetic*. That is, it relates to movement and the feeling of movement – moving and turning. Next, a child moves into a stage of *visual* learning that enables him or her to better grasp the spatial relationships of objects. The child can understand that all points on the perimeter of a circle are at the same distance from the circle's center. Finally, most adults move into an *abstract* mode of learning in which symbols can represent physical objects and their relationships. Each successive learning mode adds to, rather than replaces, the earlier modes.

A few further propositions complete the picture of the theory and how it relates to technology marketing:

- *Not all people grow into the later cognitive stages with complete success or comfort.* I was a math major in college, supposedly a very abstract pursuit; and believe me, nothing stops a cocktail party conversation in its tracks faster than when someone says "I was a math major." Yet even the use of ordinary language can be highly abstract, as words must represent people, places, things, relationships and actions.

- *Instruction that is presented to adults in terms of one of the earlier learning modes (kinesthetic or visual) is usually accepted more quickly, completely and enthusiastically than instruction represented in an abstract mode.* Alan Kay showed a film based on the book *The Inner Game of Tennis*. The film contrasted the speed with which two students learned to play tennis. The first student was told to meet the ball at a certain angle at a certain stage of its bounce – in other words, the full abstract teaching treatment. The second student was told to "make the ball go over the net." The first student experienced frustration and a serious case of clumsies. The second student progressed rapidly and evidently enjoyed himself.

- *Learning that is presented and obtained through kinesthetic or visual means is retained longer.*

- Kinesthetic learning is even more appealing and memorable than visual learning, and both are superior in these regards to abstract learning.

Kay naturally tied these concepts to computer interfaces, which are a key to computer usability and hence to computer marketing. The most obvious application of kinesthetic learning to computer interface is the mouse and the joystick. (If you imagine a fighter pilot trying to control his jet using a keyboard or voice commands rather than a joystick, the truth of these arguments about learning modes should be pretty apparent.) Seymour Papert invented the successful Logo language that enables young children to manipulate events on a computer screen by controlling a "turtle" that moves and turns. Kay himself was part of the research at the Xerox PARC laboratory that led to iconic interfaces and windows on computer screens. The mouse and visual interface were key to the market success of the Macintosh computer, differentiating it from the abstract, command-line interface of early IBM-compatible computers. It is important to realize that the choice between the MacOS and DOS was not simply "a matter of taste." The work of Brunner and Piaget[4] shows that the appeal of a kinesthetic/visual mode versus an abstract mode is deep in human psychology, and the later appearance of Windows OS was thus no surprise.

This psychological research, reinforcing Rogers' notion of the *visibility* characteristic of innovations, also shows that our preference for things we can feel (kinesthetic) and see (visual) is a deep one. Marketing examples that support the theory include geographic information systems (GIS) and the growth of document imaging and its integration with Electronic Data Interchange. Harking back to the case of water boiling in Peru, we see why selling the intangible, invisible (i.e., abstract) theory of disease was so difficult.

It is now common to see **clothes dryers** in private homes. At one time, though, most households dried clothes on clotheslines. In our terms, the home market was a new target segment for makers of drying machines. The machines seemed to be an extravagance, given that the only relative advantages were being able to dry clothes on rainy days, and not having to pin items of clothing individually to a line.

In early 1995, PBS reported the death of the well-known industrial designer who had first put glass doors on the fronts of clothes dryers. Watching the clothes spin

---

[4]   See also Hauert C A (Ed) (1990) Developmental Psychology: Cognitive, Perceptuo-Motor and Neuropsychological Perspectives. North-Holland Amsterdam;   Halford G S (1982) The Development of Thought. Lawrence Erlbaum Associates Hillsdale NJ

made early buyers feel comfortable that the expensive machine was doing a lot of work. The windows, of course, added to the cost but not to the functionality of the machine. Now that home dryers are a commonplace, it is rare to see windows in them. Windows are, however, still seen in laundromat machines, for reasons you can deduce.

Art Carroll, a pioneer in touch screen technology and later the founder of Carroll Touch, introduced this mode of input to the PLATO educational programming team at the University of Illinois many years ago. Carroll had noted that pointing is an infant's earliest method of indicating a specific desired object. Pointing is more elementary than mouse manipulation, as the latter requires more advanced hand-eye coordination. Studies at Stanford University and elsewhere prove touchscreen users retain the information gained during a computer session better than users of other input modes.

However, there are limits to the applicability of this early learning mode to human-machine interaction. A person cannot point and touch for extended periods of time, especially if it is necessary to keep the forearm in a raised position. Touchscreens have proven effective for casual information searches (for example, at retail and tourist information kiosks), for short-duration office tasks (telling the photocopying machine whether to copy on one or two sides of the paper and whether to enlarge the image or not), and for short, overtly spatial kinds of control tasks (controlling videoconferences). Touchscreens are superior to mechanical buttons when the task is of sufficient complexity that a flexible, menu-driven visual presentation is needed. But in all too many cases, the menus are not well designed, and the frustrated user turns away from the system.

The more advanced (from the point of view of human developmental learning) mouse and icon interface of the Macintosh has been effective for opening and rearranging files, selecting text in word processor documents, and so on. These are not highly abstract tasks. But can point and point-drag-click interfaces make highly abstract tasks easier?

I raise these limitations to suggest that it is reasonable to wonder whether kinesthetic input devices are best for purely spatial tasks. Are visual input devices best for computer analogs of manual tasks (like filing)? Should abstract command-line interfaces be used for abstract tasks, like computer programming? The answers to these questions are still unknown; look for example at the mixed reviews of Microsoft's Visual Basic programming language. *(Infoworld,* 11/14/94: "This may be 'visual' Basic, but you still have to write your own code.") But it seems pretty certain that users find command lines awkward for spatial or manual tasks, and interface designers should strive for control panels that match users' learning capacities and preferences.

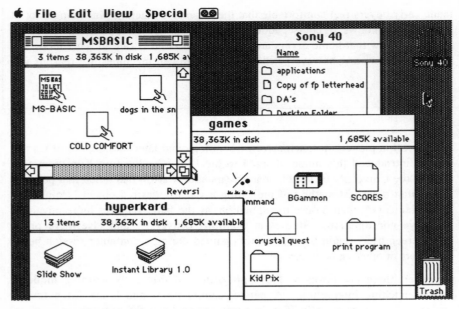

Figure 8.5(a) Mouse-driven iconic interface with windows (Macintosh)

C> PATH=C:\;C:\WORD5;C:DOS;C:\GAMS205

Figure 8.5(b) Actual command line I had to use repeatedly until it drove me bonkers (DOS)

This is all the more important for products that are overtly learning-oriented – *and* for training materials that accompany other innovative and high-tech products. Object International Corp. of Austin sells programming environments and training in object-oriented programming. They have designed and now distribute a game called The Object Game. This board game uses markers, whistles, etc. to engage all the senses in an exercise in learning the basics of OOPS. "Technology," trumpeted an exaggerated but well-intentioned 1997 Sun Microsystems ad, "should not require a learner's permit."

8.9 Effecting Change in People and Organizations

Let's motivate this short section with two quotations. The first describes an organization's frustration in getting a handle on its "culture." The second relates another organization's frustration about being unable to *change* its culture.

A study was conducted to determine the effects on cost of varying levels of specifications for reliability, quality assurance, testing and documentation. Results demonstrated that changes in the specification levels did not explain all cost differences.... Organizational work habits do not readily change ... as a result of imposing a different specification level.... Residual costs were attributable only to "the way we do business," or *culture.*[5]

Management should never underestimate the constraining effects of technical folklore and deep-rooted prejudices shaped by the accumulated successes and failures of generations of technical products. These attitudes breed barriers so fundamental that they are unrecognizable except to outsiders. Because of such barriers, many companies repeatedly market products that never overcome fundamental flaws and never achieve distinction. Such organizations believe, instinctively and without debate, that certain directions are closed to them, even as other organizations proceed in exactly those directions with success.[6]

The NASA report from which the first quote is drawn is correct to note that "work habits do not readily change." But it is not correct to imply that culture is some kind of undefinable statistical residual left over after all the "hard" and easily measurable dimensions of the business fail to account for cost differences. Nor is it true that believing "certain directions" are impossible is the sum total of organizational culture. Culture does consist of habits, attitudes, and beliefs. But these are measurable too, and fairly well understood by anthropologists, organizational behavior scholars, and some managers.

This book has offered some insights about why individuals resist change, and about how social structures are defined. But it will not offer details about culture and culture change at the organizational level, that is, in the specific kind of social structure that is a business firm. There are many good books on organizational change; you are encouraged to look at them. While organizational change management is not this author's primary expertise, I submit that in the firms I have worked for, culture change was always effected primarily by personnel turnover. In each case, top management sent a strong and consistent message about what was required to meet changing business conditions. Employees who could accept and thrive under the new conditions stayed; those who could not, left; and more were hired who came in knowing the score and morally signing on to it.

To me, the classic lesson in organizational change is in the book of Exodus. Having escaped slavery in Egypt, the Israelites wandered the desert for forty years before entering the promised land of Canaan. Why spend forty years covering this very short geographical distance? The wanderers weren't stupid, and they weren't lost! The interpretation I have found most convincing is that no one who

---

5   NASA/RCA (1988) Equipment Specification Cost Study

6   Rudolph S E and W D Lee (1990)   Lessons from the Field. R&D Magazine October 119

personally knew slavery was to enter the promised land. Because of their attitudes and habits, such people would hinder the formation of a society of free persons. A forty-year walk ensured that the former slaves grew old and died before crossing the border of Canaan. It was a very long attitudinal distance.

These days, we cannot spend forty years changing a company's culture!   Staff changes, alliances and outsourcing, and skunkworks organizations are all ways of changing a company more quickly.

You would be correct to guess that organizations which contain many people are even slower to change their attitudes and behavior than individual people are. Adopting an innovation, i.e., buying and using a revolutionary product, requires a change of attitude and behavior. The second quote above can be at least as true for buyers as it is for the seller firms that Rudolph and Lee describe.

Moreover, Chapter 4 noted that today's need to rapidly acquire technologies and markets often leads to acquisitions and alliances. These imply "culture clashes." America Online, while still trying to digest Netscape (the press used the word "clash, " noting AOL, "known as an uptight group, has taken over Netscape, known for its looseness."), acquired Time Warner and bought a new and bigger set of culture gaps. As alliances and markets become global, we must attend to issues of national and ethnic cultures as well as corporate cultures. Euro-Disney bombed in France, and McDonalds *was* bombed in France. The boxed story above about the internationalization of the cranberry market suggests many of the attendant problems.   Textbooks on international management offer some safeguards and solutions.

As a Type IV technology entrepreneur (unless you can afford to wait for your customers to prepare themselves to use your product), you would be wise to take a change agent role in preparing your prospects to absorb your innovation more rapidly and effectively.

8.10 Transferring Outside Technologies into the Firm. Diffusing Adopted Technologies within the Firm

Most recipes for managing the insertion of a new technology into the firm take a straightforward project management approach.   Now that we have studied diffusion theory, we can devise a richer recipe – in the box below – that is more likely to succeed.

Table 8.4 Introducing an Innovation in Your Firm: An Augmented Project Management Approach[7]

- Make the best business case.
- Set goals.
- Establish "proof of concept."
- Benchmark what other firms have accomplished using this innovation.
- Define the critical path.
- Prepare counseling &/or outplacement for those affected negatively by the change.
- Understand who are the (potential)
  - Opinion leaders   - Change agents
  - Innovators   - Early Adopters
- Train, train, train.

- Put the best[8] of these people on the project.
- Choose a 1st application that is high-impact & visible.
- Develop a communication infrastructure.
- Identify & sell benefits ("real" and emotional).
- Prepare responses to the naysayers.
- Define accountability.
- Train, train, train.

- Don't let early milestones slip.
- Involve executive team at regular intervals. Remind them that the rank & file are sensitive to their signals.
- Manage scope, from the beginning and aggressively.
- Expand proof of concept to 1 or more prototypes.
- Value diverse skill sets - process, technical, & organizational.
- Block efforts to return to old way of doing things.
- Celebrate successes.
- Constantly recheck people's commitment.
- Plan early for interfaces, data conversion, etc.; Consider dual operation – for how long?
- Train, train, train.

- Constantly measure against the business case.
- Quantify benefits of the change & compare to cost of making the change.
- Train, train, train.

---

[7]   This list draws on material from Rich Payne of bauhaus group, and Jack Wilborn of Arthur Andersen.

[8]   If you can spare them from regular work, says Wilborn, they are not the best people!

8.11 Life Cycle Cost

8.11.1 Introduction to LC Cost

Many high tech companies now use life cycle costing to better manage component and materials acquisition. The life cycle cost concept originated with military contractors when the Pentagon realized that most of the costs of acquiring and using a weapon system were not procurement, but the price of operating, supporting and later disposing of the system.

Table 8.5 The Pentagon Realized that 80% of Weapons System Costs Were Operations/Support Costs

Operations/Support Costs	80%
Cost of Acquisition about	15%
Disposal Costs	5%
	100%

Eventually the idea of life cycle costing was applied to commercial products, especially in cases where the cost of ownership is part of the buying decision. For example, Mazda's Miata became a very successful sports car after a long stretch in which no car maker had two-seater roadsters on the market. Mazda reinvented the category, opening the door for BMW to join that market segment in 1997. The Miata was originally conceived as a small car with a very powerful engine. Mazda, consulting with insurance firms, determined that the cost of insuring such a vehicle, likely to sustain extreme damage in a high-speed collision, was more than the market would bear. Because high insurance rates would have a huge impact on the product's operating costs, the Miata was redesigned with a weaker four-cylinder engine. Would the downsized Miata continue to appeal to the targeted market segment – that is, people who wanted a high-powered roadster? The answer Mazda engineers came up with was sheer marketing genius: They removed much of the sound insulation between the engine compartment and the passenger compartment, and re-tuned the car's muffler. Now the driver could still feel the power of a sports car, hearing a loud vrrroom during acceleration and idle. The look and sound of the Miata reinforced the original roadster image, AND the product became affordable from an insurance point of view.

In the aerospace industry, Boeing faced a similar problem in 1990. The new aircraft the company was trying to sell worldwide cost more to own and operate than Boeing's older planes. Customers, especially in foreign countries, easily calculated the advantage of stepping up the maintenance of their old fleet rather than buying and maintaining a new Boeing aircraft. To succeed with their new product, Boeing had to cut the cost of ownership; so they simplified the design of the new aircraft to reduce maintenance costs.

Life cycle cost planning became even more relevant in commercial markets where increasing regulations held manufacturers responsible for disposing of their products. BMW has led the way in making automobiles easily to disassemble for separation and recycling of materials. European manufacturers, especially in Sweden, are leading the way in the so-called take-back laws, under which manufacturers must accept worn out products from consumers and be responsible for their environmentally safe disposal. It has become in manufacturers' best interest to design and manufacture producs with lower costs of separation, recycling, and disposal.

Life cycle cost planning is also relevant for governments when calculating the social costs of a public works program. In the commercial world, it is still possible for a manufacturer not to worry much about the cost of their products' operation and support once they're out the door; but governments don't have that flexibility. Stringent federal regulations require estimation of environmental impacts that affect the lives of different segments of the population.

Using life cycle cost planning is also a way to see downstream costs. How easy is the product to maintain? How easy and affordable is it to recycle? These are things that can be planned in the design stage of a new product. We can control or influence life cycle costs by paying attention to the so-called design-fors – in other words, products that are designed for reliability, manufacturability, safety, supportability, maintainability, quality, ergonomics, and marketability.

Table 8.6 LC Planning and the Design-fors

Design for quality
Design for disposability
Design for marketability
Design for reliability
Design for testability
Design for safety
Design for supportability
Design for distribution (availability)
Design for maintainability
Design for manability (ergonomics)
Design for manufacturability

While life cycle cost should encompass *all* the life cycle stages, the term is commonly used to delineate the manufacturing, logistics/support/ownership, and disposal stages – and sometimes, just the latter two.

Table 8.7 Life Cycle Cost Refers to all LC Stages

R&D
Investment/Development/Acquisition
Direct Operations & Support (O/S)
Disposal

8.11.2 Steps in LC Costing

Steps in life cycle costing include specifying the mission or the condition of usage of the product, and determining what kind of performance is expected under that mission. Part of the philosophy of life cycle costing is to pay attention to the significant costs that differ between two designs, also referred to as the materiality and incremental cost principles. The next step in life cycle costing is to identify cost elements, cost parameters, and cost relationships. Together these three elements comprise the life cycle cost estimate for each design alternative. These calculations will eliminate all but one of the design alternatives and will establish the characteristics of the final design of the product.

Table 8.8 Steps In Life Cycle Costing and Decision

- Determine the mission and performance specifications.
- Observe the materiality and incremental cost principles.
- Establish cost elements for each design alternative.
- Establish cost parameters for each design alternative.
- Estimate values of cost parameters for each design alternative.
- Specify cost relationships for each design alternative.
- Total the LC cost estimate for each design alternative
- Decide on final design.

8.11.3 Mission Scenarios

Let's consider an example, delineating the mission of a military electronics system. The number of systems to be manufactured and the expected performance of the system are specified in a product scenario. A maintenance scenario is based on the mean time between failure of the system. It specifies that support and maintenance of the system can be performed on site or at a central depot. The estimated mean time to repair is given at the field level, meaning on-site, at the nearest-base level and at the depot level. Keep in mind that there will be more complex repairs going on at the depot level and therefore the mean time to repair (MTTR) is higher at the depot level. The mission includes an operating scenario and a training scenario that describes how many hours of training are needed for the operators of the system and for the maintenance technicians.

8.11.4 Materiality

We can look at the cost of an automobile to understand materiality and the incremental cost principle. In this example, the frequency of having to empty ashtrays is probably not a cost that is going to make a bit of difference to anyone, nor is it likely to differ markedly between car models. So we ignore those costs as immaterial and focus on more relevant cost drivers.

The principle of incremental cost dictates that we think only about direct costs, because overhead or indirect costs don't vary much across the different design alternatives. For many products, these differences will not have any effect on the size of the factory to be built or rented, so the life cycle cost analysis for deciding on the proper design can ignore the issue of factory rents.

8.11.5 Cost Elements

Ideally, the term "life cycle cost" means the costs of the product from the point of view of all the owners the product may have, from the product's conception, all the way to the manufacture, launch, marketing, ownership, maintenance, and even the final disposal of the product. In practice, we are looking not just at the development, manufacturing and marketing costs of the product; but also at operations, support and disposal costs.

This expansion of our point of view is necessary. In the past, companies had one group responsible for development, manufacturing and marketing while another, separate group handled operation, support and disposal. Now, both marketing and regulatory considerations make it worthwhile for manufacturers to estimate operation support and disposal costs for any product.

We'll focus on estimating the expenses of operation and support using the family car as an example. To operate an automobile you certainly have to buy gasoline.

So money spent on fuel – gasoline expense – is a cost element in a vehicle's operation and support life cycle stage. (We'll look at other automobile cost elements in a later section.)

8.11.6 Cost Parameters

How can you estimate gasoline expense during the time you own the car? If you are the manufacturer, the question is how to estimate the total amount of money spent on gasoline during the time the customer owns the automobile. The answer requires that we decompose the cost element "gasoline expense" into cost parameters.

Table 8.9 Cost Parameters for the Element "Gasoline Expense"

gas price/gallon
inflation & other price changes
expected lifetime mileage (miles driven)
average miles/gallon
Values of these parameters are used to calculate gas expense.

Our task is complicated by the fact that different customer segments have different missions in mind when they buy an automobile. Unlike the military procurement situation, it is difficult for the auto manufacturer to figure out their customers' missions without extensive market research.

For example, one customer may use a car almost exclusively for city driving. Another may do occasional or intensive highway driving. One might live in the mountains or vacation there and do a lot of high-altitude driving. In most areas of the country, drivers count on being able to navigate in snow and icy conditions, while customers in other parts of the country have desert conditions to consider. And did I mention the other customer that pulls heavy loads in his pickup truck? You get the picture.

Table 8.10 Estimate Cost Parameters for Each Mission or Use

City Driving
Highway Driving
Mountain Driving
Snow/Ice Driving
Desert Conditions

Heavy Loads, etc

Some market research may lead us astray about the customer's real mission. Maybe the customer's self-image involves off-road driving, so he buys a four-wheel drive sport utility vehicle, and reports that he intends to drive in wilderness areas. In actuality, this SUV will be driven exclusively on city streets.

This presents a problem, because for country driving in cold weather, one design alternative may be superior in life cycle cost. If snow and ice driving were restricted to fewer than 5,000 miles per year, perhaps a different design would prove superior.

Estimating cost parameters can be a tricky problem. There are three commonly accepted methods. They are the analogy method, the statistical method (sometimes redundantly called the parametric method), and the engineering method. The analogy method utilizes similar products and looks at their cost parameters. We could use the analogy method to estimate the fuel expended during maneuvering procedures for a new spacecraft that was somewhat similar to an older spacecraft, for example.

The statistical method, or parametric method, usually involves multiple regressions of cost against different characteristics or attributes of the product. Returning to our spacecraft example, NASA has traditionally estimated the cost of a space mission as a function of the weight of the spacecraft's components. Naturally, the fuel expended at launch is proportional to the weight of the spacecraft. But, perhaps surprisingly, NASA is estimating the entire cost of the development of the spacecraft -- the launch of the spacecraft and the monitoring and execution of the mission of the spacecraft -- all on the basis of its weight. As unlikely as this sounds in principle, and as unsatisfactory as it can turn out in practice, looking at the data shows that this method of estimation is not terribly far off the mark. The results of the statistical method are actually superior to alternatives.

The third method of estimating parameters is called the engineering method. This builds detailed costs using a bottom-up method that looks at the cost of each component (estimated by the engineers who designed the components), then aggregates those estimates to form an estimation of the cost parameter for the entire product or apparatus. When engineering estimates of cost parameters are aggregated, the resulting data is often less accurate than estimates obtained by the statistical or parametric method. This is sometimes because engineers are optimistic about the performance and cost of the components they have themselves designed.

8.11.7 Cost Relationships

To estimate a cost element, tie the values of the associated cost parameters together by using a *cost relationship*. For the gasoline expense cost element, the cost relationship is (NPV stands for net present value):

Gas Expense = NPV of [fuel price x miles driven/year x gas mileage] over expected life of car.

Not all cost relationships can be expressed in such an easy quantitative formula; some must be described more loosely or even guessed.

8.11.8 An Automobile Life Cycle Cost Example

Let's pursue the example of the cost of owning a car again, and identify not just the gasoline expense but the all the other cost elements of owning and operating an automobile.

Before you buy a car, you spend time, energy, and money collecting information about the performance ratings and the characteristics of the different kinds of automobiles that you might be interested in buying. Certainly a big part of the purchase, ownership and support cost is the purchase price of the automobile.

Table 8.11 Purchasing and Owning a Car: Cost Elements

Information Collection
Purchase Price
TT&L
Finance Charges
Insurance
Routine Maintenance
Repairs
Gas, Oil
Accessories
Replace Tires, Wipers, Etc.
Washing, Cleaning
Parking, Garaging
Tolls
Selling, Trading in at End of Car's Life

Tax, title, and license – are these material costs? Well it certainly can make a difference depending on what state you buy your automobile in and what the sales tax is in that state. But unless your Oregon dealer has only pickup trucks and your Washington dealer has only sedans, the incremental cost principle says, "ignore TTL."

If you finance the car you will pay interest on the car loan, so finance charges are a cost element. You have to buy insurance. and insurance rates differ according to what kind of car you buy.

The cost of maintenance is a cost element, as is the average expected cost of repairs when something goes wrong. You will have to buy gasoline, oil, other fluids, accessories, and eventually replace the tires, the windshield wipers, brakes, and fan belts. You will spend money washing and cleaning the car. You will have to park the car if you work in an urban environment and possibly the parking charge will be lower for a compact car and higher for a bigger car. You will have tolls to pay if you travel on toll roads (though these will probably not differ according to your choice of auto), and then there is the resale value of the vehicle if you want to sell the car.

How many years will you own the car? Decide that and then think about the cost of selling or trading in the car, or disposal costs if you'll use the car until the bitter, bitter end. What's the charge to tow that car to the landfill?

Consider the design-fors and "ilities" in a luxury automobile. That's quite a different mission and quite a different market segment than a basic transportation automobile. The Lexus luxury car is advertised as being very quiet; a rider in an Lexus experiences very little road noise. One way this is accomplished is that Lexuses are sold with very soft tires. Soft tires make less noise when they hit the road then do tires made of harder rubber. The consequence for the buyer is that these tires need to be replaced more often because softer tires wear out faster than harder tires. So over the course of a given number of years of ownership, tires have to be replaced more often, and they tend to be expensive tires. So the design decision affects the life cycle cost. Another way Lexus drivers are insulated from road noise is sound insulation in the doors and floor panels. These are filled with asphalt. It's an effective insulator, but tends to be very heavy, so more energy is required to move the car from place to place. That energy, of course, comes from the gasoline, so again in terms of life cycle costs we are paying more in gasoline expense for a car that is quieter. Design decisions affect life cycle costs.

8.11.9 Aggregate to Determine LC Cost for the Design Alternative

After you estimate cost parameters and cost relationships, sum all of the costs to estimate the total for this car choice, throughout the operation and support stage of owning the automobile. Do this separately for each mission.

Then compare the cost estimates of different designs and choose the best alternative for the expected mission. Remember that design alternative A might be superior for 10,000 miles per year of desert driving, but for more than that, alternative B might the superior design for the vehicle.

8.11.10 Disposal Costs

What affects disposal costs? I have mentioned take-back laws as a factor in estimating disposal costs, and we can expect that those costs will grow as take-back laws diffuse from Europe into North America. Tax laws also affect the cost of disposing of surplus or used products, sometimes with very unfortunate consequences. For example, in September, 1989, Apple Computer buried 2,700 Lisa computers in a landfill in Utah. "Apple gets a better tax deal by destroying Lisas than by giving them away [to schools or to the needy]." (R.X. Cringely, *Infoworld,* 10/23/89)

8.11.11 LC Decisions

Would you use these considerations when you are comparing two cars that you are interested in? Or would you make your car purchase on a more emotional basis? What kinds of people would use the life cycle cost calculation to determine their car purchase choice?

Certainly the fleet purchasing manager for Avis Rent-a-Car would use these considerations, because she has to decide whether to buy 3,000 Buicks or 3,000 Fords.

An automotive designer thinking about the design-fors – reliability, manufacturability, safety, supportability, maintainability, quality, ergonomics, marketability – also needs to base decisions on LC cost. How can design decisions affect the cost parameters of the car? For example: a high performance engine will increase fuel expense because the owner is going to be urged to use high-octane fuel. This affects gasoline expense over the lifetime of the car. Would manufacturing the car with a lighter body increase the gas mileage if all other things stayed equal?

While companies feel it necessary to estimate life cycle costs, the actual process of making those estimates can be very difficult. Why is this? One issue is the problems inherent in forecasting things like energy prices, pollutant emission regulation, and so on. The cost of owning your car can change when unforeseen events occur. If you are involved in an automobile accident, dealing with the insurance company and the repair garage may involve surprising expenses. If you give birth to triplets you may need to buy a new, bigger car; or keep your present vehicle for fewer years than you had originally intended. All these unexpected things change the product's life cycle cost.

You know that life cycle costs depend on the incremental cost principal, but it isn't possible to break out all incremental costs using conventional accounting systems. Sometimes material, incremental costs are hidden in "general and administrative" overhead accounting lines. When different departments in the company are responsible for design and for marketing, bringing together all the information necessary to estimate a life cycle cost can be a challenge. This cross-departmental function can be sabotaged by poor communication between individuals and departments. Finally issues of corporate culture and product complexity can make life cycle cost estimation even more difficult.

Table 8.12 Why Life Cycle Cost Planning Can Be Difficult

Inadequate Information/ Bad Forecasts

Unforeseen Events

Accounting Methods Hide Costs in Overhead

Inadequate Communication Among Departments

Cultural Barriers

Complexity of Product

Some examples from NASA and Boeing will illustrate how easily this can happen. Unforeseen expenses in life cycle cost estimates have led NASA to take issues of corporate culture quite seriously. If the incremental cost principle were true, and design cost depended only on the differences between two alternatives, then differences between the cost of two spacecraft should depend only on the differences in the specifications of the spacecraft. An extensive study at NASA determined that this was not true. What other factors could explain such differences?

NASA concluded that the only explanation was corporate culture or "the way we do business." Daniel Goldin, current NASA Administrator, is committed to executing space missions "faster, better and cheaper." How can this be done when the entire corporate culture displays a certain stickiness? A study proved differences in product specs did not explain cost differences between NASA Projects. NASA's problems include the fact that most of their mission clusters, like the Apollo flights, last over many years, even decades. People at NASA tend to assume that all projects should be long projects even if it isn't necessarily true.

Another NASA study showed that a radio in a planetary spacecraft can costs thousands of times more than that same radio placed in army truck. Why is this?

NASA believes people assume a piece of equipment *should* cost more on a spacecraft, and the organization acts accordingly.

The same radio on a planetary spacecraft should cost more than one on an army truck – or so people believe. This is organizational culture. People think – sometimes with justification and sometimes without – "This is a spacecraft. It's important to quadruple check things that we usually only double check." And the costs of quadruple checks can add up, necessary or not.

Much of NASA's current organizational activity stems from their frank realization that changing their own organization is going to be more difficult than executing the mission to Mars.

An example from Boeing shows that product complexity injects difficulties into the life cycle cost estimation process. How complex is an aircraft? A modern commercial aircraft has more than three million parts. With luck, all of these three million parts will fly in close formation! But as complexity goes up, so rises the difficulty of communicating product specifications from designers to builders to contractors. Boeing estimates that 10% of its total current development costs are due to designers' and contractors' rework.

And reliability goes down as a product gets more complex. With compromised reliability, maintenance schedules must be accelerated, and maintenance costs go up. Unexpected breakdowns increase, and repair costs go up. Even if every one of those three million parts is highly reliable, the laws of probability make it likely that something will fail on every flight. Luckily for us all, most of these failures are not mission critical. It is a heavy burden on designers to make sure that most product failures are indeed not mission critical. As complexity goes up, costs go up and   reliability goes down. This led former Lockheed-Martin Chairman Norman Augustine to say, tongue in cheek, that "It is very expensive to achieve high unreliability."

In 1997, as corporations once again began to look longingly at cheap Net Computers as replacements for expensive PCs, they inquired about the cost of ownership of PCs. Sources such as *The Economist,* the Gartner Group, and *Business Week* turned in estimates ranging from $7,000 to $13,000. With a similar range on cost of owning an NC, a Fortune 1000 firm may save $105 million annually by switching to NCs, or may save only $34 million. But switching costs – the non-recurring costs of "extricating the company from a fully implemented client/server architecture" – would run $60 million, according to former U.S. Department of Defense CIO Paul Strassmann. This leaves the switching decision indeterminate. Strassmann attributes the variation in cost estimates to the widely differing kinds of users, degrees of software customization, levels of security, and kinds of network access needed for each station.

8.12 How to Be a Change Agent on the Seller Side: The Case of a Marketing Decision Support System

8.12.1 The Company

In 1938, a company called Industrial Surveys Inc. became one of the first to build a successful business on the then-new mathematical technology of survey sampling. The company's product was sample-based information on the distribution flows and inventories of industrial and consumer goods. Other pioneering firms of that era began to use surveys to measure radio and television audiences and for attitude and political polling.

In the 1940's Industrial Surveys changed its name to Market Research Corporation of America (MRCA for short), and concentrated on sampling consumer households, interviewing homemakers on their purchase and use of foods, drugs and other items. MRCA further specialized in market measurement by panel sampling. In this sampling technique, individual households remain in the sample for years, facilitating longitudinal studies and giving reliable estimates of trends in the marketplace.

In the 1960's, telephone and personal interviews were discontinued in favor of questionnaires administered through the mail. Although MRCA still does some local market measurement, most of its business stems from nationwide consumer panels.

The 60's and 70's saw the startup of new national diary panel companies, and new technologies of consumer measurement such as supermarket checkout scanners. MRCA's responsive strategies included market diversification, and the aggressive technical development of new ways to make its consumer-related data accessible, timely and useful.

In 1984 the company underwent another name change. The new name, MRCA Information Services, conveys the company's greater diversity as well as its orientation to the information needs of the customer. Its continuously updated databases then included, from a panel of more than 12,000 households in the lower 48 states, records of the purchase and/or use of financial services, processed foods, personal care items, home cleaning aids, textiles and home furnishings, shoes, jewelry, footwear and luggage. MRCA's clientele includes the largest manufacturers, retailers, trade associations and government agencies associated with these industries.

8.12.2 The Data

DYANA(TM), the decision support system (DSS) I will describe below, connects to the purchase database for consumer package goods, which include foods, health and beauty aids, and household products. It is based on a census-balanced subsample of about 11,500 households. The financial services database, based on a different subsample, drives a similar DSS, with one difference: Besides transactions and demographics, the financial database contains an additional file for the net worth of, and accounts held by, each sample household, as well as preferences for and attitudes about financial institutions.

The databases for food usage and for textiles have their own interactive access systems which, because of the different data structures within these DBs, are not DYANA-like.

The package goods purchase database contains one record per purchase, each record detailing the product category, date, retailoutlet, brand, price, size, coupon or deal, and other variables such as flavor and packaging. Each such record is linked to the appropriate household record, which contains demographic and socioeconomic information for the sample household, as well as reporting status information, and geographic recodes corresponding to census and client sales regions. Auxiliary files within the database contain client account and other information.

Until the mid-1970's, all these data were stored on magnetic tapes, and all jobs were processed from tape using proprietary computer programming languages which had been devised within MRCA. After the acquisition of MRCA by the present management, aggressive modernization began. The database described above was designed by MRCA software engineers and implemented on the RAMIS II mainframe database system\. The design of the database required tough decisions on the standardization of product categories and code structures. These decisions were the foundation for the later (as yet unplanned) development of DYANA.

The package goods diary (questionnaire) contains almost fifty product categories, some of which may yield three or more transactions per week from a sizeable fraction of the sample households. Multiply by 11,500 households, and consider that a rolling two to three years of purchase data are resident on disk for each coded category. (At any given time, some of the categories are not revenue generators and therefore are not key-entered to the database.) This calculation will give you some idea of the size of this database -- and in fact MRCA's package goods purchase data service was the largest private RAMIS database in existence. The sheer size of the database, in relation to the computing power available to us at reasonable cost, was one of the key design considerations for DYANA.

MRCA's data delivery system is now a dual one: Reports generated by MRCA personnel and sent to the client in hard copy; and DYANA interactive sessions, in which the user is a client at a microcomputer located at the client company's offices. Although the RAMIS database is still used for custom reporting jobs, the DYANA/DYANAgraf decision support system was written from scratch in a variety of lower-level languages. It does, though, use records which are formatted similarly to the RAMIS records I have just described.

The most important aspect of the database is that MRCA owns it. Ownership of the data is of profound importance for marketing a DSS based on it. First, no user likes to transfer data into an analytical program. Whether the transfer is a matter of key entry or involves having the mainframe operator mount a tape, it's an annoying, error-prone and time-consuming process that is eliminated when the data and the software are integrated.

Some word-processing programs for PCs come with standard business letters built in -- showing that the software houses are beginning to recognize this truth. Yet the Wall Street Journal (September 7, 1984) reports that electronic publishing has proven unprofitable for nearly everyone who has tried it. We might interpret this simply to mean that data without analysis is as bad as analysis without data. But most material published electronically is of a static or "timeless" nature, leading to no repeat buying. The fact that MRCA's databases are continually updated ties MRCA to the customer.

Second, selling a DSS jointly with a continually updated database affords flexibility in pricing. One product can subsidize the other -- which in economic terms is its complement -- when this is appropriate.   And last, owning the data means that MRCA is the unchallenged expert on the processing and analysis of it. Its delivery system (DSS) therefore has unmatched credibility in the market.

8.12.3 The Clients

MRCA's clients were characterized by three distinctive features. First, they are large companies. They have sufficient margins to absorb the cost of rather expensive market information services and remain competitive. There are a fair but limited number of such companies and institutions. Clearly with present cost structures MRCA's market growth must be based on additional services rather than additional clients.

Second, the clients use a variety of data from sources other than MRCA. In addition to their internal data on marketing costs and shipment volumes, they buy data on the movement of product through intermediate sections of the distribution pipeline; perhaps consumer attitude and preference surveys; and data on the advertising expenditures of their competitors and the magazine and newspaper readership of their target market households. They may also buy taste tests and focus groups. Consumer panel information now competes with many other services for the attention of corporate market researchers and for their budget dollars.

Third, panel data's point of entry into the client corporation is at a fairly low level. When market research was young and a novelty, it commanded the attention of top executives. Nowadays, except for unusually exciting innovations, research is routinely purchased and handled at lower levels, where it is collated and filtered upward in the organization. Similarly, in decades past, client corporations hired statisticians to evaluate and analyze market data from outside services. In modern times such data is, upon delivery, first inspected by an analyst who is often a college graduate but who may have only a rudimentary statistical background. Research results lose their "brand name" identification with the originating supplier as they are summarized and filtered upward through levels of client management. The research supplier loses visibility when this happens.

8.12.4 The Necessity Which Was the Mother of DYANA™

In addition to the problems posed by limited visibility in a limited market, MRCA in the 70's suffered from a cumbersome delivery system. Throughout the decade, the document most commonly delivered to a client was the "Market Summary Report", a matrix whose rows represented individual products (broken down by package size, flavor, etc.) and whose columns represented various market measures (brand share, penetration, average price paid, and the like).

The "MSR" was often replicated for different regions of the country and for different demographically defined buyer groups. The row/column format and replications were customized for each client and remained unchanged from month to month.

Non-MSR reports were called "special analyses" and had to be programmed on a custom basis. These included brand switching studies, new product growth extrapolations and so on.The MSR was delivered monthly, by mail. It was generally quite a few pounds of closely printed paper, and we called it the "everything by everything" crosstabulation method of market research.

There were probably some gems of market insight in every one of those MSR's, but it took lots of perseverance and not a little luck to find them.

Worse yet, if there were an error in the MSR, the turnaround time on a corrected run was on the order of six weeks. On the other hand, if a client did stumble onto a gem of information in an MRCA report, and if this discovery gave him or her an idea for further analysis, MRCA managers had necessarily mixed feelings. A special analysis meant added revenue, but MRCA had to tell the client that it couldn't deliver the analysis for six weeks. By the time the analysis reached the client's desk, he or she had lost the train of thought -- or lost all interest.

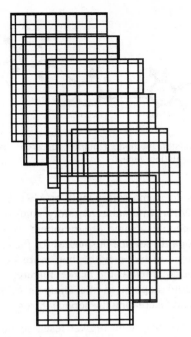

Figure 8.6 The "everything by everything" format

Special analyses were regularly ordered only by clients who could fit them into large research plans, with long lead times. There was no scope for specials done on a whim or in an exploratory manner. Since exploratory analysis could be expected to reveal interesting sales angles and conversation starters for MRCA's sales people, the lack of this capability was felt from an internal point of view also.

In the information industry, companies compete for a share of the clients' attention, or his or her time or mental space. An information product has to provide continuity (i.e. be timely and on time). It has to have attention-grabbing properties. And for the long term it has to be useful and must minimize wasted time on the user's part. The MSRs had few of these virtues.

8.12.5 Special Features of the DYANA DSS

DYANA (the name is an acronym for DYnamic ANAlysis) was conceived through a fortunate combination of events. On the computational side, the RAMIS database had been available for some time, and interactive programming had become the norm within MRCA. The time was ripe for new thrusts in the data processing area.

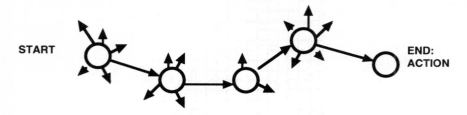

Figure 8.7 Sequential decision method of market research: Gather information; decide whether to gather more information; repeat.

On the organizational side, some salespeople at MRCA began to write special analyses directly in RAMIS. Internal exploratory analysis was now possible via RAMIS for those salespeople. Some programmers were moved into the sales department. It is traditional in service businesses for the creation of the service to occur mostly in operations to begin with, then more in marketing, and finally mostly at the point of consumption as the market matures. It began to be apparent that the same thing could happen in the analysis of market information -- ultimately the clients would be creating the service themselves.

At the same time we knew that client market research staffs tended to look at a limited range of data configurations, although different companies had different favorite or traditional ways of looking at their markets. Research suppliers, MRCA included, were burying these users under much unwanted data. At the same time, the fixed data sets were stifling the clients' need to strike out in innovative directions of inquiry. Dave Learner, MRCA's president, began to urge the development of a service which would allow the user to extract one fact from the database and then make a decision: "act now", or "request more information."

Our research department had spent some years immersed in the literature of marketing and consumer behavior models, and this experience had focused our attention on the distribution of purchase frequencies as a fundamental tool of market analysis. This frequency histogram allows for the identification of light and heavy buyer groups, average interpurchase times, and product market penetrations. Comparison of the overall distribution of purchase frequencies with conditional distributions (that is, d.p.f.'s based on a subset of the data) reveals market shares, brand loyalties, and other relative measures.

Adding a second set of bars to this graph, showing the pound volume purchased by by households buying n times, closed the circle by incorporating volumetric measures and market segmentation based on them. This double bar graph, generated interactively and parameterized by household characteristics and time delimiters, etc., was the complete market research tool.

The first commercial releases of DYANA used dumb printer terminals, with no graphic output. Later, full-screen terminal capability was added. In 1984 we introduced DYANAgraf™, DYANA's color graphics analysis enhancement, which was run on a PC with color monitor, color printer, and autodial modem. It was a hybrid mainframe-micro product which was close to state-of-the-art for its time.

In all versions, the user requests a customized working database to be drawn from the master generic database. (A catalog command subsequently shows the user all the working databases which are still active on his/her account.) All DYANA modules are run on the mainframe from this working database.

Figure 8.8 The empirical distribution of purchase frequencies

The autodial function is invoked according to the user's menu selections. A naive user is not aware of whether he or she is talking to the mainframe or to the micro. The autodial software was authored in-house, and seemed to work more reliably with MRCA's hardware than the commercial systems then available.

In DYANA and DYANAgraf, module runs are parameterized while in communication with the mainframe. In FUNDS, the PC accepts the module specifications and uploads them to the mainframe. In DYANAgraf and FUNDS, the output files from the module computations are downloaded to the PC. The graphics, analysis and reporting software (written in compiler BASIC) resides in the PC. Graphics are displayed on the screen; a function key causes the graph to be printed, and another F-key causes the corresponding tabular report to be printed.

The scope of this discussion does not allow a full description of DYANA's capabilities. The rest of this section will focus on some of the original and innovative features of DYANA. For reasons given in the next section, we expanded the original concept of a frequency distribution analysis until as of this writing DYANA consists of eight modules. These modules are the nodes in the "act now or request more information" sequential research path. The link between one node and the next is the parameterization or "filtering" that defines the next analysis.

Filtering in DYANA is of four kinds: On the demographic/geographic/ socio-economic characteristics of the household; on the attributes of the transaction (time, deals and coupons, etc.); on the definition of the product (brand, size, flavor); and on the behavior of the buying household as defined by previously run modules. An example of the latter would be a demographic profile of the households who bought brand A two or more times (or who bought both brand A and brand B). These behavioral tags alter the working database, until the tags are changed or until the database is erased; the master database is never contaminated.

A DYANA analysis, then, is a sequence of module runs and filter specifications, resulting in the answer to a specific question. The extant modules together with the filtering mechanism address the most frequent inquiries of market researchers. Looking only at the common questions allows the user to learn only a fraction of the commands necessary for use of a general DBMS. This reduction approach is very common in special-purpose software. But DYANA also contains functions that are beyond those found in a general purpose DBMS (or at least would be very awkward in such a system): Current filter sets are saved for future use and are subsequently offered to the user when he/she is most likely to need them. Also, as I've mentioned, DYANA allows a working database to be marked or altered to facilitate an analysis, without polluting the original data.

8.12.6 Problems, Progress, and Opportunities

After formulating a complete system of market analysis based on the purchase frequency distribution, we ran into the practical problems of DSS implementation, and realized we had outsmarted ourselves. There turned out to be many useful market measures (e.g. combination buying, the extent to which a buyer of item A also buys item B) which, although derivative from the purchase frequency graph, were simply too much trouble to derive that way. And once the measure was derived, the purchase frequency graph was an inconvenient way to display it. We had to design separate DYANA "modules" to display these measures and to show the demographic characteristics of the relevant households. We turned the problem into an opportunity by designing some unique interactive graphics formats (including the first areally-accurate on-screen Venn diagrams) which are specially suited to market research uses of a DSS.

Another problem, stemming from the multivariate nature of the data, prevented the "one fact at a time" philosophy from becoming a reality. Because that one fact must be teased out of the DSS with as many as 25 parameters (more or less), users felt they were getting too little return for their finger exercises. We compromised by instituting some tabular outputs, bringing the average closer to "25 numbers in, 25 numbers out", which seemed to be more acceptable to users. Facilities for saving parameter sets were also helpful in this regard. Product definitions, which commonly are used across many modules, are saved as separate files. Demographic filters used in the particular user's last run of a module are echoed back when the same module is next called.

The large number of variables also meant that the graphics software had to be written in-house. Commercial graphics packages provide neither the flexibility for special-purpose graphic forms, nor the multiple titles, headers and legends which we needed to represent data sets defined by multi-multivariate parameters.

We humans are not multivariate thinkers, and our spoken language does not allow us to deal with more than three or four variables at a time. Even when the right data are extracted from a DSS or a DBMS, a decision is not likely to be made until that number can be expressed in common language. In DYANA, filtering on demographics, geography, income, etc. is equivalent to creating a conditional probability space. Looking at the incidence of a behavior within that filter, say the proportion of households buying both brand A and brand B, may result in an expression like

$$\text{Prob } \{X=A \text{ and } Y=B \mid T=t, u_1<U<u_2, \text{ and } V=v\}.$$

How to express this in ordinary English, not to mention as a concise label on a graph, remains problematical for many DYANAgraf functions. If you have ever used U.S. Census data (most marketers do), you are familiar with this problem. Even *American Demographics,* by far the most readable of all demographics publications, suffers from awkward sentence construction, and they have all the room they need – not just a screenwidth – to say what they need to say. To take another example, the Apple Macintosh, the "computer you already know how to use", has many novel functions which, when used in combination, produce unanticipated and undocumented results. The folks at Apple must have decided that no documentation is preferable to tedious and convoluted documentation, and they may be right.

This, then, is a general problem of DSSs based on multivariate databases. We can do things with computers that we cannot discuss in ordinary language. A semester-long course of study would enable people to develop the necessary common orientation and terminology, but this is not an acceptable option for an ostensibly "friendly" commercial DSS.

In DYANA, an echo-print of the input parameters clears up most ambiguity of written labels and headers. But this is "computer language", not natural language. The report and graph menus are very easy to use, given that the analyses have already been downloaded.

The number and size of menus can also quickly become oppressive in a multivariate environment. We have managed to keep the number of menu levels to two or three. The most commonly needed options after drawing a graph are discreetly displayed at the bottom of the graph's screen. When the graph is printed, this menu is replaced by an MRCA copyright statement. Of course, MRCA hopes its name remains on the document as it is circulated through the client company.

One final problem deserves mention. Since the DYANA databases are sample information rather than complete enumerations, and since DYANA users may exercise any and as many filters as they wish to specify a report, we had to guard against the possibility of a user relying on insufficient sample sizes. This need spurred a project for exact real-time calculations of confidence intervals for the reported measures. The result was faster, friendlier and more accurate than the old system of hardcopy lookup tables.

8.12.7 The Future

DYANA, like all modular structures, is a confession that the designers did not expect to think of everything the first time around. MRCA invited members of the academic community to submit designs for DYANA modules which are either new analyses or cost-effective shortcuts to analyses which are currently possible within the system.

DYANA's philosophy and strategy dictate that the DSS will not contain general purpose statistical or spreadsheet functions, but will provide software links to selected commercially available programs that do these more general tasks. DYANA and FUNDS linked early on to the Lotus Corporation's 1-2-3 spreadsheet program.

We have seen that DYANA is both data and methodology. It addresses a real problem in the marketplace, and also incorporates a theory of marketing. Many successful systems do not include theory. That is, they allow unstructured or free-form calculations and access to data. But these free-form data mining systems can generate nonsensical results. Decades after its invention, it seems that DYANA's use of theory as a framework for data mining was the right approach.

8.13 How to Be a Change Agent on the Buyer Side: PC Insertion in a Mainframe-Centric Company

The memo below dates from the days when early adopters and early majority segments in corporations were beginning to embrace personal computers. Ours had been a mainframe-oriented company, and this heightened the tensions concerning adopting PCs on a more widespread basis. The memo's language is more daring than I would recommend for general use; I was, at the date of writing, already contemplating leaving the company, and this may explain the choice of words. Fortunately, the memo's resulting frankness makes it more valuable for learning the topic of this section.

MEMORANDUM

To: D__   date: 10/27/88

From: Fred Phillips   cc: President

Subject: MRCA PC application development policy guidelines

D__, this is in response to your request for help in designing such guidelines.

You asked with reference to using contract programmers/analysts for developing PC applications. We also ought to give thought to what's needed for applications originating within MRCA. In the end, I think the same policies and guidelines suffice for both kinds of developments, with a few exceptions.

The latter will be of two varieties. The first is programs developed by an MRCA employee with the explicit intent of making them available to other people in the company, and maybe to clients (like many that have come out of my department). The second variety is tools people hack together for their personal productivity that nonetheless find their way into circulation in an informal fashion. Although we may try to discourage this informal proliferation of wildcat programs, we have to recognize that it's going to happen regardless.

All three modes of development – contracted development, planned internal developments, and wildcat programs – are desirable because they increase the creativity, responsiveness, flexibility, and velocity of innovation of MRCA as a company. More than desirable, I'd say they are absolutely necessary for our survival. They also pose the danger of things getting too far out of control, and that's why we need a policy and development guidelines.

Specifically, the principal danger is depending on programs with hidden bugs or inadequate documentation, written by a person who then quits MRCA and/or disappears. When the program crashes, who's the user gonna call? Not

Ghostbusters – they'll call D__ and G__ because you're the guys who are supposed to know everything about computers!   And of course there will be a client deadline involved.

Policies and guidelines should cover:

- Program testing

- Completeness of code documentation

- Completeness of user manual(s)

- Standardization of user interface

- Standardization of consumer panel terminology used in I/O screens

- What MRCA considers reasonable interpretation of its data

- Definition of milestones for purposes of controlling the project

- Understanding of intellectual property and secrecy

- How this ties in with MRCA employees' acquisition of PC hardware

- Other things you may think of.

Some things not appearing in the above bullets are missing because of inadvertant omission or because I haven't yet thought it through; others are not there because they shouldn't be there. Such things as the [programming] language to be used, or how much to pay a contract programmer, should not be controlled too tightly. This is because the whole idea is to foster flexibility and creativity – PC development should in certain ways be more freewheeling than mainframe development.

Thus an important point. MRCA's mainframe people should not be the ones to formulate these policies. With all due respect, our mainframe gurus do not have the slightest notion of what user-friendliness is; their career choice involved agreeing to meet the computer more than half way. Today's PC users rightly expect the computer to bridge over 90% of the communication gap. For MRCA's part, we want to encourage creativity and leverage the intellects of our analysts by extending computer power to them even if they don't think they want it. This means devoting 75% of project man-hours to an *impeccable* user interface, and only 25% to the central algorithm and I/O (just about the reverse of mainframe projects); writing a non-scary, easily read manual that ideally the user will never actually have to open; and putting some fun into the program to market its use more effectively.

"But," says the traditional manager, "I don't want to give my people toys. They'll be distracted."   You and I are aware of the connection between play and creativity. Let's all face the fact that a little control has to be sacrificed in order to find new and better ways to satisfy our clients. There's a difference between nondirected fishing in the [National Consumer Panel] data, and playing Space Invaders on our PCs. The former will result in new insights; managers should discourage only the latter.

I strongly recommend the use of an outside consultant to help formulate these PC policies. I'm not an expert on them, but other people have amassed a great deal of expertise on microcomputer management. They can help us avoid the biases and omissions we would bring to the guidelines were we to do it ourselves. [Today, I would add, "a consultant who is not a vendor and has no financial ties to vendors." The term "microcomputer management" sounds so quaint to 21st century ears!]

In case you think I'm being to touchy-feely with this creativity stuff, I can only repeat that our current product development cycle is much too long and doesn't find room even to start on many promising submitted ideas. The marketplace is changing rapidly, and our highly capitalized competitors can afford to shorten their form development cycles and stay on the leading edge. To survive, we have to innovate, and that means opening up the development process.

Now another point, just to be sure I insult everybody. Just as our mainframe folks don't have an inkling of user friendliness or decentralized control of computing, our account executives don't have an inklling of proper software engineering. They will be offended when MRCA tries to exert even the small amount of control the company must exert over the account exec's *very own clever and original computer program.* We must have something like the following policy:

- An employee may use any software he/she wants for self-training or creative exploration in market research, especially involving original programs.

- MRCA has established guidelines for original or contracted software development. The Product Development department will be reasonable in signing off on the conformance of such projects to the company's guidelines.

- If an employee writes or receives a program and that program is signed off by Product Development, and Product Development has copies of the documentation and manuals for the program, then Product Development will support the program. If those conditions are not met and a program crashes, the user/employee understands he/she will be left flapping in the breeze.

- If an original program is used to produce numbers that will be reported to a client, and if Product Development has not signed off on that program, then the employee's supervisor should be notified in advance and will take responsibility for approving or not approving that specific use.

To get back for a moment to the intellectual property issue, here's one area where contractors have to be treated a little differently from employees. A contractor can sign all the nondisclosure forms in the world, yet if he/she pirates the idea and it show up in the marketplace elsewhere, MRCA will have a devil of a time, and much legal expense, to prove piracy. On any sensitive project, an MRCA person must break up the total development task into discrete – and discreet – subtasks, and hire a different contractor for each subtask. This not only increases security, but encourages modular programming that produces (perhaps) program segments that can be reused in other projects.

8.14 How Long Will It Take?

Beneficial innovations have spread slowly, due to bureaucratic influences, difficult conceptual underpinnings, ineffective change agents, and other reasons. For example, scurvy was already a feared killer of sailors for over 200 years when in 1601 one Captain Lancaster of the British navy discovered lemon juice would cure it. This cure was not widely used until 1747. Furthermore, the cure was not adopted for all navy ships until 1795, and scurvy continued to be a major killer until it was wiped out in 1865 upon adoption of the cure by the Merchant Marine. Why was this innovation so slow to spread? One reason is that other remedies were being proposed by more prominent individuals, with greater credibility than Lancaster. As a naval problem, scurvy cures faced unique competition: Scurvy prevention was not readily accepted, yet new ships and guns were.

In his syndicated newspaper column (October 8, 1993), Lou M. Boyd reported that "doctors didn't get around to prescribing aspirin routinely until 46 years after Carl Gerhardt discovered it in 1853." In market research, longitudinal survey sampling has not been widely used after 50 years, despite ample proof of its power relative to cross-sectional statistics.

The graph on the next page shows typical adoption times for manufacturing technologies, and one organization's goals for compressing these lag times. It should be clear that the profit implication of accelerating these revenue streams is extreme.

One commentator offers these rules of thumb for time from patent to start of sales (a different matter from time to widespread use!): Sensors, two to three years, depending on the kind of FDA approval needed; biomedical devices, three to five years, more if clinical trials are required; and research instruments and equipment, one to three years.

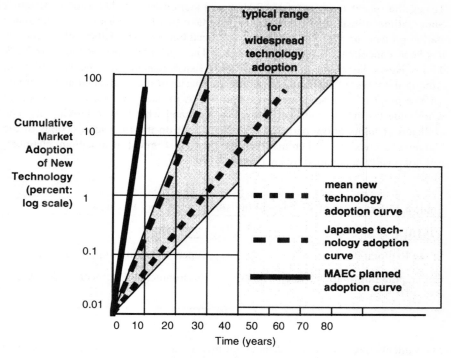

(source: Manufacturing Application and Education Centers)

Figure 8.9 Time scales for adoption of manufacturing technology

8.15 Marketing Invisibles

A mother, raising her son according to the "Cleanliness is next to Godliness" principle, finally pushed the boy to the point of rebellion. "All I hear around here," he exclaimed, "is germs and Jesus. And I can't see either one!"

One of the greatest successes in recent times, among high-tech products, is geographic information systems (GIS). The main success factor is *you can see it.* Sales regions and routes, troop deployments, retail locations, customer addresses, and the layout of subway and sewer lines spring to brilliant visibility. This chapter has established via theory and example that visibility is an important plus in the diffusion of an innovation.

Yet invisibles are important in the modern world. How can we market, for example, environmental engineering services? If you clean toxins from the soil under a construction site, how can you demonstrate the results of your work? In

Texas, the recent troubles of NASA's space station and DOE's Superconducting Supercollider illustrate the difference. To make two long stories short, the space station project survives, though under a reduced budget. The Supercollider project has been cancelled by the U.S. Congress. Dave Montgomery of the New York Times News Service (October 22, 1993) listed reasons for the cancellation. Among them were "Difficulty in selling the merits of a complicated and futuristic science project." It's no coincidence that NASA missions are seen by millions on worldwide television, and everyone can understand what they're looking at. The collision of high-energy particles that sheds light on the first nanoseconds of the Universe is invisible. When made visible on an instrument or in a cloud chamber, it is not readily understood by lay people.

Table 8.13 Examples of Visible and Invisible Technology Products

VISIBLE	INVISIBLE
Space Exploration	Superconducting Supercollider
Geographic Information Systems (GIS)	Environmental Engineering Services

My students have had these ideas about marketing invisibles. The most succinct suggestion was, "Pork"! Create jobs in a Congressional district. Second, there's always *something* to see. Show the senators the Supercollider tunnel. Show the environmental engineering client the OSHA regulations. Finally, don't promise that the innovation, e.g., the Supercollider, will produce *specific* spinoffs. Disappointments may ensue. If you are confident about the innovation, sell the unforeseen spinoffs or the general potential for spinoffs that are bound to happen.

8.16 Psychographics of the PC Market

"Technology Marketers Turn to Alcohol in Record Numbers" is a headline we haven't seen yet. But we might see it soon, given the real headlines that have been appearing: "Techno Terror Slows Info Highway Traffic"; "Fear of Technology is Phobia of the 90's"; "Marketing Technology No Easy Task".[9]

---

[9]   Wiseman P and D Enrico (1994)   Techno terror slows info highway traffic. USA Today (November 14) B1; Associated Press (1994)   Marketing technology no easy task.   Austin American-Statesman (February 17) E1-E2; Dell Computer Corporation (1993) Fear of Technology is Phobia of the 90's. July 26 (press release)

The focus of a 1994 Gallup survey: Fear of technology in America. This phobia, seemingly widespread according to Gallup, could be a hurdle for companies who want to merge onto the information highway. The Gallup survey confirms what many commentators have been hinting at: People are afraid and anxious about the technology available today, notwithstanding that technologies like movies on demand are becoming available in homes and offices. Gallup questioned 605 white-collar workers and found:

- 32% of people confess to being "cyberphobic."

- 39% of women say they fear technology compared to 27% of men. Michelle Weil, a clinical psychologist, says, "women have not traditionally been encouraged to explore technology," though "had we done the survey 20 years ago, you would have seen a vast difference" between the sexes.

- 59% of people will look at technology only after it has proven.

- 12% consider themselves "early adopters"–those first to grab new technology.

- 56% believe the advancement of computers threatens their privacy.

- 38% are overwhelmed by information available.

- 38% worry about losing face-to-face contact with other people.

- 58% have not heard of the Internet – the international web of computer networks that let users find information and send e-mail worldwide.

- 78% say that the information highway is either here or on the way.

- 54% believe their industry is growing because of the information highway.

More evidence, or at least more anecdotes, can be marshalled to support this trend. A Dell Computer Corp. study found that 55% of adults were hesitant to embrace technology. Market researchers Odyssey Ventures Inc. found that the 33-34% estimate of homes with computers is closer to 27% of US homes. When Bell Atlantic CEO Raymond Smith asked his top 50 executives how many used the Internet, he found what should be a shocking answer: four.

According to Dell Computer Corporation, fear of technology is the prominent phobia of the 1990s. Dell's survey of 1,000 adults and 500 teens across the country showed that 55 percent of Americans – despite the United States' history of leading technical innovations – remain fearful of using technology. Nearly 25% have never used a computer, programmed a VCR, or stored stations on a car radio.

These results are part of extensive studies conducted by Dell over the last eleven years. The focus of this work is to understand the "techknowledge" in America. Dell has conducted extensive market research, interviewed focus groups, and responded to feedback from Dell's toll-free customer service lines.

Teens and adults agree that computers are a valuable time saving tool, yet 32 percent of adults are afraid they will damage a machine while using it. Almost a quarter of these people are fearful of setting their digital alarm clock at night.[10]

8.16.1 An "Understanding Gap"

A main cause of this paranoia and misunderstanding is the computer industry has failed to educate the public about the simplicity and utility of personal computers. People do not know exactly what computers can do for them, and few have any way to learn. Says Dell, "most computer manuals are as difficult to understand as a graduate-level course in quantum physics." This lack of education creates an accelerating gap between what is available to them and what people can do with technology. This gap, reasons Dell, is responsible for the technophobia in America.

8.16.2 Techno-Typing

To help narrow this understanding gap, Dell employs a technology-need appraisal system called Techno-Typing. Some Techno-Type classifications are:

- Techno-Wizard: An experienced guru who wants the latest technology at a low price. Wants to have most up-to-date products.

- Techno-To-Go: A new customer who wants to pull the computer out the box and see what it can do. Not concerned with how the technology works. Values support and service.

- Techno-Boomer: Takes a proactive stance on the buying decision. Reads and seeks recommendations before purchasing. Wishes to makes the logical, correct, buying decision.

- Techno-Phobe: Refuses to have anything to do with technology.

There are two corporate Techno-Types that Dell identifies:

- Techno-Teamer: Thinks of computer as a time-saving tool to increase productivity. Computer use on network is mainly team and job oriented. Primary concern is network failure.

- Techno-Critical: Uses computers for critical job tasks–computer-aided-design or engineering. Often is a independent consultant/contractor who depends on

---

[10] At this point I feel compelled to reaffirm that I am not making this up, or even exaggerating the numbers.

technology for business.   Fears system failure because it is detrimental to business.

Table 8.14 TechnoFear in America: The Numbers

American Technophobia:

- 55% of Americans are technophobic to some degree

- Of all technophobics, 55% are female and 45% are male

- Of all technophobics, 67% are adults and 33% are teens

- 25% of adults have never used a computer, set a VCR to record a TV show, or programmed station on a car radio

- 32% of adults are intimidated and feel they will damage a computer if they use it without assistance. Of this group, 22% do not feel comfortable setting a digital alarm clock.

- 51% of adults find new technologies difficult to understand

- 58% find the rate of change in technology confusing

Adult Apprehension:

- Adults are not comfortable using a variety of electrical products:

Computer	23%
Compact Disk Player	20%
Automatic Teller Machine	22%
VCR	9%
Answering Machine	15%
Programmable Thermostat	15%
Digital Alarm Clocks	8%
Car Phone	34%

Teens are More Techno-Literate than Adults:

- Teens are more comfortable with technology:

  % Comfortable Using Technology

	Adults	Teens
Digital Alarm Clock	88%	98%
VCR	87%	98%
Compact-Disc Player	61%	90%

Answering Machine	78%	87%
Computer	70%	91%
Car Phone	40%	55%

- Ninety-two percent of teens are comfortable with VCRs, compact-disc players, answering machines, and computers; compared to only 74 percent of adults that are comfortable with those technologies.

- Almost 33% of all adults have never used a computer, compared to only 8 percent of teens.

- More teens are at ease using a computer than are comfortable with an answering machine.

- Eighty-five percent of teens enjoy figuring out what a new technology can do; compared to only seventy-one percent of adults.

American Attitudes Toward Computers:

- 27% of U.S. adults have never used a computer; 55% have never bought one.

- 25% of adults miss the days when only typewriters were used.

- 77% of teens and 70% of adults wish computer technology was easier to understand.

- More than 25% of adults would never use a computer unless they were forced to.

- 90% of teens think computers are fun, compared to 74% of adults.

Source: Dell Computer Corporation (1995)

Odyssey Ventures Inc., a market research firm, found that companies will have to appeal to customers in innovative ways to sell advanced technology. Odyssey's 1994 survey of 4,020 homes identified six core attitudes about technology; beliefs based on education, income, and other demographics.

High-tech marketers have traditionally spoken of two types of customers – "early adopters," who immediately embrace a new technology, and the rest of the population. This approach classifies all people as a potential market.

The Odyssey study emphasized the willingness of people actually to purchase technology. This showed that the current approach to tech-marketing is too elementary for the new services becoming available – such things as movies-on-demand, encyclopedias on computer, home shopping, and modernistic communications. Odyssey CEO Nick Donatiello says, "The message to companies is they have to begin to think about this as a set of markets with different needs and different attitudes."

The company has identified six different type of technology customers:

- New enthusiast – will readily spend money on new technology if the benefit is apparent.

- Hopeful – usually cannot afford the new technology they want.

- Faithful – are not opposed to new technology, but have little desire for it.

- Oldliner – have no desire to spend money or try new technology.

- Independent – although they can afford it, do not use technology now, so do not   seek improvements of new technology.

- Surfer – does not care about new technology, and is wary of companies that sell it.

Odyssey's research combines the classical life cycle segmentation concept with an "ability to pay" classification. Each of the Odyssey types compose about 14 to 18 percent of the total market. "This is not dominated by one segment so people can say 'Ah there's the target,'" said Donatiello. This is alarming to companies working hopefully on a "killer application." Technology intensive companies usually try to create a "killer app" – like the Lotus 123 spreadsheet program in the mid 1980's – that will be so popular that all people will buy it. Odyssey's research stresses that only people who can afford the new technology will buy it. The "new enthusiasts" and "hopefuls" may be the only people that support a revolutionary technology.

8.17 Notes

It was a hybrid mainframe-micro product which was close to state-of-the-art for its time. See Freedman D H (1984) Tapping the Corporate Database. High Technology April

Culture. See (1997) Schools Brief. The Economist (no byline) (November 29) 71-72; Kalish D E (1999) AOL tries to eas the clash of corporate cultures. The Philadelphia Inquirer (March 28) C2; Marer P (1998) Introducing Culture in Business Courses. International Teaching Resources for Business Indiana University CIBER 12 Autumn 1-8

...the mission of a military electronics system. Adapted from Blanchard, B S and W J Fabrycky (1990) Systems Engineering and Analysis. Prentice Hall Englewood Cliffs NJ (2nd edition)

Take-back laws. See Scarlett L (1999) Product Take-Back Systems: Mandates Reconsidered. Policy Study #153 Center for the Study of American Business Washington University St. Louis Missouri

A slightly updated scheme for psychographically segmenting technology buyers is in (no byline)...(1997) Do You Know Your Technology Type? The Futurist September-October 10-11; The article is based on "Why Consumers Buy." (1996) The Forrester Report. Forrester Research in Cambridge Massachusetts December; See also (no byline) (1998) Are Tech Buyers Different? Marketers say new consumer categories are needed. Business Week (January 26) 64-66

One commentator offers these rules of thumb for time from patent to start of sales... Pincus J (1999) Email via the techno-l listserv January 1

the non-recurring costs of "extricating the company from a fully implemented client/server architecture" – would run $60 million, according to former U.S. Department of Defense CIO Paul Strassman. See www.strassmann.com

8.18 Questions and Problems

Short Answer

S8-1. Lou Boyd's newspaper column reports that the modern technological innovation that swept the world most quickly was the sewing machine. List possible reasons for the rapid diffusion of the sewing machine. What were the business implications for Singer and Brother?

S8-2. Discuss how the classification scheme of Figure 8.1 affects marketing decisions for products with which you are familiar.

Discussion Questions (Answers should be one or two paragraphs.)

D8-6. Many technology companies are started by engineers to make products that are sold to and used by other engineers. Eventually, as in the case of Hewlett-Packard, continued demands for growth require the company to make and sell products for non-technical buyers. Comment on some of the changes that a company must make in order to make this transition successfully.

Miniprojects

M8-1. Check your understanding of "cost elements" by listing the significant cost elements relating to purchase/ownership/disposal of a new personal computer.

M8-2. Automobile purchase/ownership/disposal cost estimation exercise.

(a) Write down the "usage scenario" for your new car. This is your mission in purchasing and operating your automobile. What do you want a car for? How do you expect it to perform and under what conditions will you be driving the car? How long will you drive it (i.e., how long do you expect to own the car)?

(b) Choose one automobile cost element and list associated cost parameters in the second column of the worksheet below. Make and hand in some side notes about how you would estimate or gather data to determine numerical values for these parameters. Use the third column to show how the cost parameters you have listed can be related. This will usually be a formula. For example, here is a cost relationship for gas expense:

Gas expense = NPV of [fuel price x miles driven/year x gas mileage] over
expected life of car.

(c) Discuss whether you would use these life cycle cost considerations when car shopping. When comparing two cars you are interested in? If you wouldn't, who would?

(d) Put yourself in the place of an automobile designer and think about the "Design fors": Design for reliability, testability, safety, supportability, distribution (availability), maintainability, manability (ergonomics), quality, disposability, marketability. How can design decisions affect each cost parameter? For example, a high-performance engine increases fuel price because the owner is urged to use high-octane fuel. A lighter car body increases gas mileage, all other things being equal. Use the fourth column of the worksheet to identify a design factor that affects the corresponding cost parameter.

(e) OPTIONAL: Different automobile manufacturers, including Saturn, have web sites that allow a car shopper to calculate certain costs. Are these calculations comprehensive? Do they really provide all the information that you might need to make a well-informed purchase decision? Check out some of the web sites and make some notes about how the web sites could be improved to be more useful for a customer who wants to know the real costs of owning a car. How would you improve the interactive price calculator at http://www.saturn.com/car/ipc/index.html to make it more useful for a car buyer?

Cost elements	Cost parameters	Cost relationships	Can be improved by...
information collection	_____	_____	_____
	_____	_____	_____
	_____	_____	_____
purchase price	_____	_____	_____
	_____	_____	_____
	_____	_____	_____
tt&l	_____	_____	_____

_____   _____   _____
_____   _____   _____

finance charges   _____   _____   _____
_____   _____   _____
_____   _____   _____

insurance   _____   _____   _____
_____   _____   _____
_____   _____   _____

routine maintenance   _____   _____   _____
_____   _____   _____
_____   _____   _____

repairs   _____   _____   _____
_____   _____   _____
_____   _____   _____

gas, oil   _____   _____   _____
_____   _____   _____
_____   _____   _____

accessories   _____   _____   _____
_____   _____   _____
_____   _____   _____

replace tires, wipers, etc. _____   _____   _____
_____   _____   _____
_____   _____   _____

washing, cleaning _____   _____   _____
_____   _____   _____
_____   _____   _____

parking, garaging _____   _____   _____
_____   _____   _____
_____   _____   _____

tolls   _____   _____   _____
_____   _____   _____
_____   _____   _____

selling, trading in   _____   _____   _____

_____   _____   _____

_____   _____   _____

Problems to Solve

P8-1. The purchase cost of a machine is $30,000. The annual regular maintenance costs $1,000. The mean time between failure (MTBF) is 15 months, if regular maintenance is performed. The cost of repair when failure occurs is $3,000. What is the expected cost of purchasing and owning this machine for four years?

P8-2. Life Cycle Decision Exercise. You are to evaluate two alternative designs for a piece of equipment, based on life cycle (or more specifically, a ten-year calculation of) cost. Estimated contemporaneous costs for alternative #1 and alternative #2 are shown below, in millions of dollars.

Year	Cost of Design #1	Cost of Design #2	
1	0.65	0.5	
2	0.75	0.7	Design Activity
3	2	0.7	
4	5	3	Production Activity
5	10	6	
6	1	3	
7	1	4	
8	0.8	4	Operation/Support Activity
9	0.8	4	
10	0.8	4	

Discounting these costs at an annual interest rate of 10% yields a discounted cost flow for each design alternative. Based on a net present value calculation (see the sums at the bottoms of the right columns), alternative #1 is preferred.

Year	Discount Factor	Discounted Cost of Design #1	Discounted Cost of Design #2
1	0.9091	0.5909	0.4545
2	0.8264	0.6198	0.5785

3	0.7513	1.5026	0.5259
4	0.6830	3.4151	2.0490
5	0.6209	6.2092	3.7255
6	0.5645	0.5645	1.6934
7	0.5132	0.5132	2.0526
8	0.4665	0.3732	1.8660
9	0.4241	0.3393	1.6964
10	0.3855	0.3084	1.5422
TOTALS		14.4362	16.1842 <- Net present value of each alternative.

Which design alternative would you choose if there was a significant chance that the project would last less than 8.5 years (for example, if the equipment became obsolete two years after its deployment)? Why would you make this choice? *HINT: Graph the cumulative discounted costs of each alternative.*

"It was puny. It was underpowered. It was overhyped. It cost a bundle. It was completely incompatible with the IBM PC. It didn't run in color. And there was hardly any software available for it. But it was still pretty wonderful."

Mark Potts, writing about the Macintosh in *The Washington Post*, 1/24/94

9 Strategies and Tactics for Marketing New-to-the-World Products

It is a proverb of the Internet Age that "information wants to be free." The near-universal realization that the Internet is important for marketing – and the difficulty of devising schemes for using the Internet as a direct sales and collection tool – underscore the fact that companies must decide more carefully than ever before what it is they are selling and what they are willing to give away. Internet advertising is not delivered to couch potatoes between sitcom segments. It is sought out by potential customers who are interacting with the medium. Internet advertisers therefore must give away a lot – just to keep the interest of prospects.

This is becoming true for non-Internet producers as well. Peter Coad of Object International Inc. distinguishes his "front-end" or investment-oriented customer interactions from his "back-end" (money making) interactions:

Front End	Back End
Conference presentations	Paid admission workshops
Sponsored workshops	Programming tools
Book on OOPS	Consulting services
Shareware	
Hypertext guide to OOPS	
Customer letters	
Press releases and ads	
Distribution channel incentives	

These and other strategies are discussed in this chapter, with examples, benefits and risks noted. We tie each strategy or tactic to the theory presented earlier in the book.

Why all this talk about marketing in a book about management of technology (MOT)? First, MOT must be market-oriented! Great skill in inventing and acquiring technologies will come to nothing if the end result is not a benefit that customers value and pay for. Second, facilitating the adoption of new technologies is a part of MOT...

... when you "champion" tech adoption in your own company.

... when you encourage tech adoption in your customers' firms.

And when you do either of these, you are... marketing!

9.1 A Catalog and Workbook of Strategies

9.1.1 Comfort Factors

We have seen that a new product's compatibility with the customer's sense of self, sense of place, and habits are its most important characteristic. It turns out there are many marketing strategies and tactics that convey a sense of compatibility.

• **"Bridge" products.** Introduce a sequence of "bridge" products to encourage the customer to learn a little at a time. Elsewhere in this book, this is called the "divisibility" principle.

EXAMPLES include "starter sets," prototype models, "crippled" versions of software or microchips (e.g., the 386SX), free-time offers for online services, Monopoly Junior and Scrabble Junior. The many software companies that sell consulting services rather than a software product are using a similar principle to ease the customer's entry into a new way of doing things. To the same end, Power Macintoshes are sold with software that emulates the older 680xx chips, in order to run the software users already have. Consulting companies will sometimes solve a simpler problem than the customer actually has, "using the data we have even if we can't yet get the data we need," in order to raise the client's confidence to a level where the client is willing to obtain the data needed to solve the full problem.

POSSIBLE PITFALLS: One pitfall is that the customer may stick with the toy (starter) version. Or, a bridge product may be so grossly inadequate that customer will turn off. An example is the Microsoft Windows shell called "Bob." Bob's raison d'etre was to get the population of non-computer users to begin using. There is always a tradeoff between ease of use and the time it takes to get a task done. Bob, according to *Infoworld* (1/16/95), goes too far, requiring seventeen mouse clicks to open a file. The *Infoworld* columnist calls this intolerable, "sure to drive users crazy in no time." Furthermore, the shell does not learn the novice user's habitual requests, making Bob "not a remarkable achievement."

BEST LIFE CYCLE STAGE TO USE THIS: Any early stage.

A related strategy is the **"novice vs. advanced level"** (this may be "user configurable"). Its pitfall is that a customer may buy the full product, and still expect lots of support – though you feel the full product is only for "experts."

• Booklets on "All You Need to Know About ..."

EXAMPLES: Banyan Corp. publishes "A.Y.N.T.K.A. LAN Computing." Epson does a similar "A.Y.N.T.K.A. Laser Printers," Microsoft an "NT vs. Windows" primer, and guess-who publishes "Intel Technology."

POSSIBLE PITFALLS: These can have a ring of falsity about them; by presenting the product in an oversimplified way, prospects know they are being condescended to or even lied to.

BEST LIFE CYCLE STAGE TO USE THIS: Early take-off.

• User groups.

EXAMPLES: Early Macintosh user groups interacted with Apple evangelists for efficient dissemination of Mac skills among opinion leaders and among innovators. This works for software as well, viz., SAS user groups. User groups are a way to get direct contact with end users if you market through VARS, and a cheap way to do third-party marketing support, e.g., Apple allows its software partners to talk directly to user groups. Furthermore it is "viral marketing" in that UG members communicate new uses of the product to each other.

POSSIBLE PITFALLS: High word-of-mouth factor means any blunder by your rep in front of a user group or user group member has real consequences. Computer user groups now tend to be less critical of manufacturers than in early days; the manufacturer will learn less than it once might have.

BEST LIFE CYCLE STAGE TO USE THIS: Any stage.

• Embed it as a "feature" in (someone else's) more conventional product. This strategy has great distribution advantages.

EXAMPLES: Arthur Anderson sells third-party A.I. services to its consulting clients. The 3-D data rotation feature in in SAS's JMP and in Datadesk are also written by third parties, as are the linear and nonlinear programming add-ons in Microsoft's Excel.

POSSIBLE PITFALLS: You don't get the recognition you would get by marketing to endusers. There is the possibility of intellectual property disputes. And your revenue per item sold is likely to be relatively small, making this an attractive strategy only if your OEM has truly broad market power.

BEST LIFE CYCLE STAGE TO USE THIS: Any stage, though it is less attractive in later stages.

• **Encourage growth of complementary products** that build the market. (E.g., hardware that runs certain software, and vice versa).

EXAMPLES: Apple certified (third party) software developers, IBM partners, Nintendo game developers. The car rental industry benefits from deregulation of airlines, and would sensibly have lobbied for that deregulation. An extreme example: Sony vertically integrates by buying Columbia house, ensuring the growth of musical content to be played on Sony machines. In a way, information systems like SABRE facilitate air travel, and also turn into profit centers on their own. "Fat software" is also an example, but a marginally ethical one; as software gets bigger, customers must buy faster PCs to run it on, and the new PC will require the purchase of still more software.

POSSIBLE PITFALLS: Consumers get confused by tie-ins. (Can you name all the "air travel partners" of American Airlines, what offers they provide, and why?) There is a hierarchy here of total control (vertical integration) through tie-ins through loose control (eg. lobbying for general industry benefits). Be aware of what you're trying to achieve. Consumers have expressed pique with (to pick just one company engaging in the practice) Microsoft's "fat software" tendencies. It can be seen as blatantly manipulative.

BEST LIFE CYCLE STAGE TO USE THIS: Late growth/maturity.

• **Use homely representations.**

EXAMPLES: Cartoon characters for introducing new tech products, e.g., Japanese animé. Betty Crocker was a comforting figure who convinced us it would not be too hard to make recipes with the company's products. Similarly, MRCA Information Services used a Betty Crocker-like "Doris Reed" as the fictional sympathetic figure who explained the intricacies of filling out and mailing consumer purchase diaries. VideoTelecom (a video conferencing company now known as VTEL) publishes an "Ask Aunt Agnes" newsletter.

POSSIBLE PITFALLS: You must keep the image up to date with social norms. Betty Crocker needed several makeovers during the twentieth century!

BEST LIFE CYCLE STAGE TO USE THIS: Any.

• **Emphasize Compatibility, adherence to standards, and interoperability.** (E.g., "open systems")

EXAMPLES: Machines that read both DAT and analog audio tape. Making Excel 4.0 for Windows compatible with Excel for the Mac. Apple File Exchange (reads PC files), and the "superdrive" that reads special multi-megabyte disks as well as traditional floppies. All "backward compatibility" of software, if intentional, is an example of this strategy, as are IBM emulators (like Virtual PC) for Apple computers. This is an excellent strategy for the number-two company in a category; WordPerfect is well-known for being compatible with all other word processors.

POSSIBLE PITFALLS: Standards are often manipulated; a firm can "declare it's a standard" when it isn't ready to be a standard. There are too many UNIX "standard" versions!

BEST LIFE CYCLE STAGE TO USE THIS: Early maturity.

• **Human factors** in design of control panels, etc. Friendly design of box & package.

EXAMPLES: VideoTelecom includes exemplary control templates on its mediaconferencing machines. The NeXt machine was especially ergonomic. While earlier Macintoshes required special Allen wrenches, the Mac IIci had simple thumb tabs to open the main box. The user interface itself (Mac interface or Windows) is a human-factors development.

POSSIBLE PITFALLS: You can inadvertantly make a machine very easy to use, but hard to understand. The Mac 'Easy Access' (easy keyboard equivalents for people with serious disabilities) crosses up fast typists if they don't know it's turned on. A much-heralded invention was the bathtub with a door; now it could be easy for elderly and disabled people to get in and out of the tub. But nobody liked it because... you couldn't fill up the tub before getting in!

BEST LIFE CYCLE STAGE TO USE THIS: Any.

• **User-accessible audit trails.** These let the user track down what went right or wrong after a given use of the product.

EXAMPLES: Expert systems can tell you not only the best answer to a problem, but how they arrived at the best answer. The UNIX operating system creates error files instantly and automatically when a UNIX computer crashes. All traditional log files and core dumps are examples, even if primitive, of this strategy. Some Xerox machines will self-diagnose, and some PBXs and routers do too.

POSSIBLE PITFALLS: There is added 'overhead' expense in providing these features.

BEST LIFE CYCLE STAGE TO USE THIS: Any.

• **User-Upgradeable.**   If presented as options, these are comfort factors even if they add to the learning burden. User-reconfigurability holds open the possibility that the customer can modify the product for a second application area, without additional payments to the vendor.   This goes along with breaking up the innovation into easily accessible component pieces – "divisibility."

EXAMPLES: PBXs (private telephone branch exchanges) are among the many examples of this strategy in industry.

POSSIBLE PITFALLS: Why can't I easily put a more efficient engine in my old car? The answer is because modular designs, while appealing, may not be the *best* designs.   The next generation car, with a better total design, may obsolete the current modular scheme.   Also, last-minute design changes can endanger compatibility/upgradeability.

BEST LIFE CYCLE STAGE TO USE THIS: Early maturity.

• **Don't overpromise.**   Customers are sensitive to hype, vaporware, etc. "The history of technology shows us we overestimate what a technology can do for us in a few years and underestimate what it can do in a decade or two." (Feigenbaum, McCorduck & Nii.)

EXAMPLES: IBM's OS/2, high temperature superconductors, Windows NT.

POSSIBLE PITFALLS: Loss of credibility.

BEST LIFE CYCLE STAGE TO USE THIS: As this is a "negative" strategy – something *not* to do – don't ever do it!

• The flip side of "don't overpromise" is, **Don't overemphasize limitations** of the technology. In particular, there will be a tension between salespeople and technical support/product managers.   The latter, trying to get customers to understand the product, may be overprecise about the limits of its applicability. Indeed, having had corporate management on their case about bugs and delays in development, techies are overly conscious of what's wrong with the product. They should be encouraged to verbalize what's right about the product; to understand that even with limitations, it will bring significant benefits to the customer; and to sell benefits, not features.

• **Network.** Selling products with high perceived risk depends on trusting personal relationships. If you are an established firm with other, conventional product lines sold to the same targets as your innovative product, rely on "relationship marketing" for selling the new product. If you are a startup firm, you have not had time to build long-term relationships with potential buyers. Find advisors, directors, consultants, and salespeople who have these personal connections. See also "sell yourself" under III below.

EXAMPLES: Autogenesis Corp. introduced an unconventional bone-growth technology from Russia, to U.S. physicians. They employed a U.S. doctor, a trusted opinion leader, as spokesman.

POSSIBLE PITFALLS: Everything hinges on a few relationships Continuum Corp. sells software to the insurance industry by hiring insurance executives to sell to their old colleagues. It is an expensive salesforce, but it's expensive software. A more extreme example is the government revolving door. The perceived unfairness of taxpayers paying the salaries of company "moles" inside government agencies gives rise to the term "beltway bandits."

BEST LIFE CYCLE STAGE TO USE THIS: Early stages are where networks that were established for other purposes can be bent to the new purpose. After the product is mature, a separate network of involved individuals will have solidified.

• **Introductory low price offer.**

EXAMPLE: Credit cards with low introductory interest rates.

POSSIBLE PITFALLS: Customers suspect they're going to get taken to the cleaners after the introductory period is over.

BEST LIFE CYCLE STAGE TO USE THIS: Early growth.

• **30-day price protection.**

EXAMPLES: A number of household appliance sales, consumer electronics sales, and business-to-business contracts offer to refund the difference if a bona fide lower offer is found within a certain time period after the sale.

POSSIBLE PITFALLS: Establishing bona fides can be difficult. Changing business conditions could mean sizeable refunds.

BEST LIFE CYCLE STAGE TO USE THIS: Because it will appeal to more conservative buyers, and presupposes an abundance of competitors, it is most applicable to the late growth stage.

• **Discount coupon** for your related product. When the coupon is for the next-generation product, it can accelerate technology substitution.

EXAMPLE: Radio Shack periodically mails coupons to prior customers who have not made recent purchases.

POSSIBLE PITFALLS: Coupons can attract the most price-sensitive buyers, who do not tend to make the best repeat customers.

BEST LIFE CYCLE STAGE TO USE THIS: Any, but best for getting people to try a new product sooner than they might have without the incentive.

• **Bundling.**

EXAMPLES: Application software written by unknown companies is sold, pre-installed, on all new Dell computers. The unknown companies get exposure, and Dell gets a value-added image while paying very little if anything for the software.

POSSIBLE PITFALLS: Bugs and defects in the other producer's product can exert a halo effect on your product.

BEST LIFE CYCLE STAGE TO USE THIS: late growth, maturity. But note in the Dell example the PC is mature but the bundled software may be an entrepreneurial product.

• **Competitive Trade-In.** Offer rebates on the products your prospects currently use.

ADVANTAGES: This tactic reduces your customers' cost of switching. A rebate offer can be launched on short notice, unlike major ad campaigns.

EXAMPLES: Software AG offered rebates up to $100,000 when users bought ADABAS D and turned in their Ingres package.

POSSIBLE PITFALLS: The rebate eats your profit on the initial sale; you must have a plan in place for recovering profits on the after-sale support. Existing customers may complain that new buyers got a deal that prior buyers did not get.

BEST LIFE CYCLE STAGE TO USE THIS: This tactic can be effective when a competitor is experiencing difficulties of some sort. This can happen most frequently in the early maturation/ shakeout LC stage, but of course can happen at any time.

9.1.2 Picking Your Customers

Look for...

• **Industries with high return/risk ratios** as initial target markets

EXAMPLES: Convex sold its first minisupercomputers to a semiconductor chip design firm and the geophysical research department of a petroleum company. "GAP"/entertainment products (see chapter 10) are high-return and low-risk; an example was VCR Plus+, a product that enabled people to program their video recorders using bar codes published in the weekly newspaper.

POSSIBLE PITFALLS: These markets are ultimately limiting, and you must be ready to shift your sales model to accommodate a broader market.

BEST LIFE CYCLE STAGE TO USE THIS: Early growth.

• **A first customer whose name will cut some ice** with the second customer.

EXAMPLES: Tandem sold its first fault-tolerant computer systems to Citicorp.

POSSIBLE PITFALLS: A sales executive from Novell believes the contrary is true. He says, "Target the Inc.-100, not the Fortune 500. The latter are well-known, but not innovators. The Inc. 100 are more open to new things, if you connect with them, you can share in their growth." Not to mention that getting 100 reference calls from your next 100 prospects can turn off your first customer.

BEST LIFE CYCLE STAGE TO USE THIS: Introductory.

• **Well-demarcated application areas with easily quantifiable payoffs.** Often, this means an internal application in the customer company, which the customer may see as more quantifiable, less risky, and controllable than applications involving suppliers or customers.

EXAMPLES: CTC, a "decision room software" company, makes Dell's internal meetings more effective. The Macintosh was first marketed heavily for DTP (desktop publishing) applications. Lotus specialized first in spreadsheets, which every company needs for accounting purposes even if not for additional reasons.

POSSIBLE PITFALLS: You may be labeled a niche company when that was not your intention.

BEST LIFE CYCLE STAGE TO USE THIS: Early. Lotus later was able to diversify into groupware applications.

• A champion within the customer company.

EXAMPLE: A research manager within the ad agency BBDO pushed to have the agency adopt the first linear programming system for choosing media placements.

POSSIBLE PITFALLS: The champion may leave the company before the project's completion. And even if not, the champion is playing "bet your job" and can get cold feet at any time.

BEST LIFE CYCLE STAGE TO USE THIS: Introduction, early growth. However, it can be used also in the maturity stage. For example, SAP or Peoplesoft can usually find an I.T. manager in a company who would love to have "involved in an ERP (enterprise resource planning, a generic name for products like SAP's) installation" on his or her resume.

• A customer company whose CEO has the vision needed to see the importance and the payoff of new technologies.

EXAMPLES: Shell and Schlumberger were the first oil companies to move into PCs, due to vision at the top. In 1990, AT&T put managers into field with laptop computers, for the same reason. General McDermott at USAA first computerized the insurance industry. Paul Allaire at Xerox was similarly credited with many visionary innovations.

POSSIBLE PITFALLS: Resistance at lower levels in the company.

BEST LIFE CYCLE STAGE TO USE THIS: Introduction, early growth.

• Innovators who are opinion leaders. Although members of the "innovator" market segment are not generally communicative or influential individuals, there are exceptions. Find innovators who are opinion leaders. They may be magazine editors, columnists, user group presidents, etc.

EXAMPLES: Malcolm Forbes was said to use all the latest gadgets personally. Today, there are many opportunities for "Lego entrepreneurship," that is, buying a block from here and a block from there and making a core competency of knowing where and how to snap the blocks together. As entrepreneurship becomes not only accepted but admired in many communities, Lego entrepreneurs become opinion leaders as well as innovators, and avid customers for any technology product that can make their enterprise more successful.

POSSIBLE PITFALLS: They usually don't have a lot of money.

BEST LIFE CYCLE STAGE TO USE THIS: Introduction, early growth.

• **"Hooks."**   Make your machine connectable to other machines, and your software an inviting platform or component for other software.

EXAMPLES: The Palm Pilot personal digital assistant provides wired and infrared connection to desktop computers. Microsoft allows users to write Excel macros in Visual Basic.

POSSIBLE PITFALLS: The machines your product is connectable to may change their configurations in ways you have not foreseen, resulting in a failure in your product's function that disappoints the customer.

BEST LIFE CYCLE STAGE TO USE THIS: Introduction, early growth.

• **Showcase it in science museums and theme parks.** Here is where you will catch educated and/or affluent people in a receptive, curious, relaxed frame of mind. A related strategy is "product placement" in motion pictures.

EXAMPLES: Monsanto's "house of the future" at Disneyland.

POSSIBLE PITFALLS: My house in 1999 was nothing like what the 1964 "house of the future" led me to expect it might be. So one pitfall here is conspicuously wrong predictions. It can also be an expensive strategy.

BEST LIFE CYCLE STAGE TO USE THIS: The earliest innovators do not look for ideas in theme parks, so the best stages to use this strategy are early/late growth.

9.1.3 Marketing Mix and Positioning

• **Alliances / marketing agreements.**

EXAMPLES: VideoTelecom (VTEL)'s alliance with Intel. EASI, a company developing expert system software to help manufacturers meet Underwriters' Laboratories standards, allied with U.L. The decision room software company CTC allied with Lotus in order to better integrate Lotus Notes into its product.

POSSIBLE PITFALLS: A partner can go out of business, and a partner a lot bigger than you can beat you up.

BEST LIFE CYCLE STAGE TO USE THIS: Introduction, early growth. However, mature products can benefit from sales alliances (like Fuji-Xerox) for purposes of gaining access to overseas markets.

• **Make it cheap.** Make it so cheap it can be sold like candy. A powerful barrier to entry. Also great for your reputation as an innovator.

EXAMPLES: Freeware software. Sell Nintendos (Kodaks, razors, etc.) cheap, and make money on cartridges, film, and blades.

POSSIBLE PITFALLS: If you sell too cheap, no cash is generated to fund your next-generation product. Investors won't see an attractive profit margin, and perhaps will not invest in your company.

BEST LIFE CYCLE STAGE TO USE THIS: Introduction.

• Consider yourself in the technology education business. Give seminars, write books, feel out the market, and only then introduce a product.

EXAMPLE: Texas Instruments' artificial intelligence business started in this way.

POSSIBLE PITFALLS: Competitors will be "free riding" on your training costs.

BEST LIFE CYCLE STAGE TO USE THIS: Earliest.

• **Sell yourself.** "If any technology significantly complex cannot be distinguished from magic, and if to appreciate magic you must have faith that the performer can do wondrous things, then to have your audience appreciate [your] technology you must first develop faith, not understanding. Sell yourself. A CEO may not fully grasp the technology, but is likely a good judge of character." (Unfortunately I don't know the source of this wise advice, but the part about sufficiently complex technology being indistinguishable from magic is originally due to Arthur C. Clarke.)

EXAMPLE: FMC Corp.'s move into A.I. applications used this strategy – an appropriate one for selling an "invisible" like artificial intelligence.

POSSIBLE PITFALLS: One's personal reputation is at risk and at the mercy of possibly uncontrollable factors. Also, the customer may reason that an outstanding person like you may be in demand at other companies and may not stay with this vendor long enough to see the project through.

BEST LIFE CYCLE STAGE TO USE THIS: Early growth.

• Take advantage of pre-existing **conditions of change or shake-up**. If other forces have caused your prospect to move into restructuring mode, they may be more receptive to advanced technology products.

EXAMPLES: DEC's Exsel expert system was widely accepted only after changes in DEC's product line made it impossible for employees to keep up with pace of change without using the system. Map Resources, Inc., sold its GIS products into county governments, riding in on Texas redistricting court cases. Fusion Corp.'s accelerator boards rode in on Apple's shift to 68040 chip.

POSSIBLE PITFALLS: You may wait a long time for such a chance.

BEST LIFE CYCLE STAGE TO USE THIS: Introduction, then again in late growth, maturity.

• Remember **the customer company is not a monolith**. There will be champions and there will be enemies. Innovation diffuses through the customer company in the same way it diffuses elsewhere, i.e., not instantaneously. Your product may be well-configured to meet the needs of nine customer projects; the tenth, poorly served project may be headed by an enemy.

EXAMPLES: DEC's Xcon system was an awkward fit to one special factory. This factory happened to be headed by a vocal sceptic. It would not have been wise to conclude that the product was a bad one.

POSSIBLE PITFALLS: Be careful you don't tell conflicting stories to two employees of the same prospective customer.

BEST LIFE CYCLE STAGE TO USE THIS: Any time!

• Consider **sequential media** as an early part of the communications mix. Videotapes and audiotapes must be played in sequence, ensuring that your rather complex message is heard in total. Tapes are rarely thrown away, and will probably be passed to another potential user. These educational sources should be followed by random access media (disks, books) that can be used as reference sources.

EXAMPLE: The product introduction plan for the Macintosh used print ads for conveying information, and later, TV to generate emotional excitement.

POSSIBLE PITFALLS: Print and tape may be seen as old-fashioned.

BEST LIFE CYCLE STAGE TO USE THIS: Early.

• **Newsletters!**

EXAMPLES: The newsletter efforts of MRCA and VideoTelecom were noted above. they provide useful product use advice in palatable, bite-size chunks.

POSSIBLE PITFALLS: You lock yourself into a production schedule and a future cost. It can be hard to find filler for newsletters (these are usually recipes, child-raising tips, etc.) that are not bland and boring.

BEST LIFE CYCLE STAGE TO USE THIS: Any time.

• **Open your product** to 3rd-party add-on developers.

EXAMPLES: Lotus, Corel, IBM and many other hardware and software makers do this.

POSSIBLE PITFALLS: IBM PC suppliers encroached on what IBM considered its own product territory.   Indeed, the growth of the clone market, while it entrenched the Wintel standard, sometimes cut into IBM's own sales.

BEST LIFE CYCLE STAGE TO USE THIS: Earliest, as it does help establish your product as a standard.

• **Service bureau.**

EXAMPLES: DTM Corp. sells rapid prototyping machines that use the selective laser sintering process.   But the machines are expensive, and DTM (which originally stood for "desktop manufacturing") derives significant income from generating prototypes for customers who email CAD files to DTM.   MRCA originally intended to have its clients use its Dyana interactive market research analysis tool, but clients preferred to send MRCA the parameters of an analysis and let MRCA run the software.   Map Resources, Inc., sells Geographic Information Systems (GIS) but also generates custom maps directly for clients.

POSSIBLE PITFALLS: You must staff and maintain two organizations, one for hardware/software sales and service, and one for the service bureau.

BEST LIFE CYCLE STAGE TO USE THIS: Early/late growth, maturity.

• **Evangelism = Vision + Networking**

EXAMPLES: Apple Computer and the IC^2 Institute of the University of Texas at Austin (see Chapter 1's appendix) have been exemplary users, indeed pioneers, of evangelism.   Evangelism combines vision and networking to raise business development to a high level of user enthusiasm.

POSSIBLE PITFALLS: It can be tough to maintain high levels of excitement for long periods of time.

BEST LIFE CYCLE STAGE TO USE THIS: Early growth.

• **Targeting the luxury market** allows high mark-ups and may allow you to cash in on the cachet of "unmodified" foreign products.

EXAMPLES: Mercedes Benz sells left-hand-drive cars in Japan.   Franklin Speller, intended for the U.S. market, is unexpectedly a big seller in Japan, where it helps learners of English with their spelling. Sharper Image successfully sells high-end gadgets worldwide.

POSSIBLE PITFALLS: Your sales can be subject to faddish ups and downs. Regulations in foreign countries may complicate your export plans.

BEST LIFE CYCLE STAGE TO USE THIS: Late growth or maturity.

• The "**Killer App**"

EXAMPLES: Lotus 1-2-3 was the "killer app" for the IBM PC, and desktop publishing was the killer app for the Mac. Home photography is the killer app for the home scanner.

POSSIBLE PITFALLS: There will be cutthroat competition.

BEST LIFE CYCLE STAGE TO USE THIS: Since "killer app" means a mass-market application, the strategy only has meaning in the fast-growth stage of the life cycle.

• **Support groups**

EXAMPLE: Weight Watchers organizes support groups to help its customers understand and stick to the Weight Watchers diets.

POSSIBLE PITFALLS: Don't confuse this strategy with user groups! See above for a discussion of user groups.

BEST LIFE CYCLE STAGE TO USE THIS: All.

• "**Trojan Horse**"

EXAMPLES: In the 1960s, IBM sold two models of card sorter, at quite a price differential. The cheaper model was "upgradeable" at a steep price; it turned out that "upgrading" meant a technician would open the sorter's backpanel and flip a switch! Nintendo, under the guise of friendly games, is inserting 32-bit computers into households who may own no other computers.

POSSIBLE PITFALLS: It can be expensive to plant the extra, inactive technology in each unit. How will Nintendo leverage its Trojan horse? It will face competition from rapidly advancing Internet appliances and from desktop PCs that are now nearly free or totally free.

BEST LIFE CYCLE STAGE TO USE THIS: Late growth through maturity, as in these stages, conservative customers do not want a full range of features, but you as producer already know what some of these next-generation features will be.

• **Academic papers**/conferences can bestow credibility on an advanced technology product.

EXAMPLE: In 1985, the American Society of Orthopedists wouldn't hear a paper on the Fitzarov technique from Russia; in 1995 there were sixteen papers. Autogenesis, the bone regrowth company both instigated this and benefited from it.

POSSIBLE PITFALLS: It takes great finesse to publish academic papers and protect intellectual property at the same time. AT&T earned the disfavor of academics when it attempted to do this with some newer interior point linear programming algorithms in the 1980s. There were not enough details in the highly publicized papers to qualify as real scholarly contributions. However, the airlines have now widely adopted these LP methods for flight scheduling, and they obtained the algorithms not from academic journals but by license from AT&T.

BEST LIFE CYCLE STAGE TO USE THIS: Introduction and early growth.

• **Look ahead to the next life cycle phase.**

EXAMPLES: Dell Computer looked, and saw that customers would be better serviced by call centers than by specialty retailers. Packard Bell looked farther. They saw that as the product became even more familiar, the best distribution would be through mass and discount retailers.

POSSIBLE PITFALLS: None that I know of; foresight is always good.

BEST LIFE CYCLE STAGE TO USE THIS: Any time!

• Sell a product whose **target market is people like yourself**.

EXAMPLES: Btrieve, an Austin company spun off from Novell, produces a database product that is written by programmers for programmers. Hewlett-Packard, a company started by engineers, initially made products for other engineers.

POSSIBLE PITFALLS: Eventually, shareholder pressure may force you to broaden your target market, and this can be traumatic to your managers and employees. This was certainly the case for H-P.

BEST LIFE CYCLE STAGE TO USE THIS: Early, before product differentiation takes your product out of its initial niche.

9.1.4 The Value of the Brand

After high technology becomes familiar and ordinary technology, branding à la Procter and Gamble becomes necessary. According to consultant Chuck Pettis, a brand is "the sensory, emotive and cultural proprietary image surrounding a company or product...an enhancement of perceived value and satisfactions through associations that remind and entice customers to use the product." Despite the "soft" tenor of this definition, a successful brand is a significant part of a company's goodwill, that is, the excess of market value over the accounting value of assets. The brand is personality and positioning. A customer may associate a company more closely with its brand names than with any particular product made by that company. The brand is therefore the customer's assurance that the company stands behind its product.

Research, positioning, advertising, packaging, public relations, and electronic media are all useful for building a brand. We noted in Chapter 2 that shorter product life cycles increase the relative importance of brand marketing over product marketing. However, some brand-building strategies that are cogent for products in the mature phase are inappropriate or baffling when applied to technology products in the startup phases.

Apple attracted early adopters by positioning its brand as playful and non-corporate. In the best branding tradition, Apple put the dollars behind the image to emphasize its seriousness about living up to the image, and then lived up to it. But the image later hurt the company when it was time to transition its targeting to corporate buyers. In general, startup companies cannot have a credible brand. Branding for the early life cycle or for truly innovative products is risky or impossible. Hewlett-Packard has brand identity for the company, which enables HP to participate in mature markets while retaining its image as an innovator. In this way, "branded" does not always equal "mature" or "boring." But backing a revolutionary product with the HP brand is even more risky for HP than it is for their customers. However, even for startup entrepreneurs, it's not too soon to start thinking about branding. One possible tactic for entrepreneurs is...

• Buy a brand name.

EXAMPLE: A startup PC maker bought the Packard Bell name from its last owner, Teledyne. The brand name, well known to American households, furthered the company's strategy of making computers for the home market.

POSSIBLE PITFALLS: See "best life cycle stage" below.

BEST LIFE CYCLE STAGE TO USE THIS: Any. But Packard Bell misjudged its timing, in that the Packard Bell brand name had not been widely familiar since the 1950s. So the people most likely to remember it were now of an age that made them late adopters, and the PC market had not, at the time of Packard Bell's 1980s startup, matured to the point where late adopters were buying.

9.2 Appendix: Technology Marketing in Japan - A Bibliography[1]

9.2.1 Introduction

It has been observed (*inter alia* in many of the articles cited below) that marketing is not a distinct corporate function in Japanese firms. That is, few Japanese firms have a "Marketing Department." Yet Japanese technology products – VCRs, tractors, autos, Walkmans, etc. – have been successful in global markets, and this did not happen without marketing effort. To the extent that having a Marketing Department can imply that those employed in other departments (finance, manufacturing, etc.) need not be concerned with the customer, Japanese firms have succeeded in orienting all employees to marketing – thus creating some of the world's most thoroughly customer-driven organizations.

Japan has few business schools of the American/European type. Hence, marketing is not taught as a distinct discipline in Japanese universities; marketing has not been organized as a functional department in Japanese firms; and little has been written in Japan about marketing per se – whether technology marketing or other varieties. Our own inquiries to many Japanese contacts resulted in few references in technology marketing either in Japanese or in English.

Such material as there is usually must be culled from articles on product development, innovation, technology management, trade policy, or economic competitiveness. It is probably true that U.S. business suffers from overly emphasizing the distinction between marketing on the one hand and product development and the other activities just mentioned, on the other. Nonetheless we believe it will be of value to researchers and managers to supplement the many extant bibliographies on Japanese innovation, R&D policy, etc., by compiling a list of references that focus specifically on technology marketing as it is understood in the West.

9.2.2 Contents of the Bibliography

While it is not our intention to offer a definition of technology marketing, this list focuses on publications dealing with marketing strategy, selling, and marketing

[1]   I gratefully acknowledge the assistance of Pattie Roe and Kumi Valenty in the research for this section. The writing of this section was partially supported by AFOSR grant #F49620-92-J-0522.

mix decisions[2] (pricing, distribution channel choice, customer support, advertising, etc.). It excludes (while recognizing their relevance to marketing) works on R&D, standards, patent/copyright issues, or trade barriers resulting from national policies or bilateral government negotiations. The marketing mix decisions of U.S. firms selling technology products in Japan are represented in these articles. The U.S. firms' efforts to lobby the U.S. government for negotiated market shares, to negotiate product standards within international bodies such as ISO and ANSI, or to use patent strategy to bolster revenues, are not included. The list includes articles written by Americans as well as works by Japanese. Some general (non-technology oriented) works on marketing to or by the Japanese are included.

The works cited here are classified according to whether they deal with Japanese firms marketing to Japanese customers; Japanese firms marketing to U.S. and international customers; and U.S. firms marketing to Japanese customers. For some works, the classification is ambiguous, and some publications appear in more than one category.

9.2.3 Brief Summary of the Cited Works

Much of the marketing of Japanese products that is visible to Americans is really American marketing, in the sense that Japanese manufacturers will set up or acquire a U.S. firm to market their products in the U.S. Toyota Motor Sales U.S.A. is an example. It is staffed mainly by Americans, who hire American advertising agencies and other marketing support services. The same is true for "American" marketing in Japan, where, say, Advanced Micro Devices staffs a Tokyo marketing office with Japanese executives who market to Japanese customers in the Japanese way.

Understanding the *keiretsu* organizations is a necessary precursor to studying technology marketing in the Japanese setting. Yoshinari's (1992) article is a good overview. Johansson and Nonaka (1987) describe how market research and product planning are effected by Japanese firms under vertical integration of business functions and keiretsu member companies. Johansson and Nonaka describe the international focus of these firms and their emphasis on studying consumer needs and wants. The MITI document "Information-Oriented Lifestyle" is recommended as an illustration of how social planning and market research are integrated in Japan.

Lazer, Murata and Kosaka, in an overview of Japanese marketing that spans both high-tech and low-tech product areas, discuss such matters as price supports and

---

[2]   These decisions are characterized by the "P-words": positioning, price, promotion, placement, packaging, and persuasion.

other monopolistic practices that are common in Japan but illegal in the U.S.; the way small stores control the location and business practices of larger stores (a Japanese WalMart would not be able to drive small town mom-and-pop stores out of business); and marketing as consensus building.

In U.S. technology markets, innovators and early adopters buy at specialty retailers, where sales staff are highly knowledgeable. Later stages in the life cycle see sales through mail/phone order, with service provided through centralized call centers. At a still later, mature stage, mass retailers and discounters dominate, offering little or no service. Japanese producers dominate retail channels from the beginning of the life cycle. Through market segmentation (production of many specialized products rather than one many-function device) and by careful attention to design and ergonomic issues, they attempt to reach the mass market sooner than American producers who operate under the classical life cycle model. Thornton (1994) writes about new opportunities for Americans in Japanese retail channels.

Kokubo (1993) addresses "core technologies" as a driver of marketing strategy. It is well known that while U.S. companies have emphasized current profits, Japanese companies have sacrificed current returns in order to build market share. The market share ensures a future flow of profits. Now, however, Japanese firms are focusing on core technologies. These strategic competencies will ensure future flows of market share in the wide varieties of industries to which the core technologies are applicable.

Akio Morita (1992), co-founder of Sony, discusses Japan's image in its customer countries. This important discussion raises the possibility of Japanese products' negative brand equity stemming from non-Japanese consumer's resentment of Japan's domination of certain technology markets. Morgan and Morgan (1992) point to product quality as the key to selling American products to Japanese customers. Finally, Bowonder and Miyake (1992) present a good integrative view of how technology marketing fits into the innovation strategies of Japanese firms.

9.2.4 The Bibliography

9.2.4.1 Japanese Selling to the Japanese Market

Yoshinari M (1992) The Big Six Horizontal Keiretsu. Japan Quarterly April-June

Johansson J and I Nonaka (1987) Market Research the Japanese Way. Harvard Business Review May-June 65:16-22

Kotabe M and D Duhan (1991) The Perceived Veracity of PIMS Strategy Principles in japan: An Empirical Inquiry. Journal of Marketing January 55:26-41

Lazer W, Murata S and H Kosaka (1985) Japanese Marketing: Towards a Better Understanding. Journal of Marketing Spring 49:69-81

Turpin D (1992) Strategic Persistence of the Japanese Firm. The Journal of Business Strategy Jan-Feb

Kokubo A (1993) Core Technology Based Management: The Next Japanese Challenge. Prism / First Quarter, Arthur D Little, Boston MA 13-21.

Morita A (1992) Why Japan Must Change. Fortune March 9

Bowonder B (1992) A model of corporate innovation management: some recent high tech innovations in Japan. R&D Management 22(4)

Deshpande R, Farley J and F Webster Corporate Culture, Customer Orientation and Innovativeness in Japanese Firms: A Quadrad Analysis

Parry M and X Song Determinants of R&D-Marketing Integration in High Tech Japanese Firms

Cutts R (1992) Capitalism in Japan:Cartels and Keiretsu. Harvard Business Review July-Aug

Hamel G and C Prahalad (1991) Corporate Imagination and Expeditionary Marketing. Harvard Business Review July- Aug

Yamamoto H (1992) Technology Management in Japan. SRI International Tokyo

Kotabe M and A Lanctot (1993) The Market Orientation in Practice: A Comparison of US and Japanese Firms. University of Texas at Austin

(1993) The Information- Oriented Lifestyle/The results of a discussion between the leaders of technology in Japan held on Jan. 20

Journal of International Economics (1990) Pricing to Market in Japanese Manufacturing. November 217(20)

9.2.4.2 Japanese Selling to the American Market

Johansson J and I Nonaka (1987) Market research the Japanese way. Harvard Business Review May-June

Turpin D (1992) Strategic Persistence of the Japanese Firm. The Journal of Business Strategy Jan-Feb 49-52

Morita A (1992) Why Japan Must Change. Fortune March 9

Morita A (1986) Selling to the World. Made in Japan 74-129

Jatusripitak S, L Fahey and P Kotler (1985) Strategic Global Marketing: Lessons from the Japanese. Columbia Journal of World Business Spring 20:47-53

9.2.4.3 Americans Selling to the Japanese Market

Alden V (1987) Who Says You Can't Crack Japanese Markets? Harvard Business Review January-February 65:52-56

Ohmae K (1982) Breaking into Japan's Distribution System. Wall Street Journal June 28

Morgan J and J Morgan (1992) How Americans Can Succeed in Japan. Engineering Management Review Winter

Montgomery D (1991) Understanding the Japanese as customers, competitors, and collaborators. Japan and the World Economy 3(1)

American Electronics Association (1994) Soft landing in Japan - A market entry handbook for US software companies

March R (1991) Honoring the Customer: Marketing and Selling to the Japanese. New York: John Wiley & Sons Inc

Nakamura S (1992) Selling in Japan: Tips on Access And Distribution. Journal of Japanese Trade & Industry 4

Cutts R (1992) Capitalism in Japan: Cartels and Keiretsu. Harvard Business Review July-Aug

Simon H (1986) Market entry in Japan: Barriers, problems and strategies. International Journal of Research in Marketing Oct

Webber A (1992) Japanese-Style Entrepreneurship: An Interview with Softbank's CEO, Masayoshi Son. Harvard Business Review Jan-Feb

Thornton E (1994) Revolution in Japanese Retailing. Fortune (Feb. 7) 143-146

Harvard Business School Case Motorola's Japan Strategy. 9-387-093

9.3 Notes

Branding... Pettis C (1995) TechnoBrands: How to Create & Use "Brand Identity" to Market, Advertise & Sell Technology Products. American Management Association New York; See also Hatelstad L (2000) Briefing on Branding. Red Herring January 74:172-242. The latter is a comprehensive treatment of trends, successes and failures in high technology branding.

Aaker D A (1991) Managing Brand Equity: Capitalizing on the Value of a Brand Name. New York The Free Press

Aaker D A and A L Biel (Eds) (1993) Brand Equity and Advertising. Lawrence Erlbaum Associates Hillsdale NY

9.4 Questions and Problems

Short Answer

S9-1. Note additional examples of companies that have followed some of the strategies listed above. Your teacher will tell you how many are required.

S9-2. Add more strategies to this chapter's catalog. Support each with examples, and note its pitfalls and the best life cycle stage for utilizing it. The instructor will tell you or your team how many additional strategies you are to discover.

• STRATEGY:

EXAMPLES:

-

-

POSSIBLE PITFALLS

-

-

BEST LIFE CYCLE STAGE TO USE THIS:

-

Mini-project

M9-1. *Turning an engineering spec into a marketing pitch.* In this exercise, you will work with either (i) a product specification document, or (ii) a heavily technical advertisement. If you have brought this document to class, your instructor must approve it; otherwise, work with a document provided by your instructor. Your assignment is to pretend this product was created by you and your company, and to:

(a) Identify a feasible target market. This part of M9-1 is not a research task – you may invent a hypothetical target market, or use the target market identified in the product spec, if there is one.

(b) Change every *feature-oriented* point in the document (we will assume the targeted customer doesn't care about these) into a *benefit-oriented* point that addresses a real need of the customer.

(c) As a class presentation, pitch the product. Your classmates will pretend to be consumers or employees of a company in the target market. Are you talking about *them* and *their needs,* and not about *you* and your company's *product* and its *features*? If you focus on features, specs, or cool technology, count on your classmates to loudly inform you that you are losing their interest.

(d) Listen to the critique offered by your instructor and classmates. Return to part (b) and revise your presentation, attending to the principles of this chapter, then present again.

(e) Listen to a classmate's presentation built on his/her assigned document, and offer an honest critique.

The purpose of this mini-project is to learn to control your pride of creation and fascination with technology, and to learn to focus an equal amount of passion on your customers' need to do their job, and on your opportunity to help them.

PART V

INTO THE FUTURE

"The nice thing about standards is that there are so many to choose from."
Andy Tannenbaum

"Things are the way they are because they got that way."
Kenneth Boulding

10 Escaping the Niche Market; Moving to the Mass Market

10.1 Niche Marketing vs. Mass Marketing

Let's take time out to discuss the famous VCR incident. An early U.S. producer (Ampex) had dominated the industrial market for videocassette recorders. This market included TV networks, movie studios and professional producers of video programming. Ampex did not assault the mass market of VCRs for consumer households. Japanese companies did, with well-known results – namely that there are no longer any U.S. producers of household VCRs.

Up to this point, this book has not addressed making the transition to the mass market. Ampex targeted the professional market segment and did well in it. *Within* the professional segment, Ampex encountered innovators, early adopters, late adopters, and all the rest. We wish to make a clear distinction between success in a targeted niche market and protecting oneself by developing a mass market. Both are important.

By opening a consumer market for VCRs, Japanese competitors profited handsomely, funding R&D that put technological leadership in VCRs in Japan. This in turn had an impact on the professional segment, harming U.S. firms producing for that segment. By failing to push products for the mass market, Ampex lost technological and market leadership. There is certainly a lesson there; the transition to mass market is ultimately relevant to almost all technology products. Development of mass markets for biofeedback equipment and for expert system shells, to take two examples, are underway but still problematic.

Note that "mass marketing" is still a reality, despite all the talk of flexibility and mass customization. Aside from a few VCRs sold in designer colors, these machines are sold mostly in silvertone or black, with a fixed set of features, differentiated only by the number of record heads and the length of the timer calendar.

Strategies for developing mass markets for technology products include:

- *low, low price.*

- *standards.*

- *GAP* (games, advertising & pornography).

- *the Trojan horse.*

10.2 Standardization Strategies

The first two are related; the standard audiotape cassette, for example, enabled low-cost manufacturing, thus low price and wide adoption of cassette recorders and players. The earlier proliferation of reel-to-reel tape sizes had kept player prices high.

A dictionary might say that a *standard* is something that is accepted as a basis for comparison; a criterion by which something can be judged. For example, the American National Standard for Calibrations, ANSI/NCSL Z540-1, is a standard specific to calibration laboratories and measuring and test equipment. ANSI's definition of a national (measurement) standard is "a standard recognized by an official national decision to serve, in a country, as the basis for fixing the value of all other standards of the quantity concerned."

The International Standards Organization (ISO)'s definition of a *standard*: "A technical specification or other document available to the public, drawn up with the cooperation and consensus or general approval of all parties affected by it, based upon the consolidated results of science, technology, and experience, aimed at the promotion of optimum community benefits and approved by a standardization body."

"*Standard:* A document, or an object for physical comparison, for defining product characteristics, products, or processes, prepared by a consensus of a properly constituted group of those substantially affected and having the qualifications to prepare the standard for use." (1974 Concise Chemical and Technical Dictionary. Chemical Publishing Co. Inc. New York)

"A *standard* is a guideline documentation that reflects agreements on products, practices, or operations by nationally or internationally recognized industrial, professional, trade associations or governmental bodies." (From *www.its.bldrdoc.gov.* Note: This concept applies to formal, approved standards as contrasted to de facto standards and proprietary standards.)

Worldwide Standards and Trade: The Players and Their Networks

by Herbert Phillips, P.E. (Used with permission)

Figure 10.1 Acronyms of the U.S. and Canadian free trade agreements

An overview of the players in the global standards arena and their interactions that affect worldwide trade can best be obtained by addressing this array of acronyms.

U.S. National

Starting first at home in the U.S., ANSI, the American National Standards Institute, is the federation of standards development organizations. ANSI represents both the private sector and government in the approval of standards as American National Standards. The system in this country is unique because standards development is not under government control and is a truly voluntary system. ANSI's members are:

a) standards developers
b) manufacturers
c) individuals
d) government agencies

The U.S. system, because it is voluntary, places ANSI as the U.S. member (the U.S. body) of the international standards organizations, e.g., IEC and ISO, which will be discussed later in this article. Most countries' members in these international organizations are the governments of the countries themselves. As an example, the government of country X is the member body and that government pays the dues for that country as the member body in these international organizations. The U.S. is the exception; ANSI must pay the dues for the U.S. and currently it is in excess of a million dollars each year.

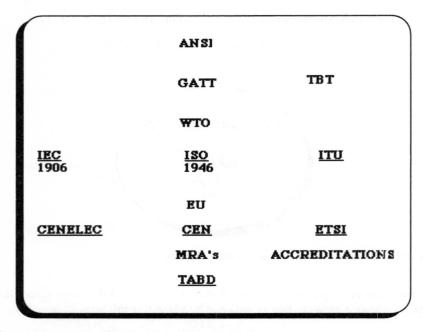

Figure 10.2 International trade and standards organizations

ANSI's role in the U.S. voluntary standards system is to be certain that there is openness in the standards development process and that standards are approved through a consensus procedure involving all interested parties. Only after these criteria are satisfied will ANSI permit a standard to become an American National Standard.

International

Expanding now to the international scene and international trade, there is an agreement called GATT, the General Agreement on Tariff and Trade, which among many other regulations, states that there shall be no technical barriers to trade (TBT). Specifically, GATT states that standards shall not be a TBT and the best approach to accomplish this is through the harmonization of the various national standards, or better, through the development of international standards.

The World Trade Organization (WTO) has been set up to carry out the GATT code. One of the WTO's functions is to hear and settle trade complaints between nations. Business reports in U.S. newspapers and magazines frequently report on these WTO actions. The European Union recently broke international trade rules by improperly changing the import status of computer network equipment, more than doubling the duty on the gear, the World Trade Organization ruled. But WTO rejected a U.S. complaint about tariffs on multimedia personal computers.

International standards are developed through three major international standards organizations.

1. The IEC (International Electrotechnical Commission) develops electrical and electronics standards. The IEC was founded in 1906 to develop international standards promoting quality, performance, safety, and reproducibility of materials, products and systems.

Its member bodies include 51 countries, and there are 87 technical committees and 93 subcommittees developing IEC standards.

In October, 1998 the IEC held its annual General Meeting in Houston, Texas – the first in the U.S. since 1970. This is rather shameful for the U.S. because other countries, including small countries and others with much smaller financial resources have hosted and continue to host these meetings each year. The uniqueness of the U.S. voluntary standards system penalizes the U.S. in this regard because meetings in the other host countries are usually fully financed by their governments.

A group under the auspices of the U.S. National Committee of the IEC, an arm of ANSI, privately raised well over a million dollars to bring the 1998 IEC General Meeting to the U.S., and this group meticulously developed plans for this meeting for over three years.

The meeting in Houston was managed for ANSI by ISA, the Instrument Society of America, in conjunction with an ISA meeting. The combined meetings was a major activity for the city of Houston.

2. The ISO (International Standards Organizations) was established in 1946. Its member bodies include 85 countries, 28 corresponding members, 185 technical committees, 600 subcommittees and 2000 working groups. ISO develops all standards not included in the scope of IEC and ITU.

Obviously the ISO is a larger organization than the IEC and, by necessity, more loosely organized. Meetings are under the administration of the chairman and secretaries of the committees rather than through the Central Office. (The IEC provides Central Office administration, has a representative at most technical committee and subcommittee meetings and provides services to these committees and their chairmen and secretaries.)

The ISO and IEC are both headquartered in the same building in Geneva, Switzerland, and until recently operated separately and independent of each other. This has recently begun to change, bringing about some economies. Ultimately the two organizations will be combined.

3.   The ITU (International Telecommunications Union) deals strictly with standards dealing with telecommunications and operates in a fashion similar to IEC and ISO.

Regional

Regional standards development groups have emerged, mainly as a result of the formations of the EU, the European Union. The EU has three such standards-developing organizations, each a mirror image of the IEC, ISO and ITU.

1. CENELEC is the European Committee for Electrical and Electronic Norms. It develops the same types of standards and is patterned to the same functions as the IEC.

2. CEN, the European Committee for Norms, is patterned to the functions of the ISO.

3. ETSI, the European Telecommunications Institute, is patterned to the functions of the ITU.

The European community wanted to be the master of its own standards. It believed that the international standards organizations were not developing standards rapidly enough to suit its needs and of course knew that it could not alone control the development of standards in the international arena. All these reasons to develop standards for the European community by the European community are justifications for the development of all regional and national standards. The goal for the European Union is to do away with the national standards of each European country once European standards through CENELEC, CEN and ETSI are completed.

Until the formation of the European Union, most U.S. industries put international standards on the back burner. International trade, for too many in the U.S., provided at the time only a minor source of business, and visions of expanding worldwide trade were retarded. For the most part, U.S. industry wanted to put its efforts into U.S. national standards and only watch what was happening in the international standards development arena to protect U.S. trade interests and spend as little as possible to do this.

The European Union or the European Community (EC), as it was known then, was developing standards at a respectable rate. It appeared that these European standards would be de facto international standards, without the participation of other countries – particularly, the U.S.

U.S. industry began to wake up and decided that this had to be resolved through serious participation in and the rapid development of international standards.

Today, most U.S. industries are providing funding for this participation, and U.S. participation is high. Today the Europeans are placing their CEN and CENELEC standards on the international standards development committees' tables. The U.S. is placing its national standards on the same table and the results are usually a harmonization of the two, with some good influence from countries other than Europe and the U.S.

CEN and CENELEC now work closely with ISO and IEC and the race has become less competitive and more cooperative as a result of what ISO and IEC are now producing through the accelerated attention and efforts provided by their member bodies and the U.S. in particular.

As an example, CENELEC standards development committees and CENELEC standards take the identical designations as the IEC committees and standards. When standards are developed by CENELEC and the IEC that appear to be acceptable to both, a simultaneous approval or vote is administered.

Until international standards are fully developed and recognized worldwide, because of different regional and national standards, more trade obstacles (not barriers) will exist. The leading example is the procedure for selling a U.S.-manufactured product in Europe. These products must meet European (CEN, CENELEC or ITSU) standards. Currently the U.S. product must be tested and approved in an European laboratory. For many reasons this takes time and resources.

For certain categories of products, e.g., electrical, marine, medical and others, MRAs, Mutual Recognition Agreements, have been initialed by the U.S. government and the EU. and implementation is underway. The implementation of these MRAs will allow U.S. laboratories to be accredited to test and approve products for sale in Europe.

Other countries such as Singapore are interested in pursuing similar MRAs with the U.S.

Although a great deal of progress has been made with regard to the EU cooperative activities on the international scene and the development of MRAs, etc., certain U.S. business executives have not been satisfied with the pace. They have formed the TABD, the TransAtlantic Business Dialog, to deal with their counterparts in Europe and with the EU in an effort to have these activities accelerated.

With the advent of the EU, there has been rapid development of other regional organizations. The U.S. has initiated some and is an active participant in the others.

First was the U.S./Canadian FTA, Free Trade Agreement. [This agreement has brought about the harmonization of U.S. and Canadian standards, the most notable those of the U.S. Underwriters Laboratories, Inc. (UL) and the Canadian Standards Association's (CSA) product safety standards.] U.S. laboratories may now be accredited by the Standards Council of Canada (SCC) to allow the sale of products in Canada that have been tested by SCC accredited U.S. laboratories. U.S. product certification organizations may also be accredited by the SCC so that the certification mark of a product certified in the U.S. is recognized in Canada without the need for Canadian re-certification.

This FTA is now expanded to its later version, NAFTA, the North American Free Trade Agreement, to include Mexico.

One more is developing; the FTAA, the Free Trade Agreement of the Americas, involving 34 countries but excluding Cuba.

An active regional body is APEC, the Asian Pacific Economics Cooperation.

Commentary

The U.S. voluntary standards system has survived government intervention because its practices have always been beyond reproach, based on openness and consensus.

In the seventies, there were two proposed legislative attempts by the Federal government to regulate the Voluntary Standards System which were defeated after lengthy hearings and testimony.

In addition, the Federal Trade Commission had proposed a trade rule which would have the Department of Commerce maintain a list of "approved standards development organizations." Products would have to be labeled with identification of the standard used for testing it, and the ratings resulting from those tests, etc.

The voluntary standards system, organized by ANSI, defeated these attempted governmental intrusions. Since then, the U.S. government has endorsed the voluntary system. One example is Circular 119, requiring government agencies to adopt voluntary standards in its procurement practices and to abolish government standards. Government agencies have since become members of ANSI and have representatives on ANSI's Board of Directors, in all cases, contributing but not intruding.

The Europeans have insisted on and succeeded in including in the World Trade Organizations (WTO) Technical Barriers to Trade (TBT) a Code of Good Practices for Preparation, Adoption and Application of Standards. Each participating country is required to sign on to this Code. The irony is that the code includes all or most of those proposed requirements of the U.S. government which were defeated by the U.S. Voluntary System. In essence, there is nothing wrong with the requirements which are in practice by the U.S. voluntary system–making them mandatory is objectionable. The U.S. resisted signing on for several years until it was one of the few holdouts. As a holdout, it became embarrassing to the U.S. which ultimately did sign on.

With fast moving technology, particularly in the field of computers and telecommunications, the procedure of involving all interested parties, openness and consensus in the standards development process wasn't working. Certain manufacturers formed consortia which fly in the face of all the voluntary standards systems' procedures which permitted it to continue to be "voluntary." Because this system of standards development by a restricted number of participants in closed meetings is growing – perhaps by necessity – is the voluntary system becoming vulnerable to the pressures of the past? Maybe, maybe not. This is no longer the seventies; much has changed, including government, but it shall be interesting to watch.

The last issue this article addresses is: "What is an international standard?" This is a question being addressed by many, including the international standards development organizations. Traditionally, the only international standards were those developed by ISO, IEC or ITU. But some national standards development organizations, whose standards have been recognized and used internationally, feel that these standards should not have to be exposed to the ISO, IEC or ITU processes; they believe that their standards are already international standards.

These organizations have been receiving growing support even by national bodies, members of ISO, IEC and ITU. The term "international standards" is used in many national, regional and international documents. It appears that redefinition is required and all the standards development organizations of the world are facing the issue.

The above article notes that in today's electronics industries, standards emerge from ad hoc producer-user consortia. They can even be adduced by single manufacturers, often as gambits or as strong-arm tactics. As this book goes to press, the Microsoft anti-trust judgment may change the picture that James Gleick painted in a 1995 *New York Times* article, summarized below.

Making Microsoft Safe for Capitalism

According to science writer James Gleick, Microsoft CEO and founder Bill Gates sets the behavioral norm for the company's aggressive, monopolistic hold on PC operating systems and its attempts to integrate more devices into its operating systems and corner the Internet Service Provider and Internet banking markets. In 1995, nearly half the world's total PC software revenue went to Microsoft. How can such a behemoth continue to grow in a way that satisfies stockholders? Gleick's answer is, ruthlessly.

Microsoft's zeal to place a computer on every desk in every office and home, all running Microsoft software, and its success in doing so up until now, lead to questions of ethics in Microsoft's dealings with developers and licensees. For example, Gleick says, many users who installed the market-leading Netscape browser, and then tried Microsoft Explorer, discovered their computers would no longer run Netscape. Users cannot delete Microsoft's MSN icon from their computers' desktops, and Microsoft was reluctant to allow other Internet gateways any screen real estate for their own icons.

He goes on to note that nothing matters more to Microsoft than the strategic management of its points of interconnection, that is, the creation, marketing, and management of standards. Microsoft has the clout to drive the creation of standards, and it is positioned to control access to its standards information and to developers' ability to operate within the standards. Microsoft has a mail standard (MAPI), a standard for the interface of software with telephone equipment (TAPI), and a proprietary online multimedia document publishing standard (Blackbird). Small software developers – the source of much useful innovation – face difficulty in getting accurate information on Microsoft's application programming interfaces (APIs), and risk having their technology swallowed up, or their company "made an offer it can't refuse" if it discloses its interfacing intentions

"The ultimate standard," Gleick notes, "is the operating system itself: the power spot in the digital ecology.... Every time Microsoft adds a new feature to the operating system, ripples flow through the software industry." This is because all application and utility programs must interface with the OS.

Gleick believes the government should not break up Microsoft, but should insist that Microsoft "open" all its standards and APIs – that is, make full information about the APIs freely available to all. "Is Windows an open standard?" asks Gleick. Unfortunately, he says, the answer is, "Yes – when and only when that suits Microsoft."

Gleick J (1995) Making Microsoft Safe for Capitalism. New York Times November 5 (The full article is also available on Gleick's personal website. Summary by Gerald Zion, Elia Freedman, Damon Webb, and Annie Leong.)

10.3 The GAP Strategy

Business Week's March 14, 1994 cover story noted the $340 billion and rapidly growing "entertainment economy." Entertainment cannot be dismissed lightly in a discussion of how a new product can be launched to the largest audience most effectively.

Even among straightlaced executives there is little argument that pornography did not create the mass market for VCRs. It may do the same for virtual reality.[1] According to *New York Times* columnist Peter Lewis, the half-dozen companies exhibiting X-rated multimedia CD-ROMs were the biggest floor draw at one Fall Comdex show, where one industrialist opined that pornography may be the "killer app" for multimedia PCs.[2]

From Hookless Fastener to Virtual Valerie

"Hookless fasteners," known now as zippers, were patented in 1893. They didn't work very well. They fell apart easily, and, at first, had to be removed from clothing before washing. Widespread adoption required two decades of functional improvement *plus* twenty additional years of marketing[3].

The zipper's relative advantage was that openings could be fastened without gaps. The zipper's early killer apps included closure of money belts and tobacco pouches, and later, closure of clothing, where it kept bulges of flesh from protruding as they might otherwise, for example, between buttons. The perceived risk was greatest, of course, with trouser flies. Fear of embarrassing zipper failure or painful injury delayed adoption for that purpose. Yet the emphasis on shapely appearance and the attention to dangers to one's privates foreshadowed the zipper's ascendance as the "symbol of futurism and eroticism" and "the instrument for allure and seduction" in the 1930s.

Early adopters in the '30s included sportswear and children's wear makers, and later, undergraduates at Princeton and Yale who were brave enough, or

---

[1] When *Scientific American* featured the "dataglove" on its cover several years ago, my faculty colleagues were quick to remark on the pornographic possibilties of this device that provides tactile input to and feedback from the computer.

[2] Lewis P (1993) Some of Show's Hits are X-rated. New York Times November 22; See also Saltzman B (1994) Now Get Your Kicks on Laser Disc: Playboy starting to embrace high-tech interactive formats. Los Angeles Times Service January 17

[3] Friedel R (1994) Zipper. Norton

sufficiently exhibitionistic, to risk zippered flies. A 1940 musical involving a striptease with zippers helped the market along, as did a scene in Rita Hayworth's movie *Gilda*. The Prince of Wales began to wear zippers. His well-publicized dalliance with Wallis Simpson, leading eventually to his abdication from the throne of England, enhanced the zipper's erotic image.

The zipper's role in literature, including Aldous Huxley's *Brave New World*, Erica Jong's "zipless f***," and Tom Robbins' 1984 description of "those little alligators of ecstasy" continued the partnership of soft porn and marketing through the present day.

The same happened with the new technology products of the 90s, though the adjective "soft" didn't always apply. New Machine Publishing Company of Santa Monica, maker of games with explicit sexual content, commenced business in June, 1993. Sales of the company's XXX-rated CD-ROMs doubled each month thereafter, resulting in sales in excess of a million dollars by the end of the year – despite the refusal of many publications to carry New Machine's ads.

Games created the market for low-capability home PCs, and helped do the same for multimedia PCs.[4] Will the kids keep the PC busy after school with games, and their parents after dark with adult CDs and websites?

Advertising made network TV a reality. It has proven to be one key to making the Internet viable when it is no longer subsidized as an academic/research network.

From Caves to Computers

The article summarized here asserts that entertainment is wedded to technology, beginning 40,000 years ago when early humans began using tools to create artwork, musical instruments, and other "non-pragmatic" uses. Working his way quickly through the more interesting technical milestones of the past five and a half centuries, Robert Toth claims that technology has fundamentally changed the way people fill their free time, leading to the thousands of computer games and other electronic gadgets available today. Commercial success for a new technology requires that it be accessible and interesting to the masses; in a word, entertaining.

---

[4]   See Business Week (1994) on Sega. February 21(cover story)

Although the Chinese had been experimenting with movable type printing for over a thousand years, it wasn't until Koster and Gutenberg invented the durable, movable metal type printing press in 1440 that printed books became a mass production item. Other developments occurred in a more condensed time frame: the first photograph in 1835 led to the creation of Kinetoscopes (moving images) in 1888 which led to public cinema in 1895. Also in 1888, radio waves were discovered and explored but remained basically a hobbyists' medium until about 1920 and the first commercial broadcast.

Accessibility to the masses takes on new meaning as the cultural and business structure takes technology from local communities to national media, creating a "common national culture." Soon after the first commercial radio broadcast, networks began to form, transforming radio into a national medium. For radio, and later television, the impetus for the network structure was the possibility of marketing the radio itself, while the importance of programming eventually became apparent. In the early 1950s, television cable was laid that connected east and west coasts, allowing people across the country to watch the same programs at the same time.

The third wave of accessibility was the invention of easily reproducible media that led to consumer ownership of a wide variety of technology-based products. Beginning with the gramophone and phonograph records, to audio and video tapes, through to the first free-standing computer video games, once the media were easily (and inexpensively) reproducible, the machines to run and create the media experienced high demand.

Toth makes the point that a lawsuit was instrumental in realizing the VCR's full potential. Walt Disney Co. and MCA Inc. sued Sony, the maker of Betamax, alleging that home videotaping infringed on copyrights. When a federal judge ruled that home taping was permissible, studios set up videotape departments, retail stores sprang up and a whole new industry was born.

Toth R J (1995)   From Caves to Computers.   Wall Street Journal September 15. Summarized by Cheryl Coupé.

10.4 The Trojan Horse

Nintendo is the best example of the Trojan Horse. By putting 16-bit then 32-bit video chips in millions of U.S. households in the guise of game players, they have readied our homes for advanced multimedia and telecommunications computing.

10.5 Conclusion

In almost every case, the "mass" market segment will have different characteristics from the innovator niche segments. The innovators regarded the product as necessary; the mass market may regard it as nice but hardly essential. The mass market will be more skeptical about your product and about your company. You will almost certainly find that some of your employees who were highly effective at selling to the early segments are totally ineffective when selling to the more conservative, harder-to-reach mass market.

The mass market will comprise the greater part of your company's lifetime and profit opportunities. So why is this chapter so short? This book's focus is on the entrepreneurial stage of technology marketing, a topic on which little has been written. Also, after making the transition to the mass market, a product is not high technology any longer, but simply technology. As such, it is subject to the standard principles of marketing and managing products that are familiar to customers. Finally, others have written on the mass marketing of technology and covered the topic at length. By all means see Geoffrey Moore's book *Crossing the Chasm* for a more detailed look at making the critical transition from selling to innovators to selling to the mass market.

10.6 Notes

For an overview of the VCR history look at Rosenbloom and Cusumano (1987) The Technological Pioneering and Competitive Advantage: The Birth of the VCR. California Management Review Summer 29:51-76. The thrust to mass market is covered in Marlow E and E Secunda (1991) Shifting Time and Space. Praeger Publishers. The standards (VHS vs. Beta) issue is addressed in Cusumano M A and Y Mylonadis et al (1992) Strategic maneuvering and mass-market dynamics: The triumph of VHS over Beta. Business History Review Spring 66:51-94.

Japan now leads in wireless Internet standards; see Schmit J (2000) Japan Pioneers Brave New World of Wireless: Nation sets standard for how people use Net on cellphones. USA Today (July 7) B1

Standards strategies. "Sun continues its effort to nominate its Java language as a standard." Read the article at http://cwlive.cw.com:8080/home/online9697.nsf/All/970923sun1841E.

Phillips H (1997) Worldwide Standards and Trade: The Players and Their Networks. (Used with permission)

10.7 Questions and Problems

Short Answer

S10-1. Look up and discuss the difference between a negotiated standard, a de facto standard, and a certification program. You may need to use sources beyond this text.

S10-2. Why do you think internal standards programs are viewed as one of the most difficult corporate functions to manage effectively?

Discussion Questions

D10-1. Refer to the Gleick article summary. (a) Write down your thoughts on how many industry players should sign off on a standard before it can legitimately be called a standard? Is Microsoft walking a fine line by disabling competing software that does not adhere to Microsoft's proclaimed standard? What will be the consequences of this?

(b) What are the costs and risks of claiming a standard early in the product life cycle? Late? Does this answer change if you have a large market share? A small one?

D10-2. Refer to the Toth article summary. (a) The article tells several stories about historic innovations in entertainment. Are there any common success factors?

(b) Are there any common characteristics of the innovators?

D10-3. Consider television; telephones; microwave ovens; "Sharper Image" gadgets; sewing machines; cellular phones; cordless phones; automobiles; electric razors; music synthesizers; "mace." For each, what was the initial appeal that created a home market? What industrial products, if any, were the precursors of each? Can you suggest additional mass market creation strategies?

"The real stories about successful high-technology start-ups are not about technical breakthroughs... or products. Rather, they are about unusual people with powerful ideas who are driven to do extraordinary things and are confident in their abilities to succeed."
Ed Zchau, *Wall Street Journal*, (12/10/1994)

"The innovator makes enemies of all those who prospered under the old order, and only lukewarm support is forthcoming from those who would prosper under the new."
Niccolò Machiavelli, *The Prince* (1513)

"Concern for man and his fate must always form the chief interest of all technical endeavors. Never forget this in the midst of your diagrams and equations."
Albert Einstein

"The seed stage VC looks for a 10x productivity advantage in the entrepreneur's product."
Tom Winter of Onset Ventures

11 The Future of Technology Commercialization

11.1 Vision. Changing the World

Why does our healthcare system favor, for example, artificial heart valves (which address a narrowly specific medical problem at great expense) over changes in diet, which can provide inexpensive, systemic improvement in health? The easy answers are that venture capital will flow to the higher-technology, higher-expense medical regime, and doctors will prefer the heart valve because its installation is highly billable, and reimburseable by insurance. We consumers have been conditioned to take a passive role in the healthcare marketplace, and don't have the "discipline" to eat what's good for us. Finally, the patient that is already hospitalized feels a sense of urgency that he/she did not feel while healthy. These answers are true but superficial.

Our model of technology marketing provides a deeper answer. Changing one's diet involves a major change in attitude and behavior. People's resistance to change is so extreme that *it is harder to get someone to "be good" than to get them to undergo open-heart surgery!* Why? It's simply because there is little or no social reinforcement available to a person who might be interested in making beneficial changes in his or her behavior.

This seems to conflict with the conventional wisdom (CW) of economics. This CW says that if I can provide a cheaper, superior solution to a problem, I should succeed. Yet, while the heart-valve entrepreneur becomes a millionaire, the dietician and the teacher of stress management remain impecunious and marginal characters in the business world. The word "injustice" is often applied here, but I think it is really, to use entrepreneur Paul Hawkin's phrase, a "failure of imagination."

The purveyor of a high-behavior-change product must also sell the needed social support structure. Before you decide this assertion is radical or crazy, consider the (admittedly few) examples where it has been done. The most successful has been Alcoholics Anonymous, perhaps because its members feel an urgency comparable to that of the acute heart patient. The many other "12-step programs" modeled after AA have been less successful, probably because each addresses only one negative behavior (codependency, credit card excesses, etc.) Weight Watchers succeeded in its original segment by building what amounted to a "support group" for buyers of its frozen meals, but faltered when trying to expand its clientele to all health-conscious eaters.

Other success stories exemplify a more rounded approach. Challenges for the organization of the future include globalization, diversity in the workplace, environmental responsibility, and so on. Executives who are actively concerned about these issues remain a minority. Groups of such executives from noncompeting companies gather periodically at roundtables sponsored by WorldTen, The Executive Committee, or other such nonprofit and for-profit clubs. Each delicately refrains from calling itself a "support group." Bruce Shotkin's Panopolis Foundation in Austin, Texas will build an ongoing community (both a physical one and a virtual, computer-networked one) where people can be surrounded by others who are concerned with systemic approaches to health and environment, and where research can be done on how the systemic approaches can be made economically viable.

Note that the "support group" strategy is for high-behavior-change products. This is distinct from the "user group" strategy, which is most applicable to high-learning products.

The conventional wisdom on technology marketing holds that technology is easily duplicated. Therefore, the CW goes, the first entrant must prolong his/her pioneer's advantage by offering superior service, and the late entrant must differentiate his/her product by offering even better service. This is a dangerous half-truth.

You are in the market because your product will change the world. Your technically capable competitors delay entering the market because they do not believe the world is ready for your product, at least at this price. Your first-entrant advantage, if you have any at all, lies in the superior force of your vision, in your ability to change the world one customer at a time.

So where does this leave Lou Gerstner, who as IBM's new CEO was roundly criticized for saying IBM doesn't need a vision? His full quote was, "There's been a lot of speculation as to when I'm going to deliver a vision. The last thing IBM needs right now is a vision. What IBM needs right now is a series of very tough-minded, market driven, highly effective strategies in each of its businesses." This statement was intended to reassure stockholders that IBM will focus on making money rather than on composing elaborate plan documents. And, in fact, IBM employees have heard and participated in vision discussions since Gerstner's statement, thus keeping them motivated.

But how can "each of [IBM's] businesses" have a strategy without a vision to give shape to the strategy? The stockholders and employees are feeling good, but what about IBM's customers? My belief is that IBM's customers have always looked to Big Blue for technological leadership. They have relied on IBM to sell their companies the most competitive information processing products. Not only that, but they have paid IBM premium prices for these products. I believe IBM's customers pay the premium prices not for the product *per se,* but for the expectation of a steady stream of high quality future upgrades and service. In other words, *a vision that includes staying on the leading technological edge can result in positive brand equity.* How can the customers believe IBM will continue to provide the most competitive products if IBM does not have a vision that articulates that goal? Will IBM be able to maintain its high margins if it doesn't formulate and publicize a vision? It will be difficult.

11.2 Systematizing Commercialization

The innovation process shown in Figure 11.1 illustrates the many stages between the laboratory and the store shelf, and also the many groups that have an interest in innovation and commercialization. Except for a few organizations that sell competing products, or that espouse an ideology that excludes your product (see "The Stealth Operation to Market RU-486" in §11.3), all these groups want you, the entrepreneur, to succeed. If you build a successful new technology company, all these interest groups will benefit due to the jobs and exports you create, the fees and investment opportunities you offer, and the greater efficiency and productivity your products make possible.

Much of what new entrepreneurs must learn is simply how to find these people and groups, and how to ask for their help. You will find they are helping entrepreneurs not only individually, but are combining in new kinds of organizations (see §11.3) to more effectively build entrepreneurial communities, and to make enterprise formation less of a black art and more of a disciplined process.

These interest groups – which include angel networks, technology brokers, incubators, university-chamber of commerce partnerships and others – appreciate and admire your technology knowledge and entrepreneurial acceptance of risk. As entrepreneur, you want and need their money and expertise. Both parties must take care that your relationship is mutually beneficial and not exploitative.

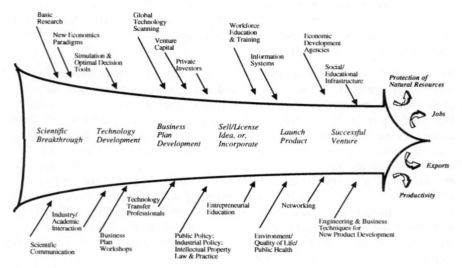

Figure 11.1 The innovation arrow

Commercializing Technology: What the Best Companies Do

This study of companies that were good at commercializing technologies versus those that were not showed definite similarities among the successful companies. Furthermore, the characteristics found in the companies that were successful at technology commercialization were not found in the companies that were not so successful. Commercialization capability is becoming more and more important in the current business environment. Shrinking life cycles, the cost of developing technology and the fragmentation of markets are driving the importance of a successful technology commercialization capability.

Companies that can successfully manage commercialization differ from other companies in four ways:

1. Time to market. Companies good at commercializing technology have much shorter development cycles than other companies.

2. Range of markets. Due to the high cost of developing technology, successful companies are able to leverage technologies across a range of markets.

3. Number of products. Successful companies are able easily to customize their products to meet market niches.

4. Breadth of technologies. The ability to understand and integrate several technologies is another characteristic of successful companies.

The characteristics of high-performing technology commercializers are not easily achieved. How do companies develop these characteristics?

- Management made commercialization a priority.

- Goals were set to measure performance and benchmarks were done to measure goals.

- There was a strong emphasis on building cross-functional skills.

- Management involvement to help speed decisions and shorten development.

Successful commercializers commercialize successfully over and over again. The authors claim the success factor is not inspiration, but discipline.

Nevens M T, Summe G L and B Uttal (1990)   Commercializing Technology: What the Best Companies Do. Harvard Business Review May-June 154-163. Summary by David Blankenbeckler.

Bringing New Materials to Market

Materials are very likely to be basic technologies, causing the birth of new industries. Examples include silicon semiconductors and optical fibers. Twenty-year cycles (invention to widespread adoption) are typical for materials:

- vulcanized rubber: 1839-late 1850s
- low-cost aluminum: 1886-early 1900s
- teflon: 1938-early 1960s
- structural titanium: mid 1940s-mid 1960s
- velcro: early 1950s-early 1970s
- polycarbonates: 1953-1970
- gallium arsenide chips: mid 1960s-mid 1980s
- diamond-like thin films: early 1970s-early 1990s

• amorphous soft magnetic materials: early 1970s- early 1990s

There are a lot of inhibitors to the adoption of new materials, and few facilitators. Inhibitors include:

• Designers are used to designing machinery using certain materials, and take a long time to learn to use new materials effectively.

• Early designs rarely live up to their full potential, and turn off potential customers.

• Materials scientists tend to develop materials that maximize a single characteristic, like strength/weight. If the new material is strong and light but not durable, it doesn't turn into good products.

• A 20-year application cycle for a material with a 17-year patent is a clear disincentive for a producer.

• Inventor/entrepreneurs hope to capture monopolistic profits with an innovative material. Customers don't want to pay monopolistic prices, and will stick with an older, inferior product until the benefits of switching are unarguable.

• Or, customers may want to hedge their price risk by requiring multiple suppliers of the new material. But this prevents any of the suppliers from realizing the economies of scale that will make the product affordable for anybody.

• Codes and standards (e.g., for building bridges) are inflexible, and often specify the specific material to be used, rather than the properties of the material.

• Codes require much in-use performance data before new materials are acceptable. But the codes also prohibit the use of the new material, thus preventing the acquisition of the data.

• The customer may have to cultivate a whole new set of suppliers. GM scientists discovered a new neodymium compound could make lighter, cheaper small motors for windshield wipers, etc., in cars. But GM had never dealt with suppliers of neodymium.

Eagar suggests these strategies to better leverage new materials: Product designers should communicate closely with materials suppliers' "application teams." Materials suppliers and users should consider partnerships for sharing risks and profits of new materials. Trade associations should write new standards and codes based on product performance rather than specific materials. Codes should recognize different categories of risk, allowing, e.g., a pipeline made of a new material in less populated areas, but require the older, proven kind of pipe within city limits.

Materials suppliers should invest at least as much in production methods as in developing their new materials. This will allow prices to drop more rapidly as manufacturing quickly becomes more efficient. Eagar notes that in the US, 70%

of R&D funding goes to product development and basic research, and 30% to manufacturing process development, and that in Japan these percentages are reversed.

Eagar T W (1995) Bringing New Materials to Market. Technology Review Feb-Mar 43-49

Table 11.1 Why Study Entrepreneurship in a Technology Management Course?

- You may want to become a technology entrepreneur.

- You may have to manage relations with a small supplier.

- You may become involved in your company's "new venture" arm, administering "intrapreneurial" and spin-off endeavors.

- You may become involved in the formation of federal or local policy regarding new business startups for creating jobs and a strong supplier base.

11.3 New Institutions for Promoting Technology Entrepreneurship

The boxed story below relates how a non-profit health research organization took the lead on commercializing a drug in the United States. While this was an unusual and controversial case, we may expect to see more such commercialization initiatives on the part of non-profits, including university foundations, public-private new business incubation partnerships, and others.

The Stealth Operation to Market RU-486

The abortifacent drug mifepristone, also known as RU-486, has been legally used in France, England, Sweden, and China for some years. The quest to have its use approved in those countries and in the U.S. has been tragic and dramatic.

Amid speculation that the Pope had pressed the devoutly Catholic president of Hoechst AG, the largest manufacturer of RU-486, to do so, Hoechst attempted to withdraw the drug from the French market in 1988. Hoechst is the successor company to I.G. Farben, which manufactured the Zyklon B gas used in Nazi extermination camps; as a result, Hoechst is also sensitive to any implication that their products could be used selectively to control population groups. But the French government took the opposite view, pleading the importance of RU-486 for women's rights and health. France's ministry of health ordered Hoecht's French subsidiary to continue to make the drug available in France, where it is now used in approximately half of all legal abortions (the balance are surgical abortions).

The U.S. is politically polarized on the matter of abortion. Hoechst and potential U.S. licensee companies feared violence, boycotts, and negative press should they introduce RU-486 here. "Hoechst took the unusual step," writes the author of this article, "of donating the U.S. patent rights to the [non-profit] Population Council, which had designed the copper T380A IUD and the contraceptive implant Norplant." This did not stop anti-abortion groups from boycotting Hoechst anyway. (Hoerchst does $7 billion of business in the U.S. in other products.) That the drug is not yet on the U.S. market is due to the Population Council's "striking lack of business sense" and the surrounding controversy and fear.

RU-486, if approved, could be prescribed by any M.D., removing the need for women to run a gauntlet of protesters to get to a surgical abortion clinic. Market research indicates most doctors would be willing to offer RU-486 in lieu of surgical abortion; the treatment would cost $300, about the same as a first-trimester surgical abortion.

Mifepristone blocks progesterone, a hormone needed to sustain pregnancy. A second drug, taken a few days after the RU-486, causes uterine contractions and the expulsion of the fetus. The combination "improves" on surgical abortion by working very early in pregnancy, and backers thus hope it will help defuse the argument that abortion is murder. It can, in fact, be used as a "morning-after" pill, and also has application in treating breast cancer, fibroid tumors, and endometriosis. Its success in the U.S., say advocates, could provide a role model for women's health care in many other countries.

The Population Council's business plan called for $28 million in investment to pay for marketing, testing, and making manufacturing and distribution arrangements. The non-profit group's encounters with venture capitalists, however, were textbook cases of culture clash, and the needed funds were not raised. The Council made a deal with an entrepreneur who quietly raised private investment for a factory, but discontinued this deal when it emerged that the Council's inadequate due diligence had missed incidents of securities fraud and disbarment in this entrepreneur-attorney's past.

The Population Council could not afford to have such a sensitive product associated with questionable people or companies. Even as the Food and Drug Administration (FDA) agreed to put the mifepristone application on fast track consideration, the Council dissolved its manufacturing agreement. The Council worried not just about its image, but about its staffers' safety. Abortion clinics had been bombed by protesters, and physicians performing abortions had been murdered. [In July, 2000, a Canadian doctor was stabbed, evidently in connection with clinical trials of RU-486 in that country.] The Council feared right-wing opponents of abortion, and (because of earlier experience with the injectable contraceptive Depo Provera) left-wing activists who oppose RU-486 because of the possibility that poor women will be pressured to have abortions. The need for extreme discretion kept the Population Council from looking as widely as they might for good business advice.

Impatient with the Council's progress, another organization called Abortion Rights Mobilization (ARM) produced a bootleg version of RU-486, using French patent documents and reverse-engineering the Chinese commercial version of the drug. ARM has received permission from FDA to conduct its own clinical trials, and thousands of women are currently involved in the trials. Demand for the drug is high, according to physicians participating in the clinical trials, and many phone inquiries come in every day. [In June, 2000, the FDA stated that RU-486 will be approved in the United States, but only under "strict guidelines."]

Kirshenbaum G (1999) The Stealth Operation to Market RU-486. George April II(4):112-125

11.3.1 New Business Incubators

The new business incubator is a system for leveraging resources to help entrepreneurs accelerate the development of growth companies. An incubator may include a building where incubated companies collocate, or may be simply a bundle of business services. The latter arrangement is called a virtual incubator, and in its extreme form may deliver services solely via the Internet. Different incubators bundle their services in different packages for different target markets, for example, for companies at different stages of development. These specialized incubators may call themselves accelerators, germinators, innovation centers, or other names. In any event, the incubator's enterprise development process should bring together entrepreneurial talent, business know-how, growth markets, and investment capital within a cohesive, supportive community.

As of 2000, approximately 1,000 U.S. incubators have created 19,000 companies and 245,000 jobs. Companies graduating from these incubators have higher success rates than unincubated technology startups.

Incubators generally engage in one or all of the following activities: business assistance to tenants; networking activities; educational activities; public relations activities; infrastructure and services; and interactions with the university.

Business assistance to tenants. Professional incubator staff and/or MBA students may review monthly financials with tenant companies; provide access to a "know-how network" of service providers; or introduce the companies to experienced mentors or foreign markets and sources of supply. The incubator hosts venture fairs at which tenant companies may present to VCs and angels their brief pitches for funding.

Networking activities. The incubator introduces the tenant companies' executives to the local business "power structure" and to representatives of other technology regions and companies. This activity includes receptions and social events.

Educational activities. Tenant execs may sit in on university classes. The incubator is site of brown-bag lectures, workshops and seminars on a variety of entrepreneurial skills – benefiting students as well as incubator tenants.

Public relations activities. The incubator issues pamphlets, press releases and other literature promoting the tenant companies jointly. The incubator itself is newsworthy as a sign of the community's commitment to supporting entrepreneurship. The incubator helps members of the know-how network publicize their connections with incubator companies.

Infrastructure and services. Needless to say, the incubator maintains attractive space, shared conference rooms, refreshment rooms and reception area. Services may include shared office machines, receptionist, switchboard, and security. These days high-quality wiring and fast Internet access is a minimum requirement.

Interactions with the university. The university and the incubator provide each other with opportunities for interns, spin-off companies, exploitation of intellectual property, and educational enrichment. More about this below.

11.3.1.1 Interactions with the University

Indeed, university-incubator interaction enhances almost all the incubator activities mentioned above. In particular:

- The incubator can host business school classes. In one such instance, engineering and business students teamed for a robotics venture laboratory, combining technical design and business planning for special purpose robots, with an eye toward launching a new venture. This writer has taught a high technology marketing laboratory for MBA students, using incubator tenants as living cases. Another possibility is the "incubator operations laboratory," preparing students to run incubators and related entrepreneurship facilitating organizations.

- The incubator provides internship opportunities for science, engineering and business students. In these arrangements, interns learn the reality of entrepreneurship. Internships benefit the incubating companies also; tenants are exposed to the latest academic knowledge, and enjoy intelligent, inexpensive labor.

- Incubator entrepreneurs are valuable guest speakers in academic classes. These guest appearances add value to students' academic experience, and give the entrepreneurs a chance to field critical questions.

- Tenants provide "living cases" and student projects. Living cases are more interactive than, e.g., written Harvard cases, giving students recourse to further information and the chance to actually have an impact on the operations and success of the subject company. Of course, these also can lead to student employment.

- The incubator provides "beta sites" for student new venture competition teams. The prospect of space in the incubator is an added incentive and reward for students competing in university-sponsored business plan contests.

A 1995 Coopers & Lybrand study showed that companies availing themselves of university resources "had productivity rates 59 percent higher than peers without such relationships." According to Glenn Doell, formerly director of Rensselaer Polytechnic Institute's incubator, "Incubators should consider developing a strong partnership with their local institution of higher learning."

11.3.1.2 Private Incubator Trends

In the 1980s and '90s, most of the incubators in the U.S. were dismissively called "real estate operations." This meant that universities were not involved and that tenants received no business assistance other than a shared receptionist and photocopier. The past few years have seen the rise of investor-owned incubators in which venture and angel funds (which invest other people's money, including pension funds, in startup companies) show fiduciary responsibility by closely monitoring their investments. These investors subsidize incubator operations out of their expected capital gains; it is thus in their interest to maintain incubator services at the highest level.

However, investor-owned incubators tend to incubate new companies for a much shorter time than public and university incubators. This was primarily due to the rush to create web-based companies in order to reap the rewards (which prior to May, 2000, were unreasonably high) of dot-com valuations. It may benefit a new company to be a member of both a university incubator and a private incubator at the same time, and indeed these may be in the same location. Venture capital-owned incubators are now re-examining their roles, and it appears the "Big Six" accounting/consulting firms will also get into the incubation game, perhaps bringing their diversified management expertise to bear on the young startups. Alliances for incubation will continue to evolve along with the markets.

11.3.1.3 Case Study: Mason Enterprise Center

This case is based on an interview with Dr. Roger Stough, Director of the Mason Enterprise Center. In his capacity as Director, Dr. Stough reports to the Provost and President of George Mason University.

MEC runs a number of incubators. One of the incubators is a physical one occupying two floors (4,000-6,000 sq.ft) and designed to shelter 16 companies (of any kind). Another is an "incubator without walls," or Technology Resource Alliance that works with 30 companies/year. These companies are pre-launch or immediately post-launch, technical companies only. They enter the resource alliance by application, and only the best are accepted.

MEC also runs three federal telework centers (cubicles with internet access, videoconferencing, computer training room). These are scattered across northern Virginia, but all are in the general radius of Washington, D.C. In addition, MEC manages an international incubator for Arlington County, for foreign companies wanting to do business in the U.S.

MEC holds contracts for four Small Business Development Centers (SBDC's). The SBDCs emphasize mentor/protege programs. In contrast to the competitive Technology Resource Alliance, the mentoring program accepts anyone wishing to start a business. A $1,000 small business loan (microloan) program is also administered. These are funded by separate federal programs, mostly intended for non-technology businesses. After completing the basic mentoring program, companies contract with the SBDC for additional services.

MEC operates a "Grubstake Program" that resembles venture fairs for angels and VCs, and cooperates with the local "Venture Investor Club" (angel network) to general leads to promising companies. A particular strength of MEC has been in naming, promoting and branding their programs in this way. For example, MEC's Dry Run™ teaches CEOs how to make a pitch for funding. The executive's presentation is heard initially by internal staff, and after, by a panel of outside advisors similar to those assembled by the Oregon Marketing Business Initiative (OMBI).

MEC finds money from federal, county, fee for service (mentoring program), and state and university matching funds. MEC's funding dictates that they must service small and minority businesses. But as community leaders agree that high-growth potential businesses are usually technology-based, MEC finds it can bend/leverage all its funded programs to encourage high tech entrepreneurship. MEC does mentor/incubate direct competitors, but at such early stages that "it doesn't matter."

University involvement is not a critical success factor for MEC, according to Dr. Stough, but is good for the university, especially due to student involvement in community relations and fund raising. MEC's network provides leverage for research contracts that come into the university having to do with measuring specific aspects of the local and state economies.

MEC is linked with academic programs in entrepreneurship throughout the university (business school, engineering, public administration, art/multimedia) leading to a university-wide minor. GMU doesn't have a full-time MBA program, so MEC fielded a campus-wide "entrepreneur profile" questionnaire to find students with experience starting businesses. These students will be MEC mentors and interns.

There are cultural as well as budget tensions within university because in MEC speed, not consensus, is of the essence. The university is better for coaching launch of non-Internet companies, Stough says, because it can take the time that is customary for university activities. One MEC success was the launch of a heritage-based travel service. But even this company leverages the net for, e.g., real-time itinerary changes.

AOL and other northern Virginia companies have created 4,000 millionaires and several billionaires. MEC leverages these for a 400-strong "knowledge network," drawing on these executives for advice on fast turnaround/short window dotcom opportunities. The Internet enhances communication among these advisors – especially if they have worked together before.

MEC hired a retired Peat Marwick executive as Executive in Residence to leverage his network for the benefit of MEC companies. The executive mentors companies and mentors incubator managers. This works well; Roger may hire another such executive.

Pieces of MEC have been evolving since 1986. Not a planned from scratch effort, in fact, according to Dr. Stough, it is "something of a mess." MEC's total budget is $7 million/year. MEC worked with 1,600 companies last year in one way or another, and helped companies get money from $50k to $10m.

MEC's website is http://www.gmu.edu/departments/tipp/center/center.htm.

While MEC continues to manage incubators on contract for county, military, etc., it is moving away from in-house, physical incubators. Roger advises not building bricks-and-mortar incubator unless a big region doesn't have a private one. Only do bricks-and-mortars in rural areas if the state provides funding specifically for that purpose, he says. Otherwise, just do traditional SBDC activities in rural areas. But other arms of MEC can buy the time of SBDC staff to help with urban incubators. The University's contribution, then, is to manage incubators for private parties with low-cost student labor.

11.3.2 Technology Brokers

Technology brokers are for-profit or non-profit enterprises that match licensable technologies with licensees for a fee.   Non-profit brokers have a mission to commercialize technologies in a certain sector (like the Population Council, described above), or in a certain country.   In fact, many brokers in the Asian Pacific Rim are non-governmental organizations (NGOs) that are chartered by national governments expressly for national economic development purposes. A broader term that also includes national government agencies is "national research and development organizations" (NRDOs). These span laboratory research, licensing/brokering, and incubation activities.   Table 11.2 lists a number of NRDOs and NGOs active in the Pacific Rim.

Table 11.2 Some NRDOs and Technology NGOs Operating in the Pacific Rim

Country	Agency
ASEAN	ASEAN Committee on Science and Technology
Australia	Australian Technology Group
China	Long Han Hi-Tech Devel. & Consulting Co.
Denmark	Danish Innovation Centre
Finland	SITRA
Finland	Otaniemi S&T Park
France	FIST (France Innovation Scientifique et Transfert)
Germany	Fraunhofer Gesellschaft
Germany	Inventors' Center of Northern Germany
Japan	MITI-AIST
Japan	STA (Prime Minister's Office)
Japan	Japan Research Development Corp. (STA)
Japan	Innovation Partners International inc.
Korea	Korea Technology Banking Corp.
Malaysia	Malaysian Technology Development Corp.
Philippines	Development Bank of the Philippines
Russia	Ministry of Science & Technology
Slovenia	Technological Devel. Fund of Republika Slovenia
Spain	Centre for Technological & Industrial Devel.
Thailand	Thailand Research Fund
Thailand	National S&T Development Agency
UK	British Technology Group
USA	NIST
USA	Research Corporation Technologies
USA	IC2 Institute/Austin Technology Incubator

Going for Brokers

It isn't easy to take a promising technology out of development and into successful commercialization. There are a variety of companies willing to help you try – for a price. However, let the buyer beware.

Commercialization efforts, says one broker, usually feature "the wrong guys selling the wrong thing to the wrong audiences for the wrong purpose." The tech broker can help remedy this situation in a limited number of circumstances, for entrepreneurs and licensors who have the time and patience to work through the process with them.

The vast majority of new technologies offered by hopeful researchers is turned down due to low prospects of commercialization. If a product is adopted by a tech broker, it may take five to seven years to get to market. Broker Kiser Research spent a decade trying to get Russian space technology into the U.S. market. If a broker is to add value, s/he must have the contacts that open doors at senior levels of licensee organizations.

Variations on the broker concept can find success. Some broker joint technology venture deals, where, for example, company A is paired with company B to produce product C. Broker Technology Management and Funding takes a 20% equity share in the new companies it helps create. It currently has 28 new companies and projects 200 shortly. The market potential for each must be at least $500 million. Dawnbreakers, another matchmaking concern, has had a 43% success rate in its joint ventures. Still another variation is the database service. Dvorkovits and Associates and ARAC charge a few thousand dollars for searches to match technologies with buyers.

Brokering can thus be a successful, if expensive, option for the entrepreneur. British Technology Group generated $50 million in revenues from a portfolio of 10,000 patents and 500 licenses in 1993. In the same year, Research Corporation Technologies generated $55.5 million in licensing fees.

Barrett R (1994) Going for Brokers. Technology Transfer Business Spring. Summary by Kevin Towle and Troy Garrett.

11.4 Additional Cases: High-Tech Entrepreneurial Companies

Entrepreneurship is critical to technology commercialization for more reasons than those noted in Table 11.1. Technology fusion ensures, first, that there are ample opportunities for entrepreneurs to hook diverse technologies together into novel products; second, that the huge number of combinatorial possibilities for fusion raises the odds against each individual combination's success in the market; and third, that large corporations are glad to have small entrepreneurs shoulder these risks – and then buy them out when they show signs of success. The coming convergence of biology (genomics) and electronics (especially information technology) will generate still another plenitude of entrepreneurial opportunities.

The risk of technology entrepreneurship is illustrated by the five companies profiled in this section. All drew venture and/or angel investment in the 1990s, and at least one of them traded on a public exchange. One, Medical Polymers Technologies Inc., based itself on a basic technology with many application areas and showed extraordinary organizational flexibility. But at least two of these companies have ceased operations, and none have reached a high profile in the marketplace.

11.4.1 Autogenesis

11.4.1.1 Background

This section is based on a 1993 talk by Thomas Gary, then a manager at Autogenesis. Autogenesis is involved in the regeneration and lengthening of limbs and tissue regeneration. Headquartered in Anchorage Alaska, Autogenesis has developed the computer technology to automate the Ilizarov tissue regeneration procedure, through the motor driven Automator. The method of mechanical fixation and distraction, developed in Russia by Dr. Gavril A Ilizarov, has been in practice for over fifty years. Basically, Autogenesis develops a computer driven motor which replaces traditional hand cranking involved with the traditional distraction histogenesis process. The automation of this process enables the doctor to increase the speed of tissue regeneration, reduce the pain to the patient, and increase patient compliance.

11.4.1.2 The Product

The computerized *Automator* can be fitted to most manufacturers' circular fixator frames. Autogenesis has currently limited the product uses to bone replacement due to cancer and trauma, and general limb growth for medical purposes. Cosmetic uses for the fixator frames and the *Automator* attachment are currently being explored. Also, an implantable fixator frame utilizing the *Automator* is being developed.

Pricing of the Automator is based on the $1500 manufacturing cost and negotiations with insurance companies and doctors so that the product is priced at a level the market will bear. This pricing strategy has provided ample margin to supply revenues for research and development into other products, including the internal automated bone regeneration product targeted at cosmetic bone regeneration surgeons.

11.4.1.3 The Industry and Market Potential

Autogenesis provides the automated apparatus for the frame responsible for the bone regeneration. There are currently six major manufactures of the fixator frame. Smith-Nephew and Richards holds a 95 percent market share for the fixator frames in the U. S.. Autogenesis expects rapid market penetration, over 3,000 surgeons are certified by Smith-Nephew and Richards to practice the Ilizarov method in the U.S. These surgeons preform over 500 limb lengthening operations a year, not including cosmetic lengthening. The Russians have successfully completed over two million of these surgeries. Therefore, it is believed that the future market is substantially larger than the currently defined surgeon market.

The cosmetic lengthening market is just developing and offers a substantial growth area, as does the veterinary market. Although these markets offer future potential, Autogenesis is focusing its initial marketing efforts on its primary market of certified surgeons simply because of the availability of names and addresses of this existing market.

11.4.1.4 The Competition

Autogenesis hold patents in several other countries. At the current time, the company foresees no viable competition to The *Automator*. Recently a Japanese company obtained a patent for a similar but simpler automated process. Autogenesis feels that the Japanese firm offers no immediate threat, as its product is based on older technology and does not offer computerization.

In order to protect its competitive advantage and comply within the strict FDA requirements imposed, Autogenesis has implemented a leasing agreement that prevents reverse engineering by potential competitors and insures a quality product.

11.4.1.5 Marketing Considerations

In taking the product to market, a direct sales force was initially implemented, with the objective of having Autogenesis executives bring the product to the market of surgeons. From this a direct sales force was to be developed. The initial direct sales campaign was not effective. Two factors prevented the successful implementation of this initial plan. First, ". . Doctors don't want to deal with multiple salesmen. " Second, the development and training of a direct sales force given the nature of the product and industry proved simply too expensive. Mr. Gary stressed the importance of developing relationships with the initial targeted market prior to selling them the product.

The current market strategy being implemented utilizes the ten Autogenesis sales men across the country as teachers to train the sales staff of the various fixator frame manufactures, in order to sell the Automator as an add-on to the traditional fixator frame.

Negotiations are underway with all six of the frame manufactures. The company views an alliance with a smaller aggressive frame manufacturer crucial to the prolonged success of the product. The main objective of such an alliance would be to gain access to international distribution channels provided by the frame manufacturer.

Aside from the importance of securing the proper distribution process for the *Automator,* Autogenesis is concerned with the generation of a premium product image in the initial stages of development. In order to accomplish this task, the firm solicited the consulting services of Dr. Draver Pele, a leading authority on the Ilizarov method. Also, in order to gain acceptance both in the surgical community and the general populus the *Automator* was first tested on John Scully of the Atlanta Falcons. The much touted surgery was preformed by Dr. Pele and drew the attention of CNN, thus creating an enormous amount of positive publicity and credibility for the product. From this initial surgery a promotional video has been developed featuring Mr. Scully stating a how "the screws (that is, the traditional procedure) scared the hell out of him."

The relationship with Dr. Pele is viewed as crucial to the current and future success of Autogenesis. Dr. Pele views the *Automator* as an integral component to his dream of an implantable fixator for cosmetic surgery purposes. The company views his services as essential in selling to the initial market of surgeons.

Another aspect of the marketing strategy is to promote and encourage training courses pertaining to the Ilizarov method in medical schools. Autogenesis feels that actively promoting the procedure inherently promotes the product.

In summation, Mr. Gary pointed out that the stage of development that the product is moving through dictates the marketing strategy and the type of promotional material needed. For example, Mr. Gary pointed out the evolutionary aspects of the brochure, evolving from a descriptive technical guide of the product, targeting investors and influential potential customers, into the current glossy sales brochure.

11.4.1.6 Market Barriers

The major barrier to the successful implementation of the product lies not in the product but the painful Ilizarov method, the negative perception that the procedure evokes. In overcoming these barriers, Autogenesis aims to work with the frame manufactures and surgeons to develop more appealing frame designs for children, as well as, promoting the benefits of automating the process to overcome the negative perceptions associated with the procedure.

11.4.1.7 Update

A student contacted the current president of Autogenesis in 1998 for an update. Then in business ten years, the firm seemed to the president to be "a startup company that has the baggage of an established company," and held only 1% of the market for limb lengthening procedures. The FDA approved Autogenesis' Automator 2000 in 1998. The product was significantly redesigned with the battery, control system and motor are all in one unit. These changes make the product easier to use and improve the mobility of the patient.

Nonetheless, the cost is still high, and the frame manufacturers, Autogenesis' putative partners, have developed competing products, some with additional useful features. Autogenesis main sales points – reduced problems due to patient neglect, and lower pain levels – are difficult to quantify and demonstrate, and few clinical studies have been done that might lend proof to these claims.

The orthopedic industry as a whole grew 16% per quarter in 1998. But Autogenesis experienced a net loss of $71,081 during the first half of that year (according to Dun & Bradstreet) on revenues of $51,277.

11.4.2 Corporate Memory Systems, Inc.

Mike Begeman, CEO of Corporate Memory Systems, Inc., articulated the problems his company has encountered in achieving higher sales volume. Leveraging off the book *Crossing the Chasm* by Geoffrey Moore, Begeman presented the plight of his company in the context of a modified technology adoption life cycle. However, before introducing the class to CMSI's dilemma, he provided a brief overview of MCC and Corporate Memory Systems, Inc., a spin-off from MCC.

11.4.2.1 MCC

In 1982, twenty computer and aerospace companies, in anticipation of heightened Japanese competition, formed a consortium, MCC, to enable long-term research and development. As separate entities, each company did not possess the resources necessary to conduct anything but R&D for the short term. As these companies pooled their resources for this endeavor, other companies in the industry, suspecting antitrust violations, cried foul.

The National Cooperative Act served as an umbrella of protection for companies forming cooperatives for research and development. However, there were limitations for the beneficiaries of the act. MCC was prohibited from producing its own products for use by its membership group. The consortium could only transfer the technology, much of which was years from the marketplace, to the companies. As it turned out, this technology transfer proved to be a formidable task. With a charter composed of fifty-percent research and fifty-percent technology transfer, MCC was an effective conduit for achieving technology years ahead of the market. It proved to be less effective, though, as a vehicle for bringing that technology to the companies for use.

Begeman presented three models utilized over a period of time by member companies harvesting technology developed at MCC. The first model consisted of liaisons and associates from each company working side by side with the specialists at MCC. The liaisons would periodically visit MCC to monitor its progress in various technologies. The associates would work at MCC for a period of time and then, after being replaced by the next group of associates, they were to return to their former companies with the knowledge gained during their tenure at MCC. Problems arose when these associates were "turned over." Having enjoyed the lifestyle in Austin for a period of time, many of these associates failed to return to their companies, thus precluding the transfer of technology to the member companies.

The second technology transfer model met with the same lack of success as the first. Member companies sent groups to MCC to evaluate the achievements of the research. Unfortunately, these groups consisted of individuals who lacked the technological background necessary to evaluate "future" technology. More profit-oriented, these individuals were overwhelmed by the highly advanced technology exposed to them.

The third technology transfer model of joint products, briefly touched upon by Begeman, did meet some success. However, at this time, the threat of Japanese competition in microelectronics was waning and, as a result, member companies were cutting back on the resources each was committing to MCC.

With a declining budget, MCC possessed limited resources to conduct its research and development. Certain technologies researched at MCC were not actively being considered by its member companies. As an employee at MCC, Begeman recognized an opportunity to bring a technology, groupware, to the market himself. Having put together a business plan, Begeman solicited the approval of each member company, a challenge in itself, to spin-off a company from MCC to develop and market groupware. MCC provided little assistance to Begeman in his efforts to spin-off a company. That policy has changed somewhat since Begeman initiated his spin-off. MCC now takes a more active role in providing assistance in the commercialization of its technology via spin-offs.

11.4.2.2 CMSI

Once Begeman and his partner, Jeff Conklin, received their license agreement for CMSI, they had little residual relationship with MCC. CMSI, the company, began its bid for the groupware marketplace. They inherited, however, one of the same problems MCC had: the groupware technology had leapfrogged the marketplace. CMSI responded by spending its first year paring down the product into a form they hoped customers would be ready for.

The groupware product, dubbed CM1, that CMSI had created was essentially a software package that ties groups of people who were working on a problem together to facilitate its solution. It does so by placing all of the information and work done on the problem in one convenient reference place.

Begeman concentrated not so much on the specific aspects of CMSI's product as on the obstacles the company had in marketing its innovative technology. CMSI's initial strategy was to target the electrical/public utilities industry with CM1. This was based on the opinions of many "experts" Begeman had consulted. All agreed on the utilities companies as the perfect customers to focus upon – that is, everyone but the customers themselves. This market provided just one sale in one year. It was time to reevaluate the strategy.

CMSI had yet to deal with its true obstacle. Groupware was a foreign concept for the companies whom CMSI tried to sell to. 100% of the company's marketing was "missionary marketing," in which their tactic was to sit down with customers, educate them on their problem, and show them how to fix it. This process proved to be slow, expensive, and not very effective.

The company received its first big break in August, 1992, when Lotus set up the first Groupware conference. CM1 took the "best of show" award. Using the momentum gained from this success, CMSI was able to garner its first wave of sales. Upon trying to assess which industry seemed to comprise the customers from CMSI, Begeman noticed something. No one industry represented the target market. Rather, the common denominator among early customers was the fact that they were traditionally "early adopters" of new technology.

Crossing the Chasm presented a solution. Rather than building up a huge sales force and the accompanying overhead costs of such a move, the book recommends focusing on a toe-hold within one industry in the early majority, and directing all efforts at this target. Upon achieving a record of sales within a particular industry, an innovative company has referrals to point to for other companies, and it can establish a record of success to base sales of the rest of the early majority.

CMSI has taken this principle to heart, and now is concentrating on gaining such a toe-hold. CMSI has determined, this time with the aid of some potential customer input, that large government agencies are just the toe-hold to target. The company now plans to maintain its skeleton staff, to keep costs down, and focus on crossing the chasm to the early majority with sales to these targeted government agencies. They hope to achieve a record of sales without growing the company to enormous cost structures that have killed so many companies when they fall into the chasm. The growth will come, the logic follows, after CMSI has safely achieved a presence among the highly targeted early majority customers.

11.4.3 Medical Polymers Technologies, Inc.

Starting a traditional biomedical or pharmaceutical company is an extremely capital-intensive process. General facilities, development laboratories, testing labs, getting FDA approval, setting up FDA-approved production facilities and establishing a marketing network all add up to enough costs to intimidate even a large corporation. In addition, the long process of obtaining FDA approval can delay any income interminably while the cash outflow continues unabated. On the other hand, the strong competitive pressures from other companies and the fast pace of technological development mean that any such company has to remain flexible and innovative to remain competitive.

Lee Cooke, Chairman/President and CEO of Medical Polymers Technologies (and former mayor of Austin) has established an innovative, flexible and dynamic biomedical R&D company in the Austin Technology Incubator, and has succeeded in bringing new products to market with FDA approval in an unusually short time, without the levels of capital investment needed for a traditional biomedical company. Their first product has proven itself to be an outstanding market success in its niche, bringing in strong revenues at an early stage and paving the way for a whole series of new products currently under develoment or testing.

11.4.3.1 The Product

Medical Polymers Technologies (MPS) develops polymer-based products for the medical, dental and veterinary markets. Polymers are long-chain molecules made up of identical subunits that intertwine in ways that give different properties from the subunits. These properties can include lubrication, moisture retention, adhesion to biological membranes and the ability to transport and deliver other molecules in the polymer's matrix. MPS's first products are based on their proprietary Polymer-based Delivery System™ (PDS™) and include cleaners, lubricants and disinfectants/sterilants for medical and dental equipment.

Their first product to reach the market is PDS Clean™, a cleaner/lubricant for dental handpieces. It was submitted to the FDA for 510(k) approval March 1, 1992, and received "substantial equivalence" authorization on August 10th of that year. The first three truckloads products were shipped in May of 1993 and have generated $2 million in manufacturing-level sales ($8 million retail). Midwest Dental Products is marketing PDS Clean™ exclusively in the U.S. under the name LifeCycle™ as the recommended cleaning and maintenence system for their high-speed dental handpieces and low-speed angles. It is the first medical device being marketed in the U.S. that has FDA authorization (PDS Clean™ is classified as a Class I Medical Device). LifeCycle™ is now in 33,000 dental offices nationwide, out of the total of 105,000 dental offices in the U.S. and 13,000 in Canada. Midwest is the largest dental products company in the U.S., operating in 44% of the total dental market, and carrying over 25% of the dental handpiece market. MPS made their first presentation to Midwest on April 14, 1992; within four months, Midwest had invested $250,000 in testing it, and had it on the market after just 12 months. MPS has also reached agreement with Al Hikma Medical

Supplies & Services to market PDS Clean™ in Saudi Arabia and the Gulf Region. Market approval has been granted in six additional countries, and negotiations for distribution are underway for at least 20 additional countries.

Cleaning and sterilizing dental equipment gained widespread national attention after the Bergalis case in Florida, in which dentist Dr. David Acer transmitted AIDS to at least five patients, and one victim, Kimberly Bergalis, testified before Congress prior to her death. Subsequent studies showed that the high-speed (400,000 rpm) air-driven handpieces could suck up bits of flesh, blood, tooth, bone and saliva into the bearing area, and later throw out some of that material – into the mouth of the next patient. Autoclaving (steam cleaning) is the method of choice for sterilizing medical and dental equipment, but the procedure causes oil-based lubricants to gum-up or varnish the surfaces, and with the added biological debris shortens the useful life of the equipment. Rather than try to change dentists' method of sterilization – a tough prospect – MPS chose a product that merely cleans the handpiece bearing exceptionally well because of its low surface tension and excellent penetration (removing all the residue and biological debris) and lubricates without oil, and is used in conjunction with autoclaving. PDS Clean™ is blown through the handpiece before autoclaving and used afterwards to lubricate it. Used after each patient and first thing in the morning, a typical dentist uses about one liter, or $40 worth of the product per month. It sells for $1.18 per oz., about 55% below the price of existing cleaner/lubricants (at $2.50 to $3.00 per oz.).

Midwest Dental was extremely impressed with the product performance, with the only downside being that it worked so well that a few handpieces were destroyed (after long-term residue build-up, the dentists had to drastically increase air pressure to maintain normal speed; after cleaning with LifeCycle™, the residue was removed and the handpiece returned to original performance, but if the dentist neglected to turn down the air pressure, the handpiece speed went up so much as to be potentially destroyed). Midwest Dental had to institute a program of informing and training dentists in the proper use of the product, and had initially even considered halting product introduction because of this. The current agreement with Midwest for PDS Clean™ is for three years, and MPS has three other products in the pipeline for Midwest. Their relationship is substantial enough that Midwest is actually adapting their products to better benefit from the PDS™ chemistry.

MPS is trying to rapidly expand into other markets, especially medical and veterinary. The largest U.S. endoscope supplier, with 80% of the market, has tested PDS Clean™ against other products and found it to be the best, and is now working with corporate headquarters to make this the recommended product for their endoscopes world-wide. The Texas A&M School of Veterinary Medicine is testing the product, and reports are that while the faculty and veterinary students hate the cleaner/lubricants now on the market, they "love" MPS's product. PDS Steril™ is a cold sterilant that has a much larger potential market than medical cleaner/lubricants, and MPS has filed a submission with the EPA for marketing it as a disinfectant (approval is expected in Fall of 1994). It contains the well-known biocide glutaraldehyde, and with its combination of hydrophilic polymer and lipophilic and hydrophilic surfactants, gets much better penetration for improved effectiveness. It kills tough TB spores (under one hour at somewhat elevated temperature) while it cleans, lubricates and inhibits rust. Other PDS™ applications include opthamalic lubricants and the large market for polymer drug delivery systems.

11.4.3.2 The Market

The current market for medical cleaner/lubricants is estimated to be approximately $75 million annually, with $25 million of this in the dental market, $40 million in the medical market, and another $10 million in veterinary medicine. Current products in this market are described as 1950's technology, and have been demonstrated as inferior to PDS Clean™ in performance in current tests. With a price point 55% below current products, MPS is poised for rapid and thorough market penetration, with the only potentially significant threat coming from competing polymer-based products yet to be introduced.

The cold sterilant market is several times larger, and as a result has much more competition. This market is estimated to be approximately $300 million annually, with the main player being Johnson & Johnson, having $25 million in sales. PDS Steril™ can occupy a unique niche in this market, as it simultaneously cleans, sterilizes and lubricates, while being non-corrosive and non-carcinogenic. Its multiple advantages should give it a large market share.

The largest potential market is for polymer drug delivery systems, in which the polymer is used to transport the drug for more effective application to the required site of activity. This is estimated to become a $2 billion industry before 2000. For comparison, the entire medical market is estimated to be $1 trillion currently, and grow to $4 trillion by 2002.

11.4.3.3 The Company

Medical Polymers Technologies currently (1994) has 11 employees and has offices in the Austin Technology Incubator, occupying most of one floor on one side of the building. They also have a 1300 sq. ft. lab about two miles from the Incubator, separated due to zoning restrictions. Manpower is kept at a minimum to keep overhead low and profits suficient to stimulate growth. This is evidenced by the fact that the company was profitable in its first quarter of production, and even though their product is priced 55% below any competitors, there is still substantial margin.

Employees of MPS have significant shares in the company, even the receptionist. This provides a strong stimulus for personal performance by the individuals. It also keeps costs low, as salaries are about one-half of market value. The company is publicly traded on the Vancouver Stock Exchange (due to lower cost) and is incorporated as a Delaware Corporation based in Texas. It is the only publicly-held company in ATI.

As a method of remaining flexible and keeping costs low, MPS outsources nearly everything. Manufacturing is done by an FDA-authorized facility in San Antonio. Marketing is done by their product partners, such as Midwest Dental. Research is conducted through a contract to NewForm Development Labs, by their marketing partners (e.g. Midwest), as well as independents such as Clinical Research Associates.

11.4.3.4 Company History

Medical Polymers was founded in 1989 in Sacramento, when a chemical engineer borrowed and mortgaged $50,000 to buy the marketing rights to a polymer system for making gas-permeable contact lenses. This was a material that came in a tube and could be sliced at the office lab to make a contact lens. There were lots of orders, but quality control problems caused the curvature not to hold. Dr. Anne Crossway came in at this point after having lunch with the founder and discussing the problems. The polymer's creator, Dr. Marvin Gold, was brought in to help solve the QC problems. However, the company pulled out of this product, leaving a staff of 10 with no source of revenues. They came up with another solution, though, and decided to sell the polymer to the homosexual community as a sexual lubricant. This product was doing well until the FDA came "knocking at the door" in September of 1991. Because the material contained Nonoxynol-9, and the product did not carry warnings that it could impede conception, the FDA gave them the choice of taking the product off the market, or submitting it for FDA testing and approval (it typically takes about $300 million to bring a new drug to market). They took the product off the market, and again were a company without a product or revenues.

At this point the company was one-half million dollars in debt, when Lee Cooke came in. After taking the product around the world to check its market potential, he brought together a group of investors to buy the patents. In the 15 months since, $5 1/2 million has been infused into the company. It now has about 1400 stockholders.

11.4.3.5 Strategy

Medical Polymers aims to become a major player in the $500 million infection control market and to enter niche markets in the $4 billion drug delivery industry. The Company's strategy is to create a wide variety of polymer-based products for the dental, hospital, Pharmaceutical, industrial and veterinary markets. Two patents were filed in December 1990 and November 1991.

The North American market for dental cleaner/lubricants is dominated by two major handpiece manufacturers. Midwest Dental is the leader with over 25% of the handpiece market and KAVO America, a German-based handpiece manufacturer, has approximately a 10% share. 'The remainder of the market is divided between a number of smaller and generic brands. All current products are synthetics or petroleum based.

The medical infection-control market is highly fragmented with many companies supplying instrument processing cleaners and lubricants. Major players include S.C. Johnson, Calgon, Vestal Labs, V. Mueller and Huntington Labs. Johnson & Johnson dominates the cold sterilant/disinfectant market with its Cidex product line. Other companies include Colgate/Hoyt, Metrex Research, Sporicidin, and Huntington Labs.

Medical Polymers changed its name to U.S. Medical Systems Inc. in 1996 and, through sales of a successful denture adhesive, was narrowing annual net losses .

11.4.4 Expert Application Systems, Inc. (EASI)

11.4.4.1 History

Officers of EASI are Pat Mack, President, Margarita Ash, VP of Development, and Debbie Sallee, VP of Sales and Marketing. EASI was founded as a result of the Fall, 1991 Moot Corp® (a new venture competition for MBA students). As the winners from that semester, they went on to represent University of Texas at Austin at the International competition. Their first product is a software package that provides a solution for companies trying to get Underwriters Laboratories (UL) approval for product safety. Although UL approval is not required by law, distributors will not carry products without this seal of approval.

11.4.4.2 Definition of Need

Currently UL has 592 standards; EASI however will concentrate on Standard 1950, which is UL's highest revenue generator. Of those companies applying for this standard, 95% of submittals get rejected the first time, and 50% of submittals get rejected the second time.

In addition, updates are not distributed regularly. The submittal process can prove to be extensive because a product might have to be redesigned in order to meet this standard. The resubmittal process increases costs as well as time to market, thus making it difficult for small companies to complete if they continue to have problems in reaching approval.

Feedback from future customers who are in the Fortune 500 companies has been positive. So has feedback from UL, who has given access to resources, engineering, customers, and funding. Part of UL's receptiveness is a result of the fact that UL is beginning to get competition; EASI's product will provide a source of competitive advantage and revenue for UL

11.4.4.3 Future Plans

EASI will try to gain strategic alliances with other testing bodies in the European Community and Canada. This is because there are similar procedures in these countries and EASI has no exclusive contract with UL. Also, they will be coming out with multimedia products to meet the various standards.

11.4.4.4 Product Description

Currently the UL application process is boring and the manuals are confusing. In addition, every component within a product must be tested. With EASI's software, this process is automated. The package offers a checklist of tests that must be completed. It is organized along the same lines as the manual, thus making it easier for those aware of the process. In addition, it has graphics and animation which will make it easier for those who have never been through the process. Other features include a manual and component directory on-line and a report generator. Since the product is built upon the Windows utilities such as Write and Terminal, it will support a bulletin board so that customers can download the most recent versions of the standard and also communicate with each other.

The product design itself has been completely customer-driven, with information from their Alpha and Beta sites used in the development. Currently they have completed their Alpha testing and will begin with Beta testing in June. These Beta sites include UL, IBM, 3M, CompuAdd, National Instruments, and Dell. Throughout development, they will try to meet the needs of everyone. Large companies will use the software for their design process in order to prevent redesigns. Small companies will use it in order to design for product safety; with the software, they can keep track of parts of the standard. They will know where they are and be able to come back where they left off.

11.4.4.5 Marketing

The defined target market will be reached via trade magazines and direct mail. EASI's goal is to publish articles in magazines in order to show the benefits of the software and to educate the customer. When the products introduced, EASI plans to send literature and a demonstration version of the software, and will initially give discounts to induce Purchase. They plan to distribute through the nationally recognized testing labs and capitalize on their muscle- They plan to make the most of their sales through mail order. Mail Order was chosen because it was the most cost-efficient. This strategy would be supplemented by advertising in nationally recognized journals.

11.4.4.6 Pricing Strategy

For price benchmarks, EASI looked for comparable products but were unable to find any. They also considered the nationally recognized testing laboratories, did not want their price/positioning to be near these entities, fearing they could become a source of competition.

For further comparisons, EASI looked at large companies that have in-house staffs, and consulting firms, but these are mostly customized for individual products. EASI's software is not comparable because it can be used for all products. EASI finally decided to look at other expert systems which range in price from $2,000 to $5,000. From this analysis, they decided on a price of $3,000. For the customer, the purchase of EASI's product would imply a savings of $7,000 in the design process.

11.4.4.7 Questions

How will you price for revisions? EASI plans to charge $500 per year for UL standard updates and access to the bulletin board. An added feature would be the ability to upgrade over the phone line.

Can UL become a competitor? This is not likely because of EASI's partnership with UL. In addition they have already supplied diskettes with information.

Why is it an expert system? At this stage, the product is not technically an expert system, but more of a multimedia solution to the application process. The expert system aspect will be added when UL supplies interpretations to the manual and legal details are ironed out. This capability will be added at an additional cost to the customer.

Is there any current competition? One company puts out a CD-ROM with all 592 standards. However, this provides no value-added to the customer. EASI plans to put out one software-based standard at a time. There will not likely be any competition as UL's material is copyrighted, and EASI currently has the rights to use this material. The company who provided the expert system shell will not likely be a competitor, as they are more interested in selling the shell. (They even plan to do advertising for EASI.)

How much will UL receive in royalties? The CD people currently give about $75 per copy and EASI expect to pay a similar amount.

How will they reach their initial customers? UL has provided EASI with a list of the 5700 manufacturers who use 1950. These people will receive demonstration disks.

What about customer support? At first, they will have an 800 number. Then they will be consultant to UL later. To gauge support, their manual will be written so that someone with no previous knowledge could use the product.

What is EASI's mission? To provide systems for expert applications in compliance industry.

What about product liability? EASI plans to avoid liability by negotiating with UL so that they will accept responsibility for interpretations.

Where did the idea come from? The idea came from problem-solution brainstorming in a hot tub over the summer.

11.4.5 American Innovations, Inc.

11.4.5.1 History of the Company

Both Mr. James Helsey (our class presenter) and the president, Carl Morris, were long time employees of Motorola of Austin. Unhappy with the lack of change in a large company, they brain-stormed for ideas to capitalize on their expertise in microprocessors, searching for a $100 million market potential within five years.

Initially, their idea was replacement of household electric meters. Market research indicated the low cost of meters and resistance to change inherent in the big utilities, and they changed the idea to a retrofit type product designed to automate meter reading. The result was AlMetering, a high function, low cost system for residential meters.

11.4.5.2 Salient Product Features

Universal refit brackets. Fast, inexpensive installation. Motorola microprocessor. Durable, compact, reliable, functional, calls back daily rather than monthly. Uses existing phone lines. No extra cost, calls placed during 1-5 a.m. Multi-Functional – with PC, can handle 150,000 houses, has several features (time-of-use data, peak demand data, etc.) that can be used at a later time. Date/Time keeping computer with prescheduled callback times – low baud (800) modem for accuracy -IBM PC as host for units. Firmware in microprocessor chip represents the biggest advantage in technology.

11.4.5.3 Initial Market Strategy – Five Niches

1. Automating expensive-to-read meters, 10-11% of the total.

2. Time-Of-Use Metering: hard data for accurate rate structures/conservation efforts.

3. Demand Metering: existing requirement for all utilities.

4. Demand-Side Management Evaluation: a growing need for effective load management.

5. Commercial/Industrial Load Recorders to replace existing, expensive products with a lower cost product that has more functionality.

Their strategy was "Go after the big utilities first." Although slow and conservative, these companies control millions of meters and "when they commit, they do it big." Use growing regulator pressure about "peak demand" reading as a selling point for AlMetering. Provide up-to-date technology at the lowest cost utilizing the telecommunications network to gain competitive advantage over lagging, cumbersome products of competitors.

11.4.5.4 Significant Problems

- Major resistance to change within the large utilities. They act very conservatively and buy from existing, large firms.

- Tendency of smaller utilities to "wait and see" what the larger ones do.

- Hard to get a documented "success story."

- Lack of capital for expansion.

The company has generated a lot of initial excitement among most of the utilities, but the biggest road-block is overcoming the initial resistance to change and getting some early acquirers to commit to hard purchase orders. The crucial question concerning the company's future existence is: How long can they live without a large capital infusion or when is that infusion coming? They have "optimized their investment dollar" too early, as opposed to first securing a large capital fund to ensure survival and then optimizing.

They are having trouble attracting new investment dollars. With an infrastructure that operates at $110,000 per month, they have spent their war chest quickly. The present crisis/crunch is attributable to a lapse in attention to steady recruiting, and attrition resulting in a "dry pipeline."

11.4.5.5 Status (1993)

They have one letter of intent from one firm and a promising prospect from another. The Canadian utilities are presently customers through licensing agreements with GE Canada, and European utilities are also showing interest.

They feel they have a two-year window of opportunity until electronics firms decide to enter the market and produce an identical product. Although reverse engineering of the microprocessor would take only six months, it would be easier for competitors to write their own firmware for their own chip.

It is apparent that the company has a product with much potential, but they have underestimated the time/capital requirement to get initial market penetration on a large scale.

11.5 Notes

Corporate Memory Systems, Inc. notes by Donnellan B and A Stein, October, 1993.

A student contacted the current president of Autogenesis in 1998... This was Mark Rehley.

American Innovations, Inc. summarized by Charlie Lanman and Robert Dach.

11.6 Questions and Problems

Short Answer

S11-1. What is the one thing you would most like an investor to understand about your company?

S11-2. The innovation arrow figure points one way only, but we know feedback occurs (it's a cycle). What kinds of flows feed back to the left?

Miniprojects

M11-1. Checklists, Diagnostics, and Assessment Instruments. In this book you have studied technology cycles and the decomposition of their cost and revenue elements; technology substitution; technology scanning; technology fusion; technology transfer; adoption and diffusion of technologies; the role and problems of technology entrepreneurship; technology forecasting; life cycle cost; types of technological risk; how government policy enhances or constrains a company's range of technology-related actions; and how several companies' experiences, good and bad, have fit into the context of these topics.

In addition, you have seen some specific tools and assessment instruments used for making decisions in these contexts. The technology adoption checklist, 3M's probability assessment for R&D projects, and Bass model forecasting are among these. As your careers continue, you will be exposed to many more such instruments and rules of thumb. They can be hard to remember, and that is why this course has concentrated on giving you the model or framework that is called "context" in the above paragraph. Frameworks give you a mental place to put the tricks, rules and experiences you will acquire in the future, and a way to relate them to each other.

This exercise is to give you a chance to use your "framework" knowledge to develop your own checklist, diagnostic or assessment tool for a technology management situation that you choose. An assessment instrument is a list of questions that are answered either by an expert or by a survey of several or all of the people involved in an organizational situation. The answers are collected using a scale of "yes/no" or "1 to 5" or "check all that apply." In other words, you will ask questions in a way that lets you easily compile the answers into a quantitative score or scores that will help you make a technology decision. Diagnostics are similar, but imply that a situation is going bad or has gone bad, and you're trying to determine what's wrong and how to fix it. A checklist is simpler, as it doesn't require scaled answers. It is just a comprehensive list of things to attend to and to do.

Do the exercise in small teams, and take 45 minutes to choose a topic and work out an instrument. We will then get together and allow you to use another 15 minutes to test your instrument on another team. (You will have to describe a specific situation to the other team so they know what they're responding to. Use your 45 minutes to write this description also.) Be prepared to present your tool to the entire class.

Your tool may address:
• Forecasting a new product's growth rate using the analogy method;
• Deciding when to introduce the next generation of a product;
• Deciding the suitability of a new technology for adoption in your firm;
• Determining whether an invention is commercializable;
• Deciding whether to develop a particular technology or product in-house or to source it; or
• Any other similar issue facing you in your work now or recently.

M11-2. Research and write a 1-2 page update on the status of a startup technology company you may find in a search of the press. Analyze their situation using the principles presented in this book, and present a SWOT (Strengths, Weaknesses, Opportunities and Threats) summary.

M11-3. In various parts of this book, technologies have been classified as basic/non-basic; continuous/discontinuous; core/non-core; and so on. Go back through the book and note each such instance of classification. Compile a catalog or glossary of the way people classify technologies.

Discussion Questions

D11-1. The "Commercializing Technology: What the Best Companies Do" article proves companies are more successful when they are *systematic* (rather than haphazard or opportunistic) about commercialization. Rate your own company's sophistication with regard to systematic commercialization. Give reasons for your answer.

D11-2. In 2160 B.C., Queen Semiramis ordered the construction of the first known underwater tunnel. Built of brick and tar, the tunnel joined the two fortified halves of the ancient city of Babylon. Nearly 4,000 years would pass before the next known underwater tunnel was constructed, under England's Thames River. Originally for carts, the British tunnel is still in use – for the subway. A tunnel linking France and England was first proposed in 1802, but Napoleon expressed no interest in it. Later proposals were ignored until 1882, when a British engineer built a machine that could cut through 39 feet of chalk every 24 hours. After one mile of tunneling, though, the project was halted when fears of a French invasion swept through Britain. The current "Chunnel," the channel tunnel linking Coquelles, France, and Folkestone, England, opened in May 1994. (Source: "The Facts." *Popular Science,* December 1995, p. 102, unsigned article.) Write a few paragraphs speculating on how and why such an anachronism as Queen Semiramis' tunnel might come to exist, and (ii) how a technology can be "lost" for 4,000 years.

D11-3. Write a short SWOT (strengths, weaknesses, opportunities and threats) analysis of one of the companies described in this chapter.

Subject Index

Name Index

3M, 191, 388, 392

A.C. Nielsen Co., 57-58, 61-62, 65-66, 72-73

A.T. Kearny Co., 122

ABC, 62-63, 72

Advanced Micro Devices, Inc., 123

Agricultural Extension Service (U.S.), 134

Alcoholics Anonymous, 362

Allaire, P, 328

Amazon.com, 197, 219

America Online, 109, 280

American Airlines, 322

American Association for the Advancement of Science, 205

American Chemical Society, 205

American Demographics, 301

American Electronics Association, 16, 340

American Express Corp., 16

American Innovations Corp., 39, 390, 392

American Society of Orthopedists, 334

American Supplier Institute, 155-156

Ampex Corp., 347

ANSI, 267, 337, 346-349, 352

Apple Computer Corp., 18, 97, 99, 151-152, 253, 274, 290, 301, 321-323, 331-332, 335

ARAC, 376

Asahi-Kasei, 23

ASEAN, 375

Asian Technology Information Program (ATIP), 121, 140

AT&T, 36, 39, 42, 69, 90, 201, 328, 334

Ateq Corp., 89, 120

Augustine, N, 122, 292

Austin Software Council, 32, 213-214

Austin Technology Incubator, 19, 31-32, 213, 241, 251, 375, 383, 385

Australian Technology Group, 375